Fruits and Plains

Fruits and Plains

THE HORTICULTURAL TRANSFORMATION
OF AMERICA

PHILIP J. PAULY

HARVARD UNIVERSITY PRESS
Cambridge, Massachusetts, and London, England   2007

Copyright © 2007 by the President and Fellows of Harvard College
All rights reserved
Printed in the United States of America

Library of Congress Cataloging-in-Publication Data

Pauly, Philip J.
   Fruits and plains : the horticultural transformation of America /
Philip J. Pauly.
      p.   cm.
   Includes bibliographical references and index.
   ISBN-13: 978-0-674-02663-6 (alk. paper)
   ISBN-10: 0-674-02663-2 (alk. paper)
    1. Horticulture—United States—History. I. Title.
SB319.P38 2007
635.0973—dc22   2007025274

This book is for Vincent A. Pauly:
worker, father, craftsman, and conifer enthusiast

Contents

List of Illustrations   *ix*

Introduction: Taking the History of American Horticulture Seriously   *1*

1 Culture and Degeneracy: Failures in Jefferson's Garden   *9*

2 The United States' First Invasive Species: The Hessian Fly as a National and International Issue   *33*

3 The Development of American Culture, with Special Reference to Fruit   *51*

4 Fixing the Accidents of American Natural History: Tree Culture and the Problem of the Prairie   *80*

5 Immigrant Aid: Naturalizing Plants in the Nineteenth Century   *99*

6 Mixed Borders: A Political History of Plant Quarantine   *131*

7 Gardening American Landscapes: From Hyde Park to Curtis Prairie   *165*

8 The Horticultural Construction of Florida   *195*

9 Culturing Nature in the Twentieth Century   *230*

10 America the Beautiful   259

   Notes   269

   Acknowledgments   327

   Index   330

Illustrations

1.1. Thomas Jefferson's orchard plan for Monticello   25
2.1. George Morgan's perception of recent and current threats to Princeton, New Jersey, 1786   35
2.2. Developmental stages of the Hessian fly   36
3.1. Plan of Mount Auburn Cemetery and Experimental Garden, 1831   57
3.2. Hovey's Seedling   59
3.3. Charles Mason Hovey   61
3.4. Horticultural Hall, Massachusetts Horticultural Society, Boston, 1870   62
3.5. Ephraim Bull   77
3.6. A "native" caricatured in *American Garden*, 1890   78
4.1. The giant sequoia "growing naturally in Massachusetts"   88
4.2. Map, "Forest and Prairie Lands of the United States," 1857   91
5.1. Map of tea regions in China and India, 1858   108
5.2. Map of potential tea regions in the United States, 1858   109

5.3. U.S. Propagating Garden, National Mall, Washington, D.C., 1859   *110*

5.4. Plan of grounds, U.S. Department of Agriculture, 1870   *114*

5.5. Johnson grass on the U.S. Department of Agriculture grounds, 1902   *120*

5.6. George Vasey   *121*

5.7. David G. Fairchild   *127*

6.1. Official seal of the Quarantine Division, California State Commission of Horticulture, 1914   *132*

6.2. Massachusetts Gypsy Moth Commission map of infestation, 1893   *138*

6.3. Commission employees hand cleaning the Dexter elm   *139*

6.4. Sterilizing a landscape near Georgetown, Massachusetts, with flamethrowers   *140*

6.5. Charles Marlatt   *144*

6.6. Burning Japanese cherry trees on the Washington Monument grounds, 1910   *150*

6.7. "The Landing of the Pilgrim Fathers," cartoon, 1919   *156*

6.8. "What we may expect if the health of the people as well as plants is given in charge of the Federal Horticultural Board," cartoon, 1919   *157*

6.9. J. Horace McFarland   *160*

7.1. Scenic view of the town of Hudson, New York, around 1820   *167*

7.2. Idlewild, Nathaniel P. Willis's country home, around 1855   *172*

7.3. Arboretum design, Central Park, 1858   *175*

7.4. Arboretum design, Central Park, 1861   *176*

7.5. Detail of the Ramble, Central Park, late 1860s   *178*

7.6. Taxonomic planting sequence, Arnold Arboretum, 1880   *180*

7.7. Dredging the Back Bay Fens, 1882   *181*

7.8. Vegetation of the Back Bay Fens, 1895   *183*

7.9. Transplanting sod at the University of Wisconsin Arboretum prairie, late 1930s   *192*

8.1. East Florida, with locations discussed in the text emphasized   *196*

8.2. Plantings, Hotel Ponce De Leon, St. Augustine, late 1880s   *209*

8.3. Plantings, Royal Poinciana Hotel, Palm Beach, around 1900   *211*

8.4. Charles T. Simpson in his garden, Lemon City, Florida, 1920s   *214*

8.5. Naturalistic vegetation on Charles Deering's estate at Cutler, 1921   *217*

8.6. Water hyacinths in the St. Johns River, Palatka, Florida, around 1895   *220*

8.7. National Guard roadblock inspection for Mediterranean fruit flies, Florida, April 1929   *225*

9.1. American Horticultural Council's Tree of Plant Knowledge, 1947   *236*

9.2. "Kudzu for Erosion Control in the Southeast," 1939   *248*

9.3. Burning the University of Wisconsin Arboretum prairie, late 1940s   *250*

9.4. Woody plants on Konza Prairie, 2003   *256*

10.1. Wellesley College students in H. H. Hunnewell's garden, 1876   *260*

Fruits and Plains

Introduction
Taking the History of American Horticulture Seriously

This book explains how horticulturists have transformed America during the last 250 years. A network of amateurs, tradesmen, and bureaucrats introduced thousands of plant species from other continents, selected and hybridized varieties so that they would thrive in American settings, and moved plants found on this continent into regions where they had not previously grown. In addition, they sought to exclude or exterminate undesirable plants and plant pests. These activities fundamentally altered not only the vegetation, but also the economic activities, social relations, and common experiences of Americans. They shaped the identity of the United States. Our understanding of contemporary environments increases substantially when the values of past horticulturists are appreciated. Management of vegetation in North America in the present day depends on grasping the consequences of horticulturists' activities.

I begin so emphatically because only a few Americans would ordinarily think of horticulture as more than an ornamental subject. In general conversation it is an upmarket synonym for gardening, involving primarily the selection, arrangement, and maintenance of flowering perennials and shrubs on suburban properties and metropolitan parklands. For the most part its development is examined in books where full-page photographs predominate over text. In the 1800s, however, horticulture was equivalent to what is now called plant biotechnology. Horticulturists focused primarily on fruits, but they were also passionate about tree planting, and their reach extended to lawn, pasture, and range grasses. Their aims were quality, productivity, and diversity. Ornament was important but secondary.[1]

Horticulturists' technologies were, for the most part, old ones. Travelers had long altered the geographic distributions of species by carrying seeds back home and planting them. Gardeners managed plants' local environments through siting, fencing, fertilizing, and irrigating. They cut and bent stems; more dramatically, they grafted different individuals together to form chimeras. Desirable plants were routinely cloned, either by dividing rootstocks or by rooting scions. In the three centuries after 1500, European collectors in eastern Asia, southern Africa, and the Americas dramatically increased the number of species in circulation; plants gathered together in gardens hybridized spontaneously and were, with varying degrees of awareness by their caretakers, selected and propagated. In the eighteenth century, plantsmen began consciously to hybridize both related varieties and different species.[2]

At the beginning of the nineteenth century, English gentlemen, gardeners, and botanists began to organize these activities as a new enterprise called horticulture. The Horticultural Society of London (1804, rechartered as the Royal Horticultural Society in 1861) promoted meetings, nomenclatural projects, a lavish journal, a propagating garden, and a seed and plant exchange system. It linked wealthy amateurs, commercial culturists, and government-run botanical gardens. Its leadership emphasized fruit trees and ornamental species (notably orchids), but they also valued trees and grasses.[3]

From both institutional and material perspectives, American horticulture was a scion of the European system. The Massachusetts Horticultural Society (1829) was created, appropriately, to distribute pear twig cuttings that Thomas A. Knight, president of the Horticultural Society of London, had given to Boston Brahmin attorney John Lowell. American horticulturists soon developed networks linking wealthy and middle-class amateurs, commercial nurserymen, professors, and bureaucrats. The populist folklore that emphasizes individual untutored American plantsmen (whether Johnny Appleseed or Luther Burbank) is seriously misleading regarding both tradition and technology.

At a deeper level, however, the relationships among people, territory, and plants in the United States were dramatically different from those in Europe. Some twenty years ago, Alfred Crosby's *Ecological Imperialism* displayed the extent to which the entry of Europeans into North America involved the replacement of New World plant species with the "portmanteau biota" that had come to western Europe from Asia and Africa during millennia of agricultural activity and commercial intercourse. English colonists created "neo-European" landscapes along the eastern seaboard that were filled with Old World plants including wheat, Kentucky bluegrass, apple trees, and agricultural weeds.[4] Crosby, however, did not consider what happened next—as the descendants of English colonists became colonials, and then, during the 1700s, Americans. They became increasingly aware of the range of climatic and biotic differences between their lands and England. They recognized that American prosperity depended on the productivity and diversity of plantings. Finally, they understood that improvements to vegetation in the United States would depend on their own initiatives. They envisioned a future of horticultural independence, and the development of a specifically American horticulture.[5]

American horticulture had four aspects. The first and most visible was the interest in finding and bringing "new" species and varieties from other parts of the world to the United States. Thomas Jefferson, justifying his own activities over three decades, emphasized in 1800 that "the greatest service which can be rendered any country is to add an useful plant to it's culture." This sentiment was cited frequently for the next century and more; plants were sought by amateurs, societies, nurserymen, and, most influentially, by agents of the national government.[6]

The second issue was "naturalizing" foreign plant lines—getting them to survive in American settings. Climates in most of the country were more extreme, in both daily and seasonal temperature changes, than those of western Europe (particularly England), and hence could be fatal to varieties that were accustomed to more even temperatures. The deeper difficulty was that, unbeknownst to Euro-Americans, certain Old World domesticates could not survive exposure to plant parasites that had evolved in North America. Successful naturalization depended on diligent selection, dogged garden work, and, frequently, evolutionary change.

The third element of American horticulture was the identification of the potentials hidden within plants already growing in North America. While settlers generally viewed American vegetation as timber to be cleared or weeds to be plowed, they intermittently hoped for the discovery of usable plants. The possibilities of the American flora could be glimpsed in domesticated tropical annuals such as maize and cultivated tobacco (the latter was brought to Virginia from the Caribbean in the early 1610s). North American ginseng, an ordinary woodland perennial, suddenly gained significance in the 1710s as an export to China. There were hopes that value could be found in grasses, plums, or berries located in unexplored parts of the continent, and that improving these species would expand the number of useful plant types.

The fourth part of horticulture was the exclusion, extermination, or suppression of undesirables. Euro-Americans, understandably if unrealistically, wanted foreign crop plants but not their associated pests: wheat but not the Hessian fly, apples but not codling moths, and cotton but not the Mexican boll weevil. Other undesirables were more generalized Old World herbivores (for example, gypsy moths and Japanese beetles), parasites (pine blister rust and Dutch elm disease), and weeds (quack grass and Russian thistle). An additional class of exclusion efforts was intracontinental: stopping the eastward spread of the Colorado potato beetle, keeping eastern phylloxera aphids from California's Old World grapes, or, more recently, fighting Atlantic coast spartina grass that was spreading in Puget Sound.

The elements of American horticulture, while placed and praised together, quite frequently came into opposition. Proponents of naturalized grapes debated advocates of native varieties, for example; or the exclusion of pests from an area prevented the introduction and propagation of new plant species. In addition, what counted as foreign or native changed as the national boundaries of the United States expanded, as land transitioned from frontier to farm to rural backwater, and as places for planting extended beyond the humid temperate landscapes of the eastern seaboard to regions that were, to varying degrees, arid or tropical.

The constant desires were to enhance plants and to manage their placement. This "gardenification" impulse, drawing on a range of naturalistic tastes, operated from the level of individual specimens to cultivars, species, assemblages, and what came to be called ecosystems. Realization of these goals depended on the development of botanical and sometimes entomological knowledge, on garden craft and soil science, on the development of transport systems, on the shape of nationalism, and on the realities of federalism and the scope for individual initiative.[7]

In order to convey the historical importance of American horticulturists, this book emphasizes, on the one hand, the coherence and continuity in their visions and activities over the last two centuries, and on the other, the details (and challenges) of practice. I focus on a selection of ideas, problems, and places that were strategic or representative. The book is not an exhaustive survey; a platitude that shines luminously from my research is how much remains to be learned about this undervalued and understudied area.

A sketch of the book's expository pathway and main topics provides a preliminary orientation. My backdrop is the emphasis in environmental history on the biotic dimensions of the entry of Europeans and Africans into North America.[8] The narrative proper begins, however, with the efforts of the first generation of self-consciously "American" naturalists in the mid-1700s to understand North American climates and plants, and to realize their territories' vegetational potentials. I follow these concerns into the Revolutionary-era "dispute over the New World" and into the horticultural and intellectual activities of Thomas Jefferson. My aim is to convey the depth and breadth of early national leaders' experiences with plants. The more immediately practical problem of insect invasion is the subject of Chapter 2—"postcolonial" Americans struggled in the 1780s with the logistics of fighting the alien Hessian fly, and with the implications embodied in this prejudicially named organism.

The next five chapters, focused on the states east of the Great Plains and north of Florida during the period from 1800 to 1930, together comprise the core of the book. I first sketch the development of organized horticulture in New England during the antebellum period. My emphasis is on fruit and on the challenges posed by tradition, novelty, and naturalization. I then move in Chapter 4 to the broader geographic canvas for arboriculture. Tree planting was particularly important in the emblematic nineteenth-century American landscape—the prairie region of the Midwest. Chapter 5 examines the processes that culminated at the end of the nineteenth century in the development of federally run "systematic plant introduction." The next chapter sketches the opposing impulse to control the movements of pests. Finally, I examine the history of landscape gardening, emphasizing plant choice rather than the usual issues of design. My goals are to convey the breadth of horticulturists' activities and their struggles to come to grips with landscape, with nationality, and with species having agendas of their own.

The two last chapters selectively extend this history in space and in time. Chapter 8 studies peninsular Florida as a concentrated recapitulation of American horticultural development in an area with a different climate and minimal preexisting culture. The final chapter outlines how horticulture declined but horticultural perspectives have become pervasive during the last seventy-five years. It sketches how pest control became the battle against invasive species, and how landscape gardening evolved into ecological restoration.

Explicating at the start three of my more idiosyncratic expository practices should prevent confusion and highlight some ways this subject is particularly important. The first concerns language. While horticulturists emphasized their ability to exercise rational control over their plants, they were also aware that they were immersed in an old and rich linguistic compost redolent with sexuality, primitive religion, and primal group consciousness. This was evident in their central concept of "culture." We need to set aside the century or more of discussions about this keyword among literary critics and anthropologists to recover a world where culture was an umbrella term for efforts at biotic improvement. Horticulturists could speak unself-consciously about strawberry culture and pear culture. For them, high culture meant neither Plato nor fancy table settings, but rather well-rotted manure and hand weeding. Yet they understood that their usages shaded smoothly into others: that pear culturists were themselves a culture; that geography, law, and race were, with climate and seeds, elements that made cotton culture possible; that culture happened not only on farms, but in schools, churches, and their own clubs. From the early nineteenth century onward, horticulturists reasonably argued that high culture, producing exquisitely flavored fruits, fancy camellias, and smooth greenswards, would lead to higher culture—to the refinement of public taste.[9]

We can see these tendencies also in politico-horticultural concepts such as native, naturalized, and alien. These words originally developed as names for the overlap between humans as legal subjects and as animal objects. From the mid-1500s, a native of England, for example, was a person whose status derived from some combination of parentage, birth locale, and education, but who was distinguishable from an alien; naturalization gave an alien the same legal status as a native. The terms were first applied to plants more than two centuries later. American horticulturists had no difficulties in discussing native and naturalized plants, and alien pests, in the nineteenth and early twentieth centuries. But usage was very difficult to specify. Was native with reference to a locale, a region, or the Americas as a whole? Given the changes in the land during the previous two centuries, could one know what was native? Did naturalized plants become, like people, native through generation? And what about hybrids? Botanists' usage now is more careful than in the 1840s, but legal and political elements have not been exterminated from their terminology; and given the conceptual and rhetorical fecundity of these terms, linguistic sterilization may not be desirable. Awareness of the historically and geographically specific arrangements that existed one and two centuries ago can, however, induce productive questioning of contemporary usages and can generate conceptual improvement.

My second basic emphasis is common among evolutionary agronomists and historians of the environment, but not beyond: the extent to which organisms have been evolutionary actors adapting themselves to the conditions created by people.[10] Horticulturists guided the forms and placements of organisms, but they controlled few, if any, completely. Plants, insects, and fungi pursued their own reproductive and evolutionary strategies, and some moved successfully into and then out of cultured settings. Many of the agricultural weeds brought to North America from England had coevolved with farming during previous millennia and had migrated to western Europe through commerce in seeds, fibers, and animals. Grasses, including important genera such as *Triticum* (wheat) and *Sorghum* (Johnson grass and both Asian and African sorghums), had evolved substantially in historic times and had repeatedly jumped the borders that distinguished crops, forage grasses, and weeds. Collectors, with their passion to juxtapose related plants in their gardens, wittingly and unwittingly created the conditions for spontaneous hybridization and the appearance of new species, notably in *Plantanus* (the London plane tree) and *Fragaria* (strawberries). Horticulturists recognized the likelihood of these developments, sought to capitalize on them when they could, and bemoaned them when crops were impacted negatively; but their actions were opportunistic rather than directive.

The third caution and invitation I offer at the start concerns the extent to which the meanings and boundaries of America and North America have fluctuated. America (the New World) was larger than North America, which was larger than America (the United States). Americans identified themselves variably with these national and continental entities. National boundaries changed repeatedly between 1750 and 1900, and the imagined scope of North America varied, from all the land between the oceans north of the equator down to the temperate humid region of the mainland. While recognizing the importance of marginal places such as Florida and California, I follow my subjects in emphasizing the extent to which the region stretching from Massachusetts to Virginia and from the Atlantic to eastern Kansas and Nebraska was the template for America.[11]

Like many of the people I discuss, I have become a rational enthusiast for my subject. Attention to the history of horticulture advances our aspirations to see the past as a whole, or more attainably, to see together the histories of the environment, agriculture, science, art, and national development. In so doing, it alters perspectives within these different fields. Historians of the environment have turned to the garden concept as a way to overcome dichotomous thinking about the relations of humans and other organisms.[12] In the history of agriculture, a focus on science, marketing, and aesthetics has emerged among scholars examining California fruit growing in the twentieth century; I suggest that studying the history of American horticulture more broadly can reintroduce ideas, leadership, and international communication into our understanding of that massive industrial enterprise.[13] The history of science, my own field of origin, looks very different when researchers at elite institutions such as the New York Botanical Garden and Harvard University are studied in their relations with the amateurs associated with the Massachusetts Horticultural Society, the American Rose Society, or the government-subsidized horticultural organizations operating in nearly every state around 1900. In addition, as historians of landscape gardening have emphasized, horticulturists have blended biological, commercial, and aesthetic interests in constructing private grounds, public parks, and, as following chapters show, regional landscapes.[14] Finally, attention to the history of horticulture leads to new interpretations of some of the themes that have long formed the foundations for general histories of the United States of America: exceptionalism, tensions between local and national interests, the problem of sectionalism, and the roles of social and technical elites in a democracy. These scholarly chestnuts regenerate creatively when their setting includes the familiar but radically nonhuman shapes and activities of fruit trees, plant rusts, and insect pests.

My goals extend, however, beyond appreciation of the richness of the past toward usefulness in the present. This outlook was pioneered more than a century ago by Cornell University horticulturist, landscape gardener, and vegetation historian Liberty Hyde Bailey. His vision of a humanized and cultured landscape integrated evolutionary and human history with social and aesthetic choices. Contemporary vegetation scientists such as the Arnold Arboretum's Peter Del Tredici have similarly asserted that realistic and progressive human interaction with vegetation in the decades to come depends on awareness of the history of those relations. I hope to give depth, color, and richness to the bare contours of that argument.[15]

CHAPTER ONE

Culture and Degeneracy
Failures in Jefferson's Garden

Thomas Jefferson devoted nearly one-third of his *Notes on the State of Virginia* to the climate and the organisms of the eastern half of North America. It was natural history written by a lawyer. Most of Jefferson's brief was directed against the French naturalist Buffon's "new theory of the tendency of nature to belittle her productions on this side the Atlantic"—Buffon's assertion that New World animals were smaller, degenerate variations of Old World types, and that American Indians were weak and lacked interest in sex. Jefferson concluded his work, however, by turning from the claims in Buffon's twenty-year-old encyclopedic *Histoire naturelle* to Guillame Raynal's current best seller in four languages, the *Histoire de deux Indes*. This book, Jefferson explained, applied Buffon's theory to "the race of whites." He denounced Raynal's "wretched philosophy" that Europeans and their descendants had declined in America over the preceding centuries; for Jefferson, this implied that George Washington should be classified among "the degeneracies of nature." He proclaimed that while Raynal's ideas would soon be forgotten, Washington's memory would "triumph over time, and will in future ages assume its just station among the most celebrated worthies of the world."[1]

Jefferson was making an easy bet in predicting that the military leader of the new United States would be more famous than a historian. On substance, however, his case against Raynal was weak. The basic relevant issue raised in the *Histoire de deux Indes* was whether American climates had negative effects on cultured plants, on domesticated animals, and on the descendants of Europeans, both immediately and over the course of generations. These concerns were neither new nor implausible. Jefferson in fact had direct experience, as both a husbandman and as a husband, with the difficulties Raynal emphasized. Yet he did not address these issues of gardening and breeding in *Notes on the State of Virginia*. His emphasis on Buffon's decades-old "new theory" was, to a significant degree, a red herring.

This chapter shows how important cultured plants were, both practically and intellectually, within the context of Americans' declaration of independence from Britain. I begin by sketching the tensions between culture and degeneracy as they were experienced in America during the first centuries of European colonization. I then discuss the ideas about climate and culture developed by John Mitchell, the most important American naturalist prior to Jefferson. This Virginia physician and imperial geographer and horticulturist was uniquely positioned to conceptualize and explain the environmental peculiarities of the "North American continent" (the United States east of the Mississippi). Raynal's history is examined to show how, during the period of the Revolution, the problems Mitchell addressed were placed within the tradition of geographic learning, within a modern approach to environmental and political history, and within the debates in France over the wisdom and value of aiding the American states.

These foundations clarify how important and difficult fruit growing was for Thomas Jefferson, and how consequential his gardening failures potentially were. Visions of cosmopolitanism, rural happiness, and profitable transatlantic trade were entwined with grapevines, olives, and oranges being planted at Monticello in the first half of the 1770s. But by 1782, when Jefferson was writing *Notes on the State of Virginia*, these plants had withered, his European neighbors were gone, his wife was dying, and his community was becoming darker. Jefferson's book was a struggle with the possibility that Raynal was right about the American climate and its degenerative consequences, and it was an effort to establish the basis for an American culture. Jefferson's solutions were not very successful, both on their own terms and in relation to the events of the 1780s. His search for alternatives displays the amount and difficulty of the cultural work to be done.

Great Promise but Slender Performance

The paired concepts of culture and degeneracy were integral to classical writings about both rural matters and morals. *Cultura* was the intelligent and attentive care of soil, plants, domesticated animals, and, by extension, children, the self, and the public. Living things that were *degeneris* had changed from the optimal cultured state in either quality or productivity, generally for the worse. Plants, animals, and people could degenerate from age or other internal causes, but they could also decline due to the relaxation of culture, or from their placement in an unfavorable or merely unfamiliar setting. The creation of *coloniae,* or new rural communities on the frontiers of the Roman Empire, gave particular significance to the tension between culture and degeneracy. In cold, arid, or wild lands, colonizing people, plants, and animals were particularly liable to degenerate. Some perished; others became uncultured—rank, rough, and tasteless.[2]

This conceptual framework gained renewed significance for western Europeans in the sixteenth and seventeenth centuries in the context of overseas colonization. One of the major purposes of European expansion was to gain control over places that differed climatically and botanically from home—to obtain products from plants such as cloves, coffee, sugarcane, and tobacco that could not be made to thrive in cool temperate latitudes.[3] The expected consequence of colonization, however, was degeneracy: both French and British commentators drew on the classical tradition to explain that people, like plants, declined when "transplanted into another soil." The phenomenon was evident in its starkest form in the tropical East Indies, where mortality was so great that Europeans were unable to establish anything larger or more permanent than outposts for traders and soldiers. Decline was less dramatic around the Caribbean, but was a continuing issue. Creoles—American-born descendants of Spaniards—comprised a small but recognizable segment within a population that was overwhelmingly American Indian and African. The stereotypical Creole was degenerate: less intelligent, less ambitious, less disciplined, and less white than his European forebears or contemporaries. The precise nature, extent, and degree of preventability of Creole degeneracy was a matter of continuing, albeit shallow and unfocused, discussion.[4]

Eastern North America was one more potential setting for colonies. The ways in which the region east of the Appalachians, north of Florida, and south of Nova Scotia differed climatically from more tropical areas resulted, however, in more complex and open-ended possibilities. A number of historians have emphasized that early English colonizers reasonably anticipated that seaboard settlements could support the species that were grown around the Mediterranean and were desired in England. Jamestown, after all, was at the same latitude as Seville, and

Boston lay south of Florence; with intelligent effort, the right kinds of oranges, olives, pomegranates, mulberries and silkworms, and—above all—Mediterranean grape varieties ought to have grown there.[5] It took time to learn that while summers on the seaboard were as hot as in Italy, the winters could be as cold as in Sweden. Which season determined what would grow and what would happen to colonists?

The successes of tobacco and rice reinforced the idea that the seaboard had a fundamentally hot climate. *Nicotiana tabacum* was a Brazilian-Caribbean species, and lowland rice had been one of the great successes in northern Italy from the 1400s onward. Both of these crops had locally degenerative consequences. Tobacco was notorious for "using up" soil, making it agriculturally worthless within five or ten years; by the 1700s, Virginia was filled with derelict tobacco land from which the poor barely scraped a living. Tidewater areas flooded seasonally for rice were soon filled with malaria and other fevers that debilitated both slaves and masters. These outcomes were unfortunate, but were the expected corollaries of culture in areas whose productivity was linked to heat and moisture. The often-noted prevalence of pestiferous animals along the seaboard was part of this picture—not just mosquitoes, but also ticks, biting flies, cockroaches, poisonous snakes, and, from Carolina southward, alligators.[6]

If the seaboard had developed the same hard calculus of culture and degeneracy as other warm areas, British colonial projectors would have been satisfied. The problem was that tobacco and rice were exceptional—the most desired Mediterranean-climate plants actually did not flourish. Citrus, olives, apricots, and wine grapes would grow a few years and then die or languish. Sometimes the cause of failure was obvious, as in the case of winter kill of citrus in Georgia. The late spring freezes that regularly destroyed blossoming apricots were more frustrating. Olives grew but did not fruit. The wine grape *(Vitis vinifera)*, most perplexingly, would grow and bear for a few years, but would then produce shriveled fruit on ever-punier growth. Colonizers believed that these problems could be overcome—that these Old World plants could be naturalized in America; but they were uncertain whether the crucial issue was to find the right locale, to introduce particular varieties that would settle in, or to import gardeners with more skill. The variables were sufficiently complex that continuing faith could coexist with an accumulating record of failure.[7]

The alternative path, less lucrative but cheaper and easier, was to emphasize replication of the familiar. Colonizers transplanted people, plants, and animals that already grew together in England. Income would come to promoters not from exports but from increasing population and appreciating land values. This approach predominated in New England and, later, south to Pennsylvania. A few species—most notably apples—naturalized easily and permanently; by the late 1700s, Americans could assert that "native" varieties like the Newtown pippin were better than those of England.[8] Some English pasture grasses grew, but in the New England climate, they would thrive only in spring and fall; with little greening in summer and none in winter, maintaining livestock was more difficult than back home. The least expected and most problematic trajectory was that of wheat. Yields of spring wheat in New England initially were very good, and this English grain became a staple crop. After 1660, however, the spread of "blast," or stem rust, during the hot, humid summers forced Yankee farmers to abandon wheat. Rye and barley were less-than-satisfactory substitutes.[9]

The remaining option was to use plants already growing in the area being settled. Colonials were unhappy with how few plants had been cultured successfully by their Indian predecessors. Crab apples, blackberries, and the common fox grape *(Vitis labrusca)* were inedible; strawberries bore poorly. Huckleberries grew widely, but the fruits were minuscule. The Indians, having no livestock, had done nothing to "improve" pasture grasses. Pumpkins were worthwhile but of limited use; beans were similar to those in Europe. The most prominent American Indian crop was maize, which, over the previous millennium, had been gradually brought north from the tropics and bred to mature during the short and intense American summers. Lacking gluten, however, it could not be made to rise; the gritty meal did not form the basis for a proper European daily bread.

Both geographic theorizers and colonial promoters advanced the argument that while the seaboard climate was problematic, settlement could improve it. Clearing forests and draining swamps led locally to drier land, purer air, fewer pests, and less disease. Some believed that in the long run, these improvements would moderate the climate—there would be warmer winters, cooler summers, and fewer storms. Such changes required healthy and diligent laborers. Colonization was thus a high-stakes race between culture and degeneracy: would colonists improve their environments enough to make family life viable and continued immigration attractive before they became too stupid and slovenly to care?

Assessments about the trajectory of a colony could vary widely. But in the decades around 1700, cultures in both New England and the

Chesapeake were widely considered to be in decline. Farmlands in longer-settled areas were less productive. Weeds, blights, and insects were increasing in variety and number. Horses were smaller and cattle were rangier. Farmers were less educated and less ambitious. Learning was less visible and less honored. Not everyone would have seen the situation from the religious perspective advanced in 1689 by Puritan minister, man of learning, and later witch hunter Cotton Mather, but they would have appreciated his assertion that New England was lapsing into "Criolian degeneracy."[10] A generation later, the Quaker merchant, man of learning, and suburban London gardener Peter Collinson expressed the common view that the problem was inherent to the continent. Discouraged by his and others' inability to grow and improve American plants, he concluded sweepingly that "great promise but slender performance" were qualities that were "hereditary to America."[11]

John Mitchell's Geographic Realism

From a biogeographic perspective, a significant transition occurred on the seaboard during the second third of the 1700s as a small group of organisms, descended from Old World invaders, began to alter the composition of species around them. Stated more concretely, a few colonials became naturalists.[12] A small network, centered in Philadelphia but reaching south to Virginia and north to New England, began to investigate plants and other natural productions, not just locally but throughout the territory east of the Mississippi. The naturalists shared knowledge and specimens among themselves and also with English correspondents. These individuals included Philadelphia nurseryman John Bartram, the polymathic Benjamin Franklin, New York royal official Cadwallader Colden, Connecticut minister and agricultural writer Jared Elliot, and, in Virginia, lawyer John Clayton and physician John Mitchell. The intermittent collecting and occasional writing these naturalists did, beginning in the 1730s, were on one level the ordinary activities of scientific amateurs. Yet as the naturalists sought out and compared plants from New England to Georgia and westward to the Mississippi, they gathered the materials and created an experience that would enable them to conceive of the territories they were striving to occupy as a geographically coherent place with particular boundaries, inhabitants, and cultural potentials.

John Mitchell is not as well known as John Bartram, whose house and garden, correspondence, and literary offspring all survived him.[13]

But Mitchell was a more significant historical actor. Operating at the intersection of science, empire, and national identity, he had a comprehensive understanding of eastern North America, its political potential, and its geographic peculiarities. While his vision of a British American future disappeared in the turmoil of the decade following his death in 1768, his realistic analysis of North American climate and culture formed a baseline for those who followed. His life and writings, therefore, deserve detailed analysis.

Mitchell was born into an upwardly mobile Virginia family in 1711. His mother died when he was an infant; his father, who remarried and produced a second (and later a third) family, sent him away to Scotland when he was about eleven. In Edinburgh, Mitchell studied under major figures of the early Enlightenment including philosopher Colin Drummond (who also taught David Hume), anatomist Alexander Monro, and botanist Charles Alston. Mitchell returned to Virginia in 1731 to establish a medical practice on the tidal Rapahannock River. Working in the spirit of his mentors' ostentatiously clearheaded and empirical *Medical Essays and Observations,* he seized the opportunities resulting from his unique geographic situation to advance learning. He prepared the first careful study of marsupial reproductive anatomy (a paper on the Virginia possum), and he reported the symptoms and postmortem appearances (with autopsy) of a new epidemic disease, probably yellow fever, which arrived in Virginia in 1737.[14]

Mitchell's enlightened perspective was particularly evident in his prominent study of the causes of differences in human skin color. In 1740 the Royal Academy of Bordeaux, guided by the *philosophe* Montesquieu, had offered a prize for the best account of "the physical cause of the Blackness of the Skin of Negroes; the Nature, Kind, or Quality of their Hair, with the Cause of their Change or Degeneration." Mitchell's entry (too late for submission to the academy, but sent to Peter Collinson and published in the Royal Society's *Transactions*) was an extended critique of Montesquieu's premises. Reporting on dissections of skin from dead slaves, Mitchell argued that their dark color was not due to the accumulation of a "black Humour" (whose effects influenced both body and mind) but rather was limited to a thin pigmented layer. More broadly he suggested that black skin was not a Hamitic curse but a "blessing" that had developed gradually to protect Africans from heat and sun. Finally, he argued that black skin was not derived from an original white, but rather that both were "degenerate" from an original tawny color. Europeans were in fact more degenerate than Africans due to their "luxurious and effeminate lives."[15]

Mitchell's career ultimately revolved, however, not around humans or opossums, but around plants. Primed by his botanical study with Alston, he collected specimens as he traveled along the Rapahannock to visit patients. By 1738 he had begun to correspond about plants with Peter Collinson, who was by this time the scientific and horticultural middleman between London and the North American colonies, and with John Bartram. He used these contacts in 1746 when, debilitated by malaria and uncomfortable with slavery, he decided to leave Virginia for England. Collinson helped him to become the primary representative of Anglo-American nature in London. He shared information with the aged naturalist Mark Catesby, and he advised Peter Kalm, the Swedish botanical traveler on his way to Pennsylvania, about things and people to seek out. Aristocratic plant enthusiasts—most notably the influential Duke of Argyll and his nephew, the Earl of Bute—relied on Mitchell to help them naturalize American plants on their suburban London estates, and to extract botanical novelties from Bartram back in Pennsylvania. Mitchell arranged for Cadwallader Colden to ship Argyll a live bison from New York.[16]

This status enabled Mitchell in the 1750s to expand his influence from the botanical to the geopolitical. For the first half of the decade he worked under patronage from the royal government's Board of Trade and Plantations and prepared what became the standard map of the British and French colonies in North America.[17] His 1757 book, *The Contest in America between Great Britain and France*, argued that the French should be driven from the continent. It also emphasized that the imperial policy limiting colonial manufacturing frustrated Americans' desires for goods. "Rupture" between England and the colonies could be prevented by encouraging American production of agricultural and mineral products that would give them the money to buy manufactures. Mitchell articulated old visions with a long and miscellaneous list of products (hemp, flax, silk, wine, oil, raisins, currants, almonds, indigo, madder, saltpeter, potash, iron, pitch, tar, and timber) whose production and export he hoped could be fostered.[18]

In the late 1750s Mitchell gained a unique position from which he could see plants and places from a global imperial perspective. His patron, the Earl of Bute, was "finishing tutor" of the Prince of Wales (soon to be George III) and confidante (and rumored lover) of the prince's mother. More particularly, Bute was the guiding force in transforming the grounds of the royal palace at Kew into a botanic garden. Between 1758 and 1761 Mitchell was Bute's personal assistant in this work. Their goal was to gather a permanent and expandable collection of as many of the world's useful and ornamental plants as possible. Such a display would provide imperial leaders with reliable and synoptic information about plants that might, through culture or transplantation, contribute to the welfare of the empire.[19]

Mitchell's connections and his perspective became particularly relevant in the aftermath of the British victory over France in the Seven Years' War. Bute negotiated the Treaty of Paris, which transformed British America from a collection of isolated western Atlantic enclaves, some of which occupied the strip of territory extending from Georgia to New Hampshire, into a single landmass that stretched from the Arctic Circle to the Tropic of Cancer (the land depicted on Mitchell's map), with outlying islands and entrepots in the Caribbean. He also developed the initial plans to help restore the government's depleted finances by generating revenue from the colonials who had benefited so greatly from victory. The intense controversies generated by the Stamp Act, the major product of this policy, prompted Mitchell in 1767 to offer an "impartial" assessment of the American situation in *The Present State of Great Britain and North America.*

Mitchell accepted the American contention that colonials could not pay the Stamp Tax because they had no cash. The underlying issue, however, was to understand why money was so scarce: why three million Americans produced so little from their huge and diverse territory that could be sold profitably in Britain. Mitchell found the answer in "the singular and peculiar soil and climate of *North America.*"[20] Weather made farming much more difficult in North America than in western Europe. Winter temperatures in New York City could go twenty degrees lower than in London. Summer heat and drought frequently killed crops. The greatest problem, however, was the weather's changeability. Mitchell described "violent north-west winds, blowing from the frozen regions of Hudson's Bay, which rage with such fury all over that continent, that they bring the climate of Hudson's Bay even to Virginia and Carolina by one blast." These late spring and early autumn freezes made the growing season in the northern colonies too brief and uncertain for common European cultured plants. Mitchell rejected the idea that the seaboard climate had been moderating during the previous century; his picture of the place he had left behind was resolutely unsentimental.[21]

The main problems in the middle and southern colonies involved land. From New Jersey to Virginia the sandy soils had been exhausted by the weedy crops—tobacco, flax, and hemp—grown during the last century. In theory, manuring could restore fertility, but forage plants did not grow rapidly and evenly enough throughout the year to support an adequate number of animals to produce sufficient dung to sustain enough production of these crops to be worthwhile. The Carolinas and Georgia were composed of either desolate pine barrens where grasses shriveled from drought or rust, or marshlands that would grow marketable rice but were so filled with fevers that they killed the people who worked on them.[22]

Mitchell was particularly antagonistic to British proposals that colonials seeking land should move to the newly acquired and even more climatically extreme territories of Canada or Florida. The former, apart from the narrow St. Lawrence Valley, consisted of "frozen lakes, drowned morasses, and sandy plains, fit only for the habitation of Beavers; or bare rocks and mountains covered with snow throughout the whole year," where the people subsisted largely on frozen eels. The latter was an unhealthy combination of "scorching sands and swamps" that had resisted settlement for two centuries. Promoters of St. Augustine as a place to grow sugarcane were unrealistic: while summers were hot, winters were punctuated by the northwest-wind freezes that killed this and other tropical crops. Canada and Florida could support Indians, but could not be parts of a truly British North America.[23]

What was the solution? In 1757 Mitchell had presented a laundry list of products that Americans might profitably grow, harvest, or mine for export. Now, with experience developing the Royal Botanical Garden, he offered a more sophisticated analysis. He argued that Anglo-American tensions would dissipate if colonials could occupy the trans-Appalachian region. The new soils would immediately support production of flax and hemp that could be exchanged in Liverpool for cash; then, after the initial fertility was gone, farmers could "for ever" produce cotton, silk, wine, and oil for sale in London.[24] Managing the transition from settlement to stability would, however, require a difficult mix of initiative and patience. Settlers would need specially adapted varieties. But while plant characteristics (like the human pigmentations Mitchell had discussed earlier) changed over time in response to climate, they did not necessarily respond with the speed that planters desired. Attempts to grow wine grapes in North America, for example, had been unsuccessful for more than a century. The alternative, Mitchell suggested, was for colonials to utilize the "natural productions of the soil and climate"—local plants that were tough enough to stand the weather. They should work to improve the grapevines, mulberry trees, and silk-producing caterpillars that already lived there. But American improvements in culture were stymied by inadequacies in American culture: "mulberry orchards and vineyards require time to be brought to perfection, which the indigent circumstances of Planters will hardly admit of." Twenty years distant from American residency, Mitchell did not know whether any Americans were financially or intellectually capable of this kind of public service.[25]

There was an alternative to the two unsatisfactory options of importing European plants or improving local types. In a long footnote Mitchell indicated that the climate of eastern North America was similar to that of "East Tartary, China, Corea, and Japan," and that "most of the staple commodities of America came from the East, as Sugar, Rice, Cotton, Coffee, Indigo, &c." He therefore suggested that some of the other "rich and valuable" plants of the Far East might flourish in Carolina or New York. Moreover, since these crops were "so totally different from any thing that Britain produces, they might for ever keep the colonies from interfering with their Mother Country, and preserve a lasting connection and correspondence between them." Mitchell's emphasis on the climatic similarity of eastern North America and eastern Asia raised questions, however, about his basic argument that North America was "singular and peculiar." Civilization had flourished for thousands of years in China. Why then was American development so slow and difficult? Mitchell extricated himself from his dilemmas by noting that "these things would require a more particular consideration."[26]

John Mitchell was a unique American-English naturalist, but his ideas were not peculiar to himself. A young coterie in Philadelphia proposed creation of the American Society for the Promotion of Useful Knowledge on the same platform in 1768.[27] Their manifesto, written largely by Charles Thomson, emphasized that European plants did not thrive in America because "the Soil and Climate of this Country" were different. But Philadelphia was at the same latitude as Beijing and climatically "this Country . . . very nearly resembles China." Introducing the "produce" of China would thus greatly improve America. Thomson's more subtle hope was that, because many useful plants—notably mulberry, persimmon, ginseng, and tobacco—grew spontaneously in both

20 · Fruits and Plains

east Asia and eastern North America but not in Europe, there might be undiscovered "native" American analogues of Asian plants such as tea. Finding such species, identifying the virtues of American medicinals such as sassafras and sumac, and culturing native silk-producing worms and grapes would result in products that would be profitable and "render us more useful to our Mother Country." How could this be done? The Philadelphians called on farmers to report on their work, but appealed particularly to the "many Gentlemen in different Parts of the Country, whom Providence hath blessed with Affluence, and whose Understanding is improved by a liberal Education," to institute inquiries and experiments.[28]

Guillame Raynal and American Mediocrity

In the early 1770s, Guillame T. F. Raynal's *Philosophical and Political History of the Settlements and Trade of the Europeans in the East and West Indies* moved discussions of continents, climates, and cultures from provincial practice to the realms of global geography and international polemics. Americans in both the eighteenth and twentieth centuries dismissed Raynal as an epigone of the great naturalist Buffon and as one of several polemicists engaged in "the dispute of the New World." Recent scholarly work conveys how limited this assessment was. Raynal's history was a collaborative project involving some of the most important *philosophes* (notably Denis Diderot). Building on longstanding intellectual traditions regarding place, climate, soil, and migration, it broke new ground in linking the natural and human history of European expansion—in advancing what is now called environmental history. Raynal's view that the new United States of America would experience a future of mediocrity was both intellectually well-grounded and empirically plausible.[29]

While Raynal's history included text supplied by different authors with distinct perspectives, its major theme was clear: the superiority of commerce over conquest in the history of European expansion around the globe during the preceding three centuries. Commercial intercourse between Europeans and others was modest, friendly, mutually profitable, and enlightening; it resulted in changes that were gradual and progressive. Territorial empires, by contrast, disrupted the lives of the conquered and resulted in the degeneration of the colonizers who tried to settle areas that were climatically alien; those baneful effects reverberated in the imperial states themselves.

Raynal explained that imperial activity had been most intense, and its problems were most evident, in America. The colonization of the New World had been, at its worst, a pageant of brutality; at best, colonists faced an uphill race to drain swamps, straighten rivers, thin forests, and manage indigenous peoples before harsh conditions, disease, mind-numbing labor, and the cruelty of slavery pushed them to barbarism or extinction. Degeneration through colonization was most evident in Spanish America. In sexually tinged language, Raynal explained that Hispanic Creoles had lost their progenitors' "firmness" due to both "the heat of the climate" and the "indolence" that resulted from their exclusion by European officials from positions of responsibility. The "barbarous luxury, shameful pleasures, and romantic intrigues" that occupied them had, over generations, "enervated all the vigour of their minds." Horses, olives, and grapevines had also degenerated in Mexico. Moreover, the deleterious consequences of conquest had spread back to Spain: supported by New World resources, the governing elites were becoming "a degenerate race" due to their disdain for work, desire for courtesans, and preference for status over industry.[30]

Degeneration also explained many of the French failures in the New World. In Canada the long winters led to both habits of idleness and too much interest in feasting and fighting. In Louisiana, the royal government had allowed settlers to disperse throughout the territory, resulting in isolation, a slide toward subsistence farming, and too-close contact with barbarous Indians. The result was that the highlands along the Mississippi, which climatically could have supported dense civilized French settlements producing fruit, tobacco, cotton, corn, livestock, and olives, had become a "great desert," stupidly given away to the Spanish.[31]

The English, Raynal understood, were at least trying to grow civilization in North America. Yet even the best colonizers faced a fundamental problem. Raynal explained that there was a "law of climates, which wills every people, every animal and vegetable species, to grow and flourish in its native soil. The love of their own country seems an ordinance of nature prescribed to all beings, like the desire of preserving their existence." North America was an alien place for Europeans. The basic problems were impure air, dank forests, and swamps. The swamps bred insects that "destroyed without opposition all the productions of nature." Both animals and humans were "attacked by epidemical disorders." Imported animals reproduced with difficulty: "each generation fell short of the last; and as it happens to American plants in Europe, European cattle continually degenerated in America."[32]

Raynal noted approvingly that through clearing and drainage, New Englanders had made their land healthy enough for humans, root crops, and apples. But European plants sprouted late. While they grew rapidly during the short, hot summers, they did not mature properly; the early freezes prevented proper ripening. It would take time for "the force of culture . . . to subdue this habit fixed and confirmed by ages," and for "the influence of art" to overcome "the dispositions of nature."[33]

Raynal's crucial example was the wine grape. "Repeated experiments" all produced juice that was "too watery, too weak, and almost impossible to be preserved in a hot climate." The causes were readily identifiable: "the country was too full of woods, which attract and confine the moist and hot vapours; the seasons were too unsettled, and the insects too numerous near the forests." Raynal acknowledged that efforts might pay off sometime in the future, but as a Frenchman, he pleaded that the Americans should concentrate on other less-competitive products such as silk, olives, or cotton.[34]

Anglo-American colonials were like their plants. They were strong and well made; and, as Benjamin Franklin had emphasized, in some places their numbers were increasing rapidly. Yet they harbored peculiar weaknesses. Raynal noted that the one direct comparison between colonials and Europeans—participation of young men in the empire's military campaigns of the preceding quarter century—had shown that the "creoles" were not as tough as their English counterparts. Colonial soldiers were notorious for their susceptibility to epidemic diseases ranging from smallpox to scurvy. Raynal thus concluded that Anglo-American people, like Anglo-American vegetables, grew quickly and robustly, but did not mature properly. Women in Pennsylvania were "of an agreeable figure" and they produced children at an early age, but they also stopped breeding early and their teeth decayed when they were quite young. Americans lacked mental maturity: "in that foreign clime the mind is enervated as well as the body: endowed with a quickness and early penetration, it hath a ready conception, but wants steadiness, and is not used to continued thought." The consequence was that the colonies had never produced a mathematician, a poet, or a man of genius.[35]

Raynal concluded by pronouncing the outcome of the confrontation between American nature and English culture still uncertain. Perhaps Anglo-Americans were, like other European organisms, "degenerated by transplanting, by growth, and by mixture," punished by Nature for having crossed the ocean. Nevertheless, education might "correct the insurmountable tendency of the climate toward the enervating pleasures of luxury and sensuality." He suggested, in an ambiguous purple passage, that the British could create in America "a new Greece" with "another Homer, a Theocritus, and especially an Anacreon" (a poet who valued drinking and sex over idealism and ambition). But such a future could only happen if the British would "clear the ground, purify the air, alter the climate, improve nature," and "form a set of people fit for the creation of a new world. This is what they have not done."[36]

In 1780 Raynal brought his history up to date with a supplement, *The Revolution of America*. French support for the revolutionaries did not change his skepticism about the Anglo-Americans' prospects. He reminded readers of his contention that decades of efforts to produce linen, wine, or silk profitably in the colonies had failed in the face of natural realities. "The poverty of the soil, which would not bear flax, obstructed the first of these views; the badness of the climate, which would not agree with vines, opposed the success of the second; and the want of hands permitted not the third to take place." He also noted that in recent decades, wheat and tobacco yields had declined by two-thirds due to soil deterioration.[37] He estimated that the maximum supportable population of the United States would be about ten million, and that, with few exports, the people would have little surplus income. The new nation would "be able to suffice for itself, provided that the inhabitants know how to make themselves happy by economy and with mediocrity." While contacts with Franklin and the defeat of the English led Raynal to acknowledge in a 1783 edition of his history that Americans could attain intellectual excellence, and to suggest that their classical model could be the grand tragedian Sophocles (not the earlier edition's primitive hedonist Anacreon), his main argument about the country's future did not change.[38]

Thomas Jefferson, Cultivator

Benjamin Franklin legendarily countered Raynal's views on American degeneration at the level of salon reparteé when he proposed, at a dinner attended by a number of tall Americans and some short Frenchmen (including the "mere shrimp" Raynal), that they stand toe-to-toe to see if Americans had in fact degenerated. (Raynal graciously demurred by admitting exceptions to his thesis.) Thomas Paine pamphleteered against Raynal's history of the Revolution.[39] Thomas Jefferson, however, wanted to do more. He wrote *Notes on the State of Virginia* as a vindication of the United States from the charge of decline.

It was a hard argument to make. I convey his difficulties by juxtaposing his gardening and breeding against his efforts to articulate the natural virtues of greater Virginia and the prospects of the United States. Jefferson's gardening interests and failures provide particular help in illuminating some of his more enigmatic emphases and omissions. They also, as a bonus, offer a new perspective on his concerns about race and on his relations with Sally Hemings, who was both the half sister and the slave of his wife, Martha.

Thomas Jefferson had an early and intense interest in fruit culture. In 1767, when he was twenty-four years old and only beginning to plan the western estate he called Monticello, he recorded a patriotically meaningful if biologically wrongheaded experiment to improve the taste of native cherries by grafting wild American buds onto cultured Old World rootstocks. Two years later he planted an orchard that included not only the usual apples, peaches, and cherries, but also pears, figs, pomegranates, and stock for grafted apricots, almonds, and nectarines.[40]

Jefferson's vision expanded in the first half of the 1770s. The land and slaves that Martha Jefferson inherited upon her father's unexpected death in 1773 suddenly made Thomas one of the gentlemen "blessed with Affluence" whom Charles Thomson and the American Society for the Promotion of Useful Knowledge had urged to experiment with plants. Contact with Filippo Mazzei, a charismatic Italian wine merchant and promoter, led Jefferson to chart a definite plan of action. Imagining that he could truly emulate the culture of the northern Italian *Piemonte* (or Piedmont) in the Virginia foothills, Jefferson provided Mazzei with property a mile southeast of Monticello. With a deferential bow, Mazzei named the property *Colle,* or Hill, to his patron's "little mountain." He supplied Jefferson with fancy Italian vegetable seeds, and he imported Italian vinedressers to plant grapes, as well as olives and oranges. In the spring of 1775 Jefferson could sit on the terrace of his partially finished home with his beautiful wife and two infant daughters, look down on Mazzei's men and his own slaves toiling in the new vineyards of Monticello and Colle, and imagine that the Virginia Piedmont would soon be dotted with gentlemen's estates producing wines, oils, and figs for local tables and for export (see Figure 1.1).[41]

Issues of climate were crucial to Jefferson's plans. He hoped that his estate's topographically unusual location—on a freestanding hill, twenty

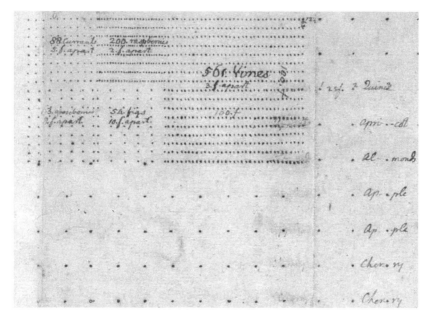

Figure 1.1 Thomas Jefferson's orchard plan for Monticello. More than five hundred grapevines were surrounded by plantings of raspberries, currants, gooseberries, and figs, and then by almonds, apricots, peaches, plums, cherries, apples, pears, and quinces. Coolidge Collection of Thomas Jefferson Manuscripts, Monticello: orchard (plat), recto, [1776–1778]. N 127; K94b. Reproduction courtesy of the Massachusetts Historical Society.

---

miles east of the Blue Ridge—would enable him to produce fruits otherwise unavailable in the region. More broadly, as a man of the Piedmont, he believed his region's climate was healthier and more pleasant than that of the sultry Tidewater region along the Chesapeake tributaries. Cooperating with Rev. James Madison in the later 1770s, Jefferson collected data comparing the climates of Charlottesville and Williamsburg. As Virginia's governor from 1779 to 1781, he used this information to support his plan to move the state capital from Williamsburg to Richmond.[42]

During the same years, however, he came to realize that weather posed major barriers to his hope that the Virginia Piedmont could support a culture and economy like that in northern Italy. Mazzei's workmen and Jefferson's slaves planted vines at Monticello in early April 1774 using the labor-intensive methods of Italian high culture.

They placed dung-dipped cuttings in a four-foot-deep trench filled halfway with old brush and an earth-dung mixture, and then filled the holes with a carefully prepared mixture of leaf mold, soil, and clay. A month later, as the precious cuttings were leafing out, Jefferson looked to the northwest and saw snow on the Blue Ridge; the next day, he wrote that widespread frost had "destroyed almost every thing," including "large saplings" and "all the shoots of vines." The following year in early March, he was planning to replant and was enthusiastic that the peach trees were already in bloom following "the most favorable winter ever known in the memory of man." Later in the month, however, he recorded "very cold weather & frosts every night for a week," which "totally killed" peaches, apples, and cherries. Mitchell would not have been surprised at these actions of the northwest wind.[43]

Mazzei left Colle in 1776 to participate in the Revolution; his workers soon dispersed, and their plantings rapidly declined. Two years later, Jefferson visited the site and could find only one olive and four orange rootstocks "out of some hundreds, the rest being killed totally." The situation with grapes was slightly less clear. In the early 1780s, Jefferson reported success with American scuppernongs, but could produce from them only vinegar. He made no further comments about the fancy varieties that he and Mazzei had planted before the war. Presumably, like all the early experiments reviewed by wine historian Thomas Pinney, they suffered the fate of other *vinifera* imports: those that did not freeze flourished for a year or two and then succumbed to American fungi and bacteria.[44]

Jefferson's marriage, like his fruit growing, was an experience with trauma and decline. In 1772 the widow Martha Wayles Skelton Jefferson was sophisticated, beautiful, and fertile, having produced her first baby four years earlier at age nineteen. During the ten years of her life with Jefferson, she bore six more children, four of whom died young. She herself became increasingly fragile from her pregnancies and possibly from diabetes. In May 1782, giving birth a seventh time, she suffered irreparable internal damage and died four months later at the age of thirty-three. Her life exemplified Raynal's stereotype of the early ripening and early fading Anglo-American woman.[45]

Notes on the State of Virginia originated in late 1780 with a series of queries posed by Francois Marbois, the French diplomatic secretary in Philadelphia, to leaders of the different American states. Marbois was seeking information that could enable the French to intervene more effectively in the war, and so his emphases were on historical, governmental, and military conditions (with afterthoughts about mining and archaeology). His secondary aim was to improve his contacts with the American national elite; his success became evident a month later, when he was named one of sixteen councillors of the American Philosophical Society (along with Jefferson and Confederation Congress Secretary Charles Thomson, who had written the American Society's 1768 manifesto).[46]

Jefferson was the only American leader who took Marbois's queries seriously. He researched and wrote responses intermittently during the chaotic last months of his governorship and afterward, and sent off his report in December 1781. Marbois received it four months later, thanked Jefferson profusely, and put the manuscript aside.[47] Jefferson, meanwhile, had announced his retirement from politics, in the wake of attacks on his performance as governor, in favor of a life that would be purely domestic and "literary." In addition to planning new flower gardens and orchards, he decided to develop his responses to Marbois's historical and military inquiry into an account of greater Virginia's natural advantages and prospects for the future. In December 1781, he asked Virginia's western military agent, George Rogers Clark, to send fossil teeth and bones from Kentucky, and he hinted to his "antient friend" Thomson that he would be happy if the American Philosophical Society were to publish his manuscript.[48]

During the relatively calm winter and spring of 1782 Jefferson collected information and marshaled his ideas. The bulk of the text was written, however, as his wife lay slowly dying during the summer.[49] The book became an effort to resolve the challenges that climate, culture, and degeneracy posed for Virginians and Americans more generally. The problem, as Jefferson presented it, was to deal with the widespread critiques of the livability of North America; the solution, as I indicated at the beginning of this chapter, was to set up the large target of Buffon's new theory, and then, by knocking it down, to undermine the broader range of doubts about American climate and culture identified with Raynal and sketched, closer to home, by Mitchell. Jefferson succeeded in countering Buffon, but was stymied in his efforts to put the larger doubts about Virginia's climate and its prospects to rest; the concerns raised by Mitchell and Raynal emerged repeatedly in Jefferson's bursts of text.

According to Buffon, New World mammals were generally smaller than comparable types in the Old World (pairings included llamas and camels, cougars and lions, and, most notably, tapirs and elephants). This fact was explained through the supposition that the Americas had risen from the ocean more recently than the Old World, and that Old World animals (including humans) had migrated there and had then degenerated in an environment that was colder, damper, and more primitive than the one to which they were accustomed.[50]

Jefferson mounted a broadside rebuttal. He collected data about the weights of American animals, highlighted the spectacular example of the moose, and speculated that claws sent to him by Clark from Kentucky were evidence for the existence of a carnivorous American "mammoth" six times as large as an elephant. He argued that the decline in size of domesticated animals in America had resulted from neglect, not climate, and he denied that Indians were physically weak or "deficient in ardor." As I noted earlier, Raynal's "wretched philosophy," reduced to an insult to Washington, was raised and dismissed within this context.[51] Jefferson's broader criticisms were that Buffon's comparison between the New World fauna and that of the entire Old World was inapt (Jefferson's tables compared the animals of all the Americas with those of Europe only), that his geological history was speculative, and that his characterization of climate in terms of heat and moisture was incoherent. Most sweepingly, Jefferson questioned whether the climate of eastern North America differed significantly from that of Europe: he complained that Buffon seemed to be writing about the earth "as if both sides were not warmed by the same genial sun."[52]

With obstacles cleared away, Jefferson took up the task of showing that his state was congenial to the culture of both plants and people. He began with a discussion of climate. Presenting five years of temperature data from Williamsburg, reduced to monthly averages, he argued that the weather there was neither that hot in summer nor that cold in winter. In addition, he reported that Virginia had both more rainfall and more sun than western Europe. The state's moderate climate was comfortable for people and would support figs, pomegranates, artichokes, and European walnuts.[53]

Jefferson was unable to carry this promotional theme far, however, because he became caught in efforts to deal with the issue of the weather's changeability. He acknowledged that severe winter freezes killed figs and other fruit trees, and that the extremes of winter cold and summer heat were "very distressing to us." In promoting his Piedmontian perspective that Charlottesville was superior to Williamsburg in having cooler and less-muggy air, he acknowledged that agricultural limits were set by freezing northwest winds.⁵⁴ Jefferson frankly admitted that a further problem was whether the consequences of improvement were, in fact, positive. He reported that the climate of Virginia had been moderating over the preceding century; like Raynal and others, he attributed this change to land clearance and swamp drainage. People enjoyed milder winters and less-sultry summers; there was, however, an "unfortunate fluctuation between heat and cold, in the spring of the year." The results were dead fruit trees and difficulties in planting.⁵⁵

Had Jefferson succeeded with grapes, the ambiguities of Virginia weather would not have mattered. *Vinifera* was the indicator species of culture; a climate that fostered fine grapes was *prima facie* a good one. But because he could not point with pride to the vintage flowing from Colle or Monticello, Jefferson passed over the subject of grapes completely in *Notes*. More tellingly, Mazzei wrote an entire volume attacking Raynal in 1788, but made no mention of his and Jefferson's hopes for Piedmont viticulture.⁵⁶ Their results had been what Raynal predicted; within the intellectual framework shared by all three writers, the most reasonable explanation was that the American environment caused this highly cultured plant to degenerate.

When Jefferson turned his discussion to humans, he struggled both implicitly and explicitly with degeneracy. As with grapes, he passed over the issues of fevers, disease among soldiers, and the early aging of women that Raynal had raised. Instead, he fell back on Benjamin Franklin's argument that the healthiness of America had been demonstrated by the dramatic increase in population over the previous 175 years. The current population of Virginia, which Jefferson estimated as greater than 560,000, was derived essentially from the few thousand colonizers who had arrived in the early 1600s.⁵⁷ The population question in Virginia differed significantly from that in Franklin's Pennsylvania, however. Jefferson worked hard to reach the conclusion that Virginia contained more free inhabitants than it did black slaves. Yet he suspected that "under the mild treatment our slaves experience, and their wholesome, though coarse, food, this blot in our country increases as fast, or faster, than the whites."⁵⁸

What did this reproductive competition between races imply for the quality and future of the state's population? Jefferson returned to this question in two places. He explicitly rejected the argument, advanced prominently by Mitchell, that the localization of the black color of Africans in a superficial skin layer showed that the variations within a fundamentally unitary human species were minor. Instead he suggested that blacks were different from and "inferior to the whites in the endowments both of body and mind," and that these inferiorities were "fixed in nature."[59] Jefferson was deeply antagonistic to slavery at this time. Echoing Raynal's prediction that the future would see a black Spartacus, he acknowledged that "a revolution of the wheel of fortune" could occur and that God would be on the side of the slaves. Freedom for blacks, however, worried him even more. On the one hand, racial antagonism would lead to convulsions and extermination; on the other, "mixture" after emancipation would "stain the blood of the master." Jefferson's solution was to expel the emancipated slaves from the United States. They would be replaced, he hoped, by an equivalent number of white laborers from other parts of the world.[60]

Could a postcolonial Virginia not built on the labor of black slaves rise above the mediocrity anticipated by Raynal? Jefferson turned to this question in his final chapters. He declined to imagine a future of cities and manufacturing; they produced dependency and "corruption of morals" that he characterized explicitly as a form of degeneracy. Virginia's pre-Revolutionary mainstay, tobacco, was not much better: this "culture productive of infinite wretchedness" impoverished the soil and demanded "a continued state of exertion" from its growers. Fortunately tobacco culture was in decline because settlement had moderated the "extraordinary degree of heat" that this tropical plant required. Jefferson believed that Virginia's future lay in wheat. He rhapsodized about a plant that, "besides cloathing the earth with herbage, and preserving its fertility . . . feeds the labourers plentifully, requires from them only a moderate toil, except in the season of harvest, raises great numbers of animals for food and service, and diffuses plenty and happiness among the whole." Jefferson's society of independent export-generating Anglo-American yeomen would arise from this soft white English mainstay.[61]

Thinking about wheat brought Jefferson back to his central desire to show that Virginia fostered excellence rather than decline. Wheat fields could double as pasture and could thus, he hoped, make "the Arabian horse an article of very considerable profit." These natural companions of gentlemen did not thrive everywhere. Jefferson echoed Raynal in noting that "animals transplanted into unfriendly climates, either change their nature" or "multiply poorly and become extinct." Virginia, in contrast to states either to the south or the north, had the "particular climate of America" where this naturally aristocratic horse could be raised "without degeneracy."[62]

After publication of *Notes on the State of Virginia* in 1785, Jefferson continued to imagine how climate, degeneracy, and culture were related. Sending a copy of his new book to his close wartime friend, the Marquis de Chastellux, he retreated from his printed views on slavery and reopened the issue of the "degeneracy of animals in America" as applied to humans. Beginning with his and the marquis' presumed agreement that "lower class" white Americans and Europeans were equally "informed" and equally susceptible to improvement through education, he emphasized that ten generations of slavery had dramatically transformed South American Indians from their "original character," which had been "on a level with Whites in the same uncultivated state." Conversely, while the "the blackman, in his present state," was not equal in "body and mind" to the "whiteman," Jefferson considered it "hazardous to affirm that, equally cultivated for a few generations," blacks would not rise to the level of whites.[63]

Jefferson's crucial term here was "cultivated." This variant on culture was not an empty metaphor, but rather a definite element within his thinking about the potentials of domesticated plants and animals, and of people. In early 1789, Jefferson gave this vision definite shape. Responding to a leading inquiry from Edward Bancroft, a Massachusetts-born London physician and fellow of the Royal Society, regarding the inability of freed black slaves to manage their affairs, Jefferson explained that on his return from France to Virginia he would initiate an "experiment":

> I shall endeavor to import as many Germans as I have grown slaves. I will settle them and my slaves, on farms of 50 acres each, intermingled, and place all on the footing of the Metayers [Medietarii] of Europe. Their children shall be brought up, as others are, in habits of property and foresight, and I have no doubt but that they will be good citizens.[64]

Jefferson knew Bancroft as a freethinking naturalist who had analyzed race relations in Guiana with scientific dispassion (Bancroft's role as a British double agent in the Revolution came to light only decades later). Both men understood that the key provision of Jefferson's proposal was placing slaves close to whites (Germans) who did not have the Virginians' ingrained racial antipathies. Within the simple sharecropping (metayer) system widespread in southern France and Italy (midway between Germany and Africa), the whites would cultivate the blacks; Jefferson, in his ambiguous reference to "their children," anticipated that intermingling would result in what he elsewhere called "mixture." The result of this intimate commerce would be good citizens.

This perspective was reinforced by Jefferson's close contact in the late 1780s with James and Sally Hemings. Serving as the United States' ambassador to France, Jefferson transplanted these slaves and half-siblings of his wife, Martha, to the exotic environment of Paris. He cultured them with lessons in French language and French cooking, and provided Sally with French clothes. If Jefferson impregnated Sally Hemings in 1789, as her son Madison claimed, this action would have been consistent with the experiment outlined a few months earlier in his letter to Bancroft. By the end of that year Jefferson was back home in Virginia and was reminded that the American environment was hostile to his foreign cultural schemes. Still, he saw in Sally Hemings a relative of his wife who was strong, available, and uncomplaining. In contrast to the pure-blooded but weak Martha Jefferson, Hemings completed at least six pregnancies without major problems and lived into her sixties. Two of her children moved to Washington, D.C., passed as white, and became good citizens.[65]

Situating Jefferson's attitudes about race mixture within the context of his experience with plants clarifies some of the more enigmatic activities of this founding father. But the deeper importance of Jefferson's garden comes from the extent to which it embodied his desires and fears about the place of the new nation in the New World. Mitchell, Raynal, and Jefferson together illuminate the degree to which American independence was—and was understood at the time to be—a biohistorical event. The issues included the real problems that the climate of North America posed for Euro-Americans, the degree to which the United States was a unique climatic region, and the extent to which cultured Old World plants could be naturalized in the United States.

The fundamental question, expressed most clearly by Raynal, was whether Americans faced a future of biotic mediocrity. Would their lives consist of cornmeal, ham, whiskey, and wool, or of white bread, wine, roast beef, linen, silk, and Arab-American riding horses? Wheat, Jefferson's great white hope, was undergoing a crisis due to foreign invasion at the moment *Notes on the State of Virginia* was being published. Wine grapes and other fruits engaged the attention of gentlemen blessed with affluence during the flowering of New England in the second third of the nineteenth century. I consider these issues in the next two chapters. Textiles make their appearance in Chapter 5.

CHAPTER TWO

The United States' First Invasive Species
The Hessian Fly as a National and International Issue

George Morgan anticipated life as a rural republican gentleman. As a member of a prominent Philadelphia family (his brother John was a signer of the Declaration of Independence and a founder of the University of Pennsylvania Medical School) he had worked before the Revolution as a mercantile agent for the British army on the far western frontier and at the same time participated in the new American Philosophical Society. He had served the cause of independence as an army supply officer in western Pennsylvania and as an emissary to the Ohio Indian tribes. Colonel Morgan, like others in the Revolutionary elite, anticipated that his postcolonial prosperity would ultimately be secured through claims to western lands. But his immediate interest as the war wound down was in exchanging his sword for a plowshare. In 1779 he purchased a 300-acre farm on the outskirts of Princeton, New Jersey. With a panoramic view of the Navesink Highlands and the distant Atlantic, the estate received the descriptive if not very imaginative name "Prospect."[1]

Morgan's property was a wreck. In December 1776, occupying imperial troops—primarily the auxiliaries from various German states who were collectively called "Hessians" by the Americans—had appropriated the food and liquor around Princeton and had burned fences to keep warm.[2] Morgan directed local workers and his slaves to rebuild fences, plant nut and fruit trees, and dig new garden plots. In addition, he participated in the Philadelphia Society for Promoting Agriculture, sharing his design for a humane beehive. Morgan's main initiative, however, involved wheat. He anticipated high yields from the fields sloping southeast from his house toward Stony Brook (now Lake Carnegie) because he had built a farmyard designed scientifically to facilitate the collection and distribution of manure. Living only a short distance from the Delaware River, he could ship competitively to the Philadelphia market, where buyers from the Caribbean and Newfoundland obtained much of their grain.[3]

In the first half of the 1780s, Prospect's prospects were good. But in spring 1786, Morgan watched his fields with foreboding. The manured wheat plants came up green and lush, but close examination found "White Worms which after a few days turn of a Chesnut Colour—They are deposited by a Fly between the Leaves & the Stalk of the green Wheat & generally at the lowermost Joint, and are inevitable Death to the Stalks they attack." These fluid-sucking larvae stunted the growth of the plants and damaged the stalks. Wind and rain knocked the plants over and made the sparse crop unharvestable. A cloud of insects was all that came from the fields later in the summer. Morgan wrote in his journal, "we call [this insect] the Hessian Fly."[4]

Whereas Chapter 1 sketched the visions that major writers advanced regarding plants and plant problems in North America between the 1740s and the 1780s, this chapter has a much tighter focus: responses in the second half of the 1780s to the appearance of the Hessian fly in the United States. The story of this wheat pest is worth telling, first of all, because it is new history: I draw on a collection of previously unstudied documents to illuminate the actions of prominent figures such as Washington, Jefferson, Thomas Paine, Samuel L. Mitchill, Joseph Banks, and (in a cameo role) Adam Smith.[5] More substantively, this chapter displays how Americans first perceived certain species as foreign invaders. It also conveys how problems that seemed local, or at most national, were actually global—in the movements of people, plants, and insects, and in responses, which were dependent on asymmetrical distributions of knowledge and of power. It shows how much problems of plant physiology and insect taxonomy could matter in international diplomacy more than two centuries ago. Finally, studying the arrival of the Hessian fly illuminates American anxieties about their future "independence"—political, economic, intellectual, and botanic.

A Useful National Prejudice

In summer 1776, General William Howe and his brother, Admiral Richard Howe, set in motion their plan to reattach the narrow strip of dissident seaboard colonies to the continental empire of British America.

Its central element was the landing of an expeditionary force composed of 33,000 British and Hessian soldiers on Staten Island in New York harbor. By mid-November, William Howe's army had occupied southern New York; in the next month, it moved west and south into New Jersey (see Figure 2.1). Howe anticipated that Jersey subjects, awed by the British military presence, would disavow the tyrannical rebels and loyally provide food and other supplies through the winter to his garrisons distributed from Hackensack to Trenton. In the next spring, the forces of order would spread out across the colonies; the rebels' demoralization and capitulation would be followed by reconciliation.[6]

Howe's hopes were threatened, on the one hand, by the inability of British leaders to induce enough of their Jersey subjects to supply food, fuel, and animal fodder, and, on the other, by the tendency of

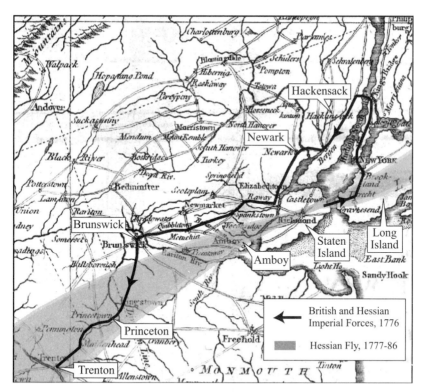

Figure 2.1 George Morgan's perception of recent and current threats to Princeton, New Jersey, 1786. Base map a detail from "A new and accurate map of New Jersey, from the best authorities," London, 1780, Library of Congress; overlay by the Rutgers University Cartography Laboratory.

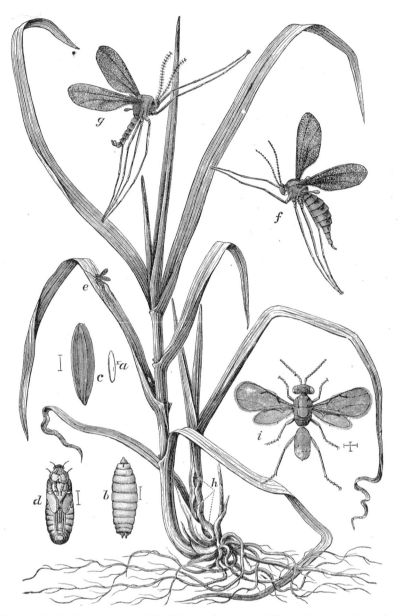

Figure 2.2 No illustrations of the Hessian fly were published in the eighteenth century. This wood engraving of developmental stages was prepared by E. L. Trouvelot (the artist-naturalist who introduced the gypsy moth to America) for A. S. Packard, *Hessian Fly: Its Ravages, Habits, and Means of Preventing Its Increase,* United States Entomological Commission, Bulletin 4, 1880. It shows (a) egg, (b) larva, (c) pupal case or "flaxseed," (d, e) emerging adult, (f) female, (g) male, (h) flaxseed inside damaged wheat stalk, and (i) parasitic wasp. Reproduction courtesy of the Rutgers University Library.

the soldiers, especially the Hessians, to alienate natives by foraging and plundering indiscriminately. The possibility of success ended on January 3, 1777, when George Washington's army, having overwhelmed the Hessian garrison in Trenton eight days earlier in a raid across the Delaware River, reappeared and pushed the imperial forces out of Princeton. For most of the next six years, the British and their allies were confined to an enclave that consisted of Manhattan, western Long Island, and lower Westchester County in New York, and a tongue of land surrounding the Raritan River in New Jersey.[7]

The inability of imperial forces to access seeds, grasses, timber, and livestock resident in North America led them to alter the basic pattern of organisms' movements across the Atlantic for the first time in more than a century. To keep the army's horses and cattle alive, supply officers contravened all the principles of economy and shipped large quantities of grass from many parts of Europe to New York. A cargo probably harvested around the Mediterranean carried the hardy puparial stages of the gall midge now called *Mayetiola destructor* (see Figure 2.2). This species had coevolved with wheat but had never become common enough in western Europe to be noticed by entomologists; on Staten Island and western Long Island, by contrast, the insects found fields filled with palatable wheat varieties and few, if any, of their predators. Over the next eight years, they spread up the Hudson and into Connecticut and (fighting prevailing winds) New Jersey.[8]

For some years, farmers discussed the new insect locally, and newspapers published a few brief notices.[9] It became a public issue only in 1786. In that year the Revolutionary elite began to shift their attention from private and local ventures—with only vague expressions of anxiety about the future of the not-very-united states—to a coherent response to a national political "crisis" and to the organization of the Constitutional Convention. Similarly, Morgan and others began to make the "Hessian Fly" an object of general awareness and action. In July, Morgan persuaded a neighbor who was visiting Long Island to interview farmers en route about the insect problem. Two months later, a letter describing the animal's life history and suggesting control methods began to circulate privately.[10] In January 1787, a new organization, the New York Society for Promoting Useful Knowledge, discussed the (still publicly unnamed) "insect destroying the wheat," and at the same time, New York governor George Clinton highlighted "the fatal ravages" that wheat had suffered "since the commencement of the late war," and asked citizens to share remedies.[11]

The *American Museum*, a new magazine created in Philadelphia to revive Revolutionary consciousness nationwide, was the venue through which the Hessian fly finally gained the attention of the general public. The inaugural issue (February 1787) warned that "that destructive insect, the *Hessian fly*," was spreading gradually from Staten Island, "where it was discovered about eight years ago" (i.e., 1778 or 1779). It claimed that unless a remedy was found, "the whole continent will be over-run—a calamity more to be lamented than the ravages of war." A collection of unsigned letters followed two months later, which discussed the animal's life history and suggested a variety of remedies (heavy fertilizing, late planting, rolling the young wheat to crush the worms, brushing fields with elderberry bushes, and spraying with salt water).[12]

In May 1787 George Morgan pushed to the fore of this discussion. He described the insect ("about the size of a small ant") and explained that it could be distinguished from similar animals under magnification by its long black legs and by its "whiskers, so small and motionless, as not to be easily perceived by the naked eye." He connected the facts reported by Americans to the world of learning by suggesting that the fly could be the same insect that the French agricultural experimenter Lullin de Chateauvieux had described in 1755. Noting that he had sent the Philadelphia Society for Promoting Agriculture samples of the fly and its "nits," he suggested that they write officially to the Agricultural Society of Geneva for more information on Chateauvieux's animal. On a more practical level, Morgan reported efforts to control the fly by manuring, rolling, and grazing the wheat. He asserted that the best long-term solution would be widespread use of a new "yellow-bearded" wheat variety (taken from an enemy sloop captured in the war and planted around Flushing, Long Island) that appeared to resist attack.[13]

If Morgan suspected that the insect was common in western Europe and had been studied thirty-two years earlier in France, why did he give it its distinctive name? The appearance of both Hessians and the insects on Staten Island and Long Island in the latter 1770s suggested that they arrived together. But what really mattered to Morgan was behavior. As he explained a year later:

> The name of *Hessian* Fly was given to this insect by myself & a Friend early after its first appearance on Long Island, as expressive of our Sentiments of the two Animals—We agreed to use some Industry in spreading the name to add, if possible, to the detestation in which the human Insect

was generally held by our yeomanry & to hand it down with all possible Infamy to the next Generation as a useful National Prejudice—It is now become the most opprobrious Term our Language affords & the greatest affront our Chimney Sweepers & even our Slaves can give or receive, is to call or be called *Hessian*.[14]

Hessian flies were, like Hessians, pitiless invaders who had advanced inexorably from New York to Princeton, ravaged peaceful farms, and left devastation behind them. The insects were worse than the humans because they had "crossed the Delaware" and showed every sign of staying.[15] For Morgan—the brother of a Signer, an army officer defending the frontier, and a gentleman building virtue and a prosperous rural life on the site of the greatest battle of the war—it was common sense both to conceptualize the insects as invaders and to mobilize Americans by associating the new threat with their most recent enemy. In his essays and in an exchange of letters with George Washington that he proudly shared, he imagined no more than a national audience. He did not consider the implications that his idiosyncratically chosen name might have beyond the United States.[16]

Worse than the Plague

In retrospect, he should have. While Americans' conversations and correspondence seemed bounded, published information readily crossed the Atlantic. Morgan's most important reader was a British agent, Phineas Bond. The son of a physician whom John Morgan had pushed aside in creating the University of Pennsylvania Medical School, the loyalist Bond fled from Philadelphia to England in 1777 to escape a treason indictment. He was able to return nine years later only because he had been appointed British consul. Bond provided secret intelligence that the British Customs Office used to stop Americans from smuggling cotton spinning machinery out of Britain or from smuggling tea in.[17]

When Bond learned in April 1788 that Philadelphia merchants were hoping to sell large amounts of wheat in England that summer, he saw value in sending the Foreign Office a brief report about the new and ruinous insect "called the Hessian Fly." Emphasizing Connecticut patrician Jeremiah Wadsworth's claim in the *American Museum* that the pests could be controlled by soaking wheat grains in elderberry juice, Bond argued that it was "reasonable to conclude that this process destroys the Egg in the Grain," and hence that grain contained Hessian fly eggs. He warned that importation of American grain infested with the fly could destroy English wheat culture.[18]

The Privy Council Committee for Trade, Bond's chief audience, was receptive to his message. Lords Carmarthen, Sheffield, and Hawkesbury, and their scientific adviser, Joseph Banks, were all hard-liners toward the new United States. They believed that the Americans should get the independence they had demanded, and more: Britain should not only eliminate all commercial ties with the former rebels but also reconfigure imperial trade to exclude them. In the short run, this meant getting Canadians to grow enough wheat so that Newfoundland fishermen and West Indies slave owners would no longer need to buy food from Americans; the long-term vision promoted by Joseph Banks was a West Indies dotted with Polynesian breadfruit trees that would provide slaves a staple food (hence the ill-fated 1787 collecting voyage of HMS *Bounty* to Tahiti).[19] In addition, higher agricultural tariffs and thus higher wheat prices would stimulate home production, thereby improving the British agricultural economy, increasing the security of the food supply, and enriching landed gentry. Parliamentary enactment of a higher tariff, which would directly increase the price of bread for urban workers, was not politically feasible in the late 1780s; but administrators were operating within their authority if they prohibited imports of potentially pestiferous grain.[20]

Banks provided both science and leadership in the government's action against the new insect. Since Bond had supplied little more than a name and a warning, Banks was (in his own words) "utterly ignorant what insect it was that the Americans meant by the Hessian Fly." The subject, however, "appeared to him so pregnant with danger to his country" that he sought to do whatever he could.[21] With only a few days to respond, he turned to the available technical literature. Finding a paper on a wheat insect in the American Philosophical Society's *Transactions,* Banks reported that the Hessian fly, "more generally called The Flying Wevil," had been known in Virginia for at least fifty years. The larval stage of this "minute moth" penetrated and fed on wheat grain. It differed from the "Wevil of Europe," by having wings and by attacking the grain in both the warehouse and in the field; as a consequence, it could easily be transported in grain and was "an Evil of a most dreadful Nature."[22] On June 25, 1788, the Privy Council and George III received Banks's recommendations and prohibited entry of American wheat until further notice.[23]

The problem with Banks's argument was that his identification of the Hessian fly with the Virginia "flying wevil" was simply incorrect. He realized his error a week later when he obtained Samuel L. Mitchill's essay on the Hessian fly, published in Noah Webster's *American Magazine,* a New York competitor of the *American Museum.* Banks did not, however, retract his earlier call to action. Instead, on July 5, he advised the Privy Council that although the Hessian fly, which "confin[ed] its Ravages to the Blade of the Corn," was not a problem, the flying weevil was a real danger. Three days later he shifted position again, arguing that there was "great reason" to believe that the Hessian fly laid its eggs on the grain, and not the straw. Warning that both the flying weevil and the Hessian fly were causing "alarming ravages" in America, he asserted that each independently would justify prohibition of imports "capable of bringing . . . the Seeds of so dreadful a Calamity."[24]

The Privy Council accepted Banks's statements without question. Issues arose, however, when the Customs Office began to set up procedures to inspect American grain already in English warehouses.[25] The economist Adam Smith, then a customs officer in Glasgow, understood that perfect paperwork that reported nothing would satisfy London. Officials in Liverpool more argumentatively informed their superiors that American grain shipments contained "wevils," but they were all the same sort, all dead, and therefore posed no problem.[26] Banks dismissed such sweeping claims from provincials who "cannot be presumed to possess so much knowledge of insects as those who have dedicated a large portion of their time to the study of natural history," but then was shocked to receive, not bureaucratic acquiescence, but a scientific challenge.[27] James Currie, a young physician who had wandered the Atlantic as an adolescent during the Revolution, critiqued both Banks's natural history and the council's policy. He argued that the flying weevil was not a problem since it had lived so long in Virginia without entering England, and emphasized that since the Hessian fly lived in the stalks, it was unlikely to also enter the grain. Noting that the "Government" had initially confused two different insects and was taking action against one after being warned against the other, Currie regretted that evidence seemed to be "biased by interest." He emphasized that "the judgment of men of science unconnected with commerce," such as himself and Banks, needed to prevail, and he boldly offered his services as a partner and consultant.[28]

Banks's response to this challenge took two forms. He directed that Currie be excluded from further service as an inspector; more significantly, he prepared a sweeping biopolitical justification for the Privy Council's course of action.[29] This careful report emphasized that, although a bug's life was generally quite predictable, under unusual environmental circumstances it could be greatly extended in length—"a Circumstance as curious perhaps as any that has been observed in the History of animated Nature." While the claims that neither the weevil nor the fly could cross the Atlantic were plausible, Banks admitted they did not "on the whole amount to that Degree of Certainty, which in Matters of such material Importance ought to be required." The weevil had gradually moved north in America over the last thirty years, and hence was now more likely to survive in England. Banks was persuaded by the argument "from general Analogy" that the Hessian fly laid its eggs on the leaves and not the grain, but he countered that eggs could mix with grain during threshing and might therefore be transported.[30]

From Banks's perspective, however, plausible scenarios about the activities of insects mattered less than the broader landscape of interactions between the natural and human economies. He explained dispassionately that the introduction of a new wheat insect into England would be "a Calamity of much more extensive and fatal Consequences than the Admission of the Plague." Plague resulted only in "the extinction of a certain proportion of the human Species, which may be, and generally is replaced in the next Generation." Insects, by contrast, could permanently affect food supply, thereby resulting in a "real Diminution of Population," and hence reducing the power of the state. Banks calculated that every English wheat farmer supplied eleven people—himself, "six manufacturers, and four of the affluent, their unproductive dependants, or the army." A one-sixth drop in domestic production would mean—since the affluent and the army would not cut consumption—that one in six workers must either "cease to eat, and consequently to labour," or would need to buy imported food. In the latter case, the benefit of that labor would go to foreigners because, from Banks's physiocratic perspective, "food, and food only is the creator, and . . . the whole of the honest gain it produces, must ultimately center in the Country that produces the food."[31]

Banks, the creator of a global botanic empire, imagined England in its agricultural character as a biogeographically isolated island. The familiar image of plague encapsulated the alternative between the potential for disaster resulting from the entry of American wheat and the indefinite continuation of the environmental present at a small but continuing cost. Framed this way, the political choice was easy. Implementation would be guided by naturalists, the new guardians of agricultural national security.

The Subsistence of the People

Joseph Banks was committed to rationality and fact. He outlined his reasons for the quarantine, he explained, so that "if by better Information than we have at present, they appear hereafter to be erroneous, it may be retracted." On the other hand, he set an extraordinary standard for acknowledging error: in a matter with such consequences, "a positive proof that no danger whatever exists should be exacted."[32] As the news about the Hessian fly circulated around Europe and, with a two-month time lag, traveled back to the United States, efforts were made to provide that proof. The different positions advanced, and the character of the eventually persuasive argument, enable a retrospective critique of this kind of practical reason in the late 1700s.

Banks, first of all, took seriously the question of the existence of a "Hessian" fly. He had read in Mitchill's essay that some Americans considered the insect "of Hessian origin, and brought to this country by the soldiers from that principality."[33] Refuting the idea that the insect lived in Germany would counter the insult that imperial troops had brought a pest across the Atlantic. Banks's larger interest, however, was in strengthening an alternative perspective advanced by Mitchill: that the insect had long existed in America, and had just discovered wheat as a food. (He ignored Mitchill's third possibility, that the fly had migrated to America from some as-yet-unidentified territory.) If Hessian flies were actually American insects, then the argument that they represented a new and alien menace to English agriculture was much stronger.

The consequence of this reasoning was a Foreign Office inquiry to embassies throughout central Europe asking whether the Hessian fly existed in their regions. All but one reported back negatively.[34] But affirming the absence of an unknown and sketchily described pest was problematic. A few respondents sought to anchor their claims in citations to the new naturalists' bible, the collected works of Linnaeus, with one suggesting that the Hessian fly was the Scandinavian *Musca secalis*. The exceptional respondent—an anonymous consultant in the Austrian Netherlands—emphasized the deeper difficulty. If the English really wanted to know whether the insect existed in the Low Countries, he argued, they would need to supply "either an exact diagram of the different stages of the insect, or the insect itself in its different stages." A merely verbal description was too vague to provide any basis for serious study.[35]

French agricultural naturalists, whose perspective was both continental and philosophic, treated the British inquiry with careful condescension. A committee of the Royal Academy of Agriculture, then the world center for practical entomology, sought to avoid nomenclatural bias by discussing Banks's insects only as #1 (the flying weevil) and #2 (the Hessian fly). They suggested that #1 was a European species investigated in Angoumois in 1755; although it periodically caused problems, it was not a major agricultural threat. Insect #2, they believed, had been found by Linnaeus in Scandinavia, and named *Phaliena graminis;* animals like it existed in France, but had not been studied in depth because they had never caused enough damage to "attract the attention of the Government." French naturalists saw Europe teeming with poorly known insects. Humans could fight these animals when they became rampant, but could not stop them from crossing the borders. To the French, the British dream of isolating their island from the rest of the world was feverishly unrealistic.[36]

Americans of different sorts responded vigorously as soon as they learned, two months after the fact, that the British government was condemning their wheat.[37] The lead was taken by the British consul-general, John Temple, a Massachusetts native married to the daughter of that state's governor, James Bowdoin. This "friend of America" interpreted the letter he received in New York in late August (with a copy of Banks's first erroneous report only) as a request for reassurance that American wheat was, in fact, safe. His inquiries quickly put him in contact with George Morgan.[38] Morgan marked up Temple's copy of Banks's report, prepared an expansive and blunt cover letter, and provided a copy of his recent exchange with George Washington regarding ways to fight the fly. Morgan's commentary included his claim that he had christened the "Hessian fly," his belief that the insect had somehow arrived in straw "at an early period of the late war," and his discovery that it had first infested fields "in the neighbourhood of Sir William Howe's debarkation [on Staten Island], and at Flat Bush."[39] He dismissed Banks's June 4 report completely, asserting that anyone who would "confound the Virginia Wheat Fly with the Hessian Fly, which are as different as a Toad from a Snake," was completely misinformed. He provided an elementary account of American wheat insects, explaining that the Virginia wheat fly (Banks's flying weevil) was a "minute moth" that fed on grain, but was long established and easily controlled; by contrast, the flying-ant-like Hessian fly "had no connection immediately with the grain," and thus would not survive in transatlantic shipments.

Morgan pressed the reconciliatory message that the English had nothing to fear from American wheat. However, the former frontier revolutionary could not resist reminding royal officials that "were a single straw containing this insect in the egg, or aurelia [pupal stage], to be carried and safely deposited in the centre of Norfolk in England, it would multiply in a few years so as to destroy all the wheat and barley crops of the whole kingdom." Americans' distress could be, at least in the imagination, a source of power.[40]

Phineas Bond received the same message as Temple and, like Morgan, saw that Banks had erred in conflating the flying weevil and the Hessian fly. He sought to reinforce his original warning about the latter, however, by cultivating uncertainties. Bond wrote the foreign secretary that Americans' claims that the animals fed and propagated only in straw were undercut by their lack of good microscopes and other "suitable instruments" for discovery. However, if the eggs were laid in the straw and, as Americans asserted "with great earnestness," the fly "was brought hitherto in the straw beds and baggage of the German troops employed in the late war," then American shippers could easily carry it to Britain. "Plain, intelligent, artless" farmers had informed Bond that Hessian flies could be found in stored wheat. This unself-conscious testimony was more trustworthy than the claims of leaders like Morgan, who were biased in favor of "the interests of the country." Bond concluded that "the works of nature are so minute, and its modes so inscrutable, as to baffle every endeavour hitherto made to form a satisfactory conclusion." Not having "any conclusive fact" from which the nontransportability of the fly followed, "the wisdom of guarding against so grievous a calamity by all due caution, must be evident."[41]

In February 1789, Joseph Banks reviewed the hundreds of pages of memoranda he had received. Believing that he now knew "as much on the subject as the Americans themselves," he considered the full range of arguments advanced and decided that Phineas Bond's analysis of dangers and uncertainties (and his own initial recommendation for action) were basically right. Contrary to American expectations, there was no indication that the Hessian fly ever lived in Hesse or anywhere else in Europe, or that it had been carried across the Atlantic during the war. In the face of Americans' "bold" assertions that the fly would not come to England in their grain, he declared that "with nearly if not exactly the same materials before me, as these gentlemen have made use of, I have not been able to draw a similar conclusion, nor indeed any certain conclusion whatever." He backed Bond's argument that until British officials could be sure that the Hessian fly could never be transported across the Atlantic, American grain shipments should be banned.

He suggested that Americans needed to further investigate both the identity and the mode of the insect's propagation, and he helpfully supplied a bibliography. In the meantime, Americans needed to control themselves: Paraphrasing Morgan's comment about the placement of eggs in Norfolk, he warned that "there cannot exist so atrocious a villain, as to commit such an act intentionally."[42]

In late March, the Privy Council (operating in the absence of George III, who was in the midst of his first attack of insanity) accepted Banks's report. Parliament ordered that the most significant documents be printed in full as testimony to the world about the reasoning behind the import ban. This pamphlet generated considerable controversy around London. Thomas Paine, always skeptical about British leaders' motives, told Banks to his face that the import prohibition "was only a political manoeuvre of the Ministry to please the landed interest, as a balance for prohibiting the exportation of Wool to please the manufacturing interest." Jefferson, in Paris, complained indirectly to Banks that the report was a "libel on our wheat." Banks ignored Paine, and he brushed aside Jefferson's "warm expressions" as unreasonable. Still concerned about Morgan's agroterrorist threat, however, he warned Jefferson that "obloquy" would fall on all Americans if "any one of them should wilfully bring over the Fly" in retaliation for British action. On the other hand, he accepted with equanimity the declaration of the Duke of Grafton, a former prime minister, that the Hessian fly was a miraculous creation—a "scourge of Heaven . . . upon such ungratefull colonies and rebellious people."[43]

Surprisingly, American newspapers were almost completely silent when the Privy Council's pamphlet arrived in the summer of 1789. To understand why, we need to look beyond the circulation of documents toward the circulation of weather currents. During the year that Banks was building his case for the continued prohibition of American wheat, storms and cold in both Europe and northern North America were producing the worst grain shortages in decades. The price of bread in Paris began to rise in August 1788, and Louis XVI's government, anxious that social instability not exacerbate the country's ongoing financial and constitutional crises, opened ports to unlimited importation of American wheat. In November the French began to offer a bounty to encourage shipments, and by the summer of 1789, Philadelphia and New York wheat prices were reaching the high end of their postwar range. (The doubling of the price of bread in Paris, meanwhile, had helped ignite the French Revolution.)[44] Within that context, the British quarantine was irrelevant.

The English also faced a food crisis, but not as quickly as the French. The thin harvest of 1788, combined with the ban on American imports and the shortage on the continent, produced exactly the result that landowners such as Banks had hoped for: the London price of wheat rose and stayed above forty-four shillings per quarter for the first time in four years. Matters became worrisome, however, when the 1789 crop failed across much of the island. London wheat prices spiked in midsummer to sixty-nine shillings, and then fluctuated for the rest of the year around fifty-five, well above the previous decades' average of forty. Privy Council leaders shared Banks's biosocial interpretation of the relations among food, prices, and population. But, while the naturalist could write philosophically that under certain circumstances one in six laborers might "cease to eat," politicians emphasized the potential for "popular commotions" in times of shortage and the government's responsibility for "the subsistence of the People."[45] During 1789 they worried increasingly about the island's wheat supply. They were equally concerned about the potential for infection from France, not by the Hessian fly, but by the Revolution. The effects that French ideas and models of mass behavior might have on a population stressed by food shortages were unknown, but the elite could readily imagine the possibilities.

British policy changed abruptly in November, just as the Privy Council was approving emergency food shipments to the Channel Islands to forestall starvation and mutiny. Foreign secretary Carmarthen (now the Duke of Leeds) sent Temple and Bond a leading inquiry: first, whether "the Evil arising from the Wevil or Hessian Fly [had] wholly ceased," and then, whether Americans had a wheat surplus, what current prices were, whether other countries had cornered the market, and whether English merchants had placed orders.[46] Three days later, the Privy Council jumped ahead of this correspondence by interviewing merchants with American connections. Hearing that "the Hessian Fly had not at all appeared, . . . that the crops have been very abundant, and that the price of wheat at New York by the last accounts was 4 shillings per bushel," the council recommended an immediate end to the quarantine and ordered the navy to dispatch a special ship to New York with the news.[47] When HMS *Echo* arrived with the new rules in February 1790, together with rumors that more than 140 vessels were on their way to buy wheat, the *New York Daily Gazette* sneered, "what avail will this be to England? None; for in the first place, the Americans have no wheat to spare; and if they had, all the French ports are ready to swallow it up at ten shillings a quarter above the price of the British market."⁴⁸ As the Hudson thawed and prices continued to rise, however, stored grain was released and shipped in large quantities to England. Temple and Bond each understood the new situation, parroting back their assurances that "the evil [is] thought to have wholly ceased."⁴⁹

A good English harvest in 1790 caused prices and imports to drop, and in the following year, Parliament amended the Corn Laws to encourage domestic production. But the issue of the Hessian fly did not reemerge. When Bond warned his superiors in July that the pest was reappearing in American fields, he received no response. Diplomatic historian Charles Ritcheson concluded that in 1789 British leaders decided that the United States would likely function in the future as their island's reserve food source. Having watched the Old Regime collapse in France, they were willing to accept the hazards posed by an insect.⁵⁰

Living with Invasions

For Americans, the Hessian fly represented a future of uncertainty and apprehension. The question of its origin was not resolved. In the early 1790s, Secretary of State Thomas Jefferson led an American Philosophical Society committee that asked the public whether the insect had existed in North America prior to the Revolution; Jefferson hoped that a negative consensus would reimplicate the imperial forces. The committee produced no report, however, because Jefferson was distracted by politics and because yellow fever invaded Philadelphia from the Caribbean in 1793.⁵¹ Samuel Mitchill set this broad lack of information about the arrival of the insect in North America against the much more specific gap in the memoirs sent to him by Banks—namely, that no European authority had described a species like the Hessian fly. He concluded that Morgan had produced a misnomer, and that the animal should properly be called "the American wheat-insect."⁵² The view that *Mayetiola* was primordially American predominated until the 1840s. Discovery of the insect in Minorca, Russia, and Austria, however, enabled American entomologists ultimately to shape a consensus that it must have been a longtime companion of wheat in the Old World and that it had reached North America for the first time in the latter 1770s.⁵³

The deeper issue was that, whatever its historical geography, the dramatic success of the new pest presaged ill for the future of wheat culture and for American culture more generally. If it was an American wheat insect, it indicated that the continent's environments were as antagonistic to wheat as they were to wine grapes (albeit in a more obscure manner). What other malignancies might be lurking farther west, waiting to pounce on cultured plants? The situation was even more discouraging if the fly really was Hessian: the devastation of wheat in America by an animal that was innocuous in Europe implied that wheat had degenerated in the New World. What other minor European pests would plague the United States in decades to come?

The Americans' practical resolution of the challenge posed by the Hessian fly was to keep wheat culture on the move. In contrast to genteel improvers like Morgan, ordinary farmers already saw migration as their optimal response to declining soil fertility in long-cultivated lands. Wheat could grow most profitably on a shifting band of lands that had untapped nutrients and were accessible to international markets but had not yet been reached by the Hessian fly. Between the 1790s and the 1840s, this strip moved westward across New York, into Ohio, and then out toward Illinois and beyond. Farmers demanded the opening of new frontier territory in part so that they would have new places to grow wheat. This dynamic accelerated the dispersal of a population focused on short-term profits, and it prevented the emergence in the East of the stable yeoman society that Jefferson had envisioned in his *Notes on the State of Virginia*. Raynal would legitimately have interpreted the Hessian fly as one more reason to predict that the United States would experience a future of mediocrity.[54]

Looking back, the most important task is to think clearly about the Hessian fly's emergence and about the perceptions it generated. The insect, arriving nearly simultaneously with independence, was the first invasive species encountered by the people of the United States of America. I have shown how completely George Morgan identified it as an alien invader threatening his new country, and how compelling this interpretation was. Yet from a biogeographic perspective, *Mayetiola* was no more alien to North America than the colonists and their wheat; moreover, given the extent to which the vegetation around New York had been transformed during the previous 150 years, it was less of an ecological disruptor than its predecessors. In its selective herbivory, it arguably enhanced the survival of North American plant species.

Insofar as the arrival of the Hessian fly was an invasion, it was more deeply nested in military policy than in natural history. In the 1770s, the species was unknown to the learned. It traveled—to New York in North America, rather than to Norfolk in England—only because of a number of singular circumstances. On the one hand, political identities were diverging to the point that British military agent George Morgan, for example, could re-create himself as a model American farmer, while his younger Philadelphia neighbor, Phineas Bond, became an English gentleman who represented his country for decades while living in Pennsylvania. On the other hand, armies were converging in circumstances that led to the unprecedented transport of huge quantities of grassy biomass to Staten Island from places such as Minorca and Poland. In contrast to champion marine stowaways like the Mediterranean fruit fly, which circumnavigated the globe in the course of the nineteenth century (see Chapters 6 and 8), *Mayetiola* had spotty success in expanding its range: while it reached New Zealand in the late nineteenth century, for example, it never established itself in Australia. The uncertainties of invasion would challenge plantsmen and scientists in the centuries to come.[55]

CHAPTER THREE

The Development of American Culture, with Special Reference to Fruit

On September 10, 1830, hundreds of Bostonians gathered at the largest hall in the city for the second annual festival of the Massachusetts Horticultural Society (MHS). Visitors listened to a formal oration on the value of horticulture, and then toured the elaborate displays of grapes, pears, peaches, apples, melons, and flowers contributed by society members. In the late afternoon, members and guests sat down to dinner, and then raised their glasses in a marathon of forty-six toasts. Pomological double entendres were prominent in these brief speeches, which ranged from comfortable self-congratulation ("*Diffusion of kinds and of kindness*—our grapes can never be sour, as they will be within reach of everybody") to commentaries about current political issues ("*The Nullificators*—South Carolina *Borers*—as nobody cares about them out of their own State, they ought to be *dug out* there").[1]

The conclusion of the day's events was communal song: "The Course of Culture." The anthem's author, Thomas Green Fessenden, had made his literary reputation a quarter century earlier with *Democracy Unveiled*, a long mock-heroic poem that attacked President Jefferson as, among other things, "A Chief who stands not shilly shally, / But is notorious for—a *Sally*." Now, however, as the editor of the *New England Farmer and Horticultural Journal*, Fessenden represented community and improvement. His poem (sung to "Auld Lang Syne") celebrated "manual culture," "mental toil," and the task of training children. It praised wives willing to be "tilled" periodically by their husbands. Its central claim, however, was that "HORTICULTURE," ordained by God and polished for millennia by man, was the source "of every art."[2]

I take these records of inebriated conviviality as an invitation, extended across nearly two centuries, to examine the development of American horticultural organization and practice. This chapter first sketches the growth of organized horticulture in the United States by examining how groups of prosperous Americans became passionate about growing and improving plants (especially those with sweet fruits); how these initiatives were linked to English activities; how Massachusetts orchardists took the lead; and how societies, nurseryman-entrepreneurs, and the gardening press interacted. I link together the elements of this story through examination of the most important plant promoter in the second third of the nineteenth century—Charles M. Hovey, a self-made nurseryman, importer, breeder, author, editor, and society manager.

I then consider what Massachusetts horticulturists tried to do. The plethora of puns at the 1830 festival was a typical expression of horticulturists' awareness of the redolence of gardening language emphasized in my introduction. Fessenden asserted explicitly that their shared keyword was culture, and he affirmed their central conviction that horticulture formed the basis for American development. These emphases were neither new nor solely American. But horticulturists in Massachusetts understood that culture was more important for them than for Europeans, and that American culture was exceptional. While Europeans and their fruits had developed together gradually, Anglo-Americans saw both themselves and the fruits they desired embedded in a context of migration and novelty. Their tasks were twofold. They sought to enable plants to make the transition that they believed their own ancestors had made. Just as colonists had settled, thrived, and reproduced in Massachusetts, gradually becoming American natives, European plants could be "naturalized"—made to stand the soils and climates, to breed, and to improve. Alternatively, they sought to culture crude but promising American natives—to take them out of the wild and make them fit for civilized tables.

These efforts to develop "American fruits" were more complex than horticulturists imagined. Four examples—apples, pears, strawberries, and grapes—convey the diversity of situations and the real, if only partially understood, results. Naturalization could be exceedingly difficult. Culturing natives entailed interactions with plants that had obscure lineages and goals of their own. In addition, success could transform human culture and American identity in unanticipated ways.

Organized Horticulture in the United States

During the first decades of the nineteenth century, a significant number of merchants, manufacturers, and professionals in northeastern cities spent time and money on suburban estates. For these prosperous and literate men, however, an integral part of their semirural existence was their ability to gather periodically in urban settings to talk about farming. Gentlemanly agricultural organizations arose in New York, Boston, and Philadelphia in the years around 1800. Then, in the 1820s, suburban "villa" owners joined with commercial nurserymen to create societies specifically devoted to horticulture, in emulation of the Englishmen who had created the Horticultural Society of London in 1804. The range of these gentlemen's interests appeared in the full titles of new rural publications: *The New England Farmer and Horticultural Journal* (1823) and the *New-York Farmer and Horticultural Repository* (1828).

What did gentlemen do on their country properties and talk about in their urban meetings? American estates north of the Mason-Dixon Line and convenient to major cities were small by comparison with those in England or the South. They were also appreciated on shorter and explicitly seasonal timescales: middle-aged businessmen sought immediate gratification during summers, not slow improvement and year-round engagement with property that had been, and would presumably remain, in a family for generations. They sought utility and economy, but were not too concerned with profits. As a consequence, almost no northeastern American estate owners developed the extensive "parks" favored by English aristocrats and their landscape designers. Few tried large-scale grain production for long, and not many pursued the expensive and time-consuming rural hobby of breeding fancy livestock.[3]

Within horticulture, there was a particular focus on fruit. Vegetables were plebeian, flowers were effete, ornamental shrubs were largely unavailable in North America, and timber trees were either too common or too slow-growing. Prior to the 1820s, "fruit" meant cider apples—a crop that was profitable but surrounded by an aura of vulgar inebriety. However, fruit culture was gradually altered by the combination of the temperance movement, steam transport, rising urban living standards, and the activities of the horticulturists themselves. The new emphasis was on table fruits, and more particularly on quality, diversity, and extended availability. Gentleman growers learned that plums, strawberries, grapes, pears, and even apples came in hundreds of varieties, whose different qualities opened opportunities for discovery, connoisseurship, and competition. Fruits looked pretty, smelled sweet, and were easily transported, displayed, and exchanged. The trees took care of themselves through the winter, and they could become, at least in the grower's imagination, a source of income.[4]

The material that American pomologists initially cultivated was limited. Indigenous "small fruits" were unsatisfactory for various reasons. Huckleberries and strawberries tasted good but grew poorly in gardens. Blackberries were sour and grapes were inedible. Eurasian table fruits—especially apples, but also pears, plums, peaches, and cherries—grew reasonably well. Apricots, on the other hand, were hard to maintain through North American winters. As emphasized in Chapter 1, European grapes were a puzzle: they grew well for a few years but then faded.

What made pomology particularly enticing to Americans was the extent to which it had advanced in western Europe in recent decades.[5] During the second half of the eighteenth century and into the nineteenth, botanists and nurserymen hunted, gathered, and began to systematically assess thousands of different varieties of fruits that had arisen under domestication in Europe and beyond. The French botanist André Michaux, for example, made his reputation in the 1780s by bringing new varieties of almonds and melons from Persia to Paris. The Horticultural Society of London, under the presidency of Thomas Andrew Knight (author of *A Treatise on the Culture of the Apple & Pear* [1797]), collaborated with botany professors and nurserymen to collect and name fruit varieties.[6]

Efforts at culture—the manipulation of growth and reproduction—also grew in prominence after 1810. European pomologists improved soil with mineral and organic manures. Pruning, training, removing bark, and grafting different species together produced trees that were monstrous but yielded larger and richer fruit. Growers harvested seeds of promising trees and selected the best individuals of the next generations; they also crossbred plants—hybridizing different varieties and sometimes different species. Horticultural gentlemen and nurserymen were amazed at the extent to which apples, pears, strawberries and other plants of culture had been altered in historic times, and were at present capable of being manipulated. But many worried whether these gains were permanent, and sought to understand how best to hold onto them or to extend them.[7]

Both the processes and results of European fruit culture were potentially available to Americans. Specialized books and magazines reported on new developments, and nurseries marketed pomological novelties. The Horticultural Society of London's *Transactions,* William Hooker's *Pomona Londinensis,* J. C. Loudon's *Gardener's Magazine,* and John Lindley's *Pomological Magazine* described new fruits and provided appetizingly vivid color illustrations. In the early 1820s, the London Horticultural Society's experimental garden in Chiswick propagated new varieties and distributed them to members and to favored correspondents who wanted to grow and show the latest fruits.[8]

American plant enthusiasts organized themselves in large part to seize these international opportunities. The leaders of the New-York Horticultural Society initiated correspondence with London counterparts in 1821 even before they had completed the process of incorporation.[9] Boston Brahmin John Lowell corresponded with the Horticultural Society of London's president, Thomas Andrew Knight; Knight's gift of scions of valuable new varieties to the "citizens of the United States," and Lowell's difficulties in propagating and distributing them, formed part of the background for the creation of the MHS in 1829. That society was justified explicitly as a way for Bostonians to acquire European stock and horticultural books; members were attracted by the promise of free and exclusive distributions of new plants.[10]

Cooperative schemes for plant importation were difficult to maintain, however. Receiving scions, grafting them onto local rootstocks, nursing the specimens for a year or two, and then distributing the hardened trees seemed to require the operation of an "experimental" garden like that of the Horticultural Society of London.[11] The New-York Horticultural Society considered the creation of such a garden to be their chief mission. David Hosack, its president in the 1820s, was a socially prominent physician and professor who had run the semipublic Elgin Botanic Garden on the present-day site of Rockefeller Center for a decade in the 1800s, and was thus well prepared for this work. He was supported by wealthy amateurs such as Pierre Lorillard; by the country's most active commercial plant importers (William R. Prince and André Parmentier); and by Hessian fly expert, Columbia College zoologist, and sometime U.S. senator Samuel L. Mitchill. Nevertheless, the society was unable to persuade Columbia, the city, the state, or the national government to fund this project. Land and upkeep were expensive; and while Manhattanites considered the benefits of a botanical garden too diffuse for public support, those from upstate or beyond deemed it a merely local amenity. In 1830, after a half decade of maneuvering, Hosack gave up and retired to Dutchess County (see Chapter 7); the society, lacking a mission and leadership, rapidly declined.[12]

Bostonians succeeded where New Yorkers failed, but their path was not what they originally envisioned. The creators of the MHS planned a practical scientific organization that would sponsor an ornamental and experimental garden modeled on Chiswick. They decided that the best way to advance this project was by combining forces with promoters of the idea that burying Bostonians in urban churchyards was unhygienic. As a consequence, MHS president H. A. S. Dearborn presented a plan in 1831 for a "Garden of Experiment and Rural Cemetery."[13] The society purchased land in Cambridge that was consecrated as Mount Auburn Cemetery. Dearborn enthusiastically supervised both the landscaping of the cemetery grounds and the construction of a thirty-acre garden where "many rare, valuable and beautiful plants may be obtained, not only from all parts of our own country, but other regions of the globe, which could be naturalized to the soil and climate of New England" (see Figure 3.1). More than 450 varieties of foreign plants were soon growing there, and both grounds and garden were opened to the public.[14]

The wealthy Bostonians who purchased cemetery lots did not share the horticulturists' values. They were enthusiastic about Mount Auburn's carriage drives and romantic landscaping, and they enjoyed picnics on their plots on Sundays (a day the public was excluded). But they were uninterested in either overseeing or looking over rows of fruit trees and radishes. A rancorous fight over design and control ended in early 1835 with the takeover of the entire property by the lot owners and the elimination of the experimental garden.[15]

John Lowell, searching for a new direction in the aftermath of this apparent defeat for horticulture, argued that "the attempt to unite the private efforts of individuals with a *vast* scheme of carrying on a *great garden* on *joint account*" had been a mistake. The MHS, he suggested, should instead foster the efforts of individual *"practical men"* whose *"latent talents"* had come to light though their participation in the society's affairs. The means for fostering talent would be a library and a system of cash prizes for superior plants. The society could take this new path because—in return for ceding control of Mount Auburn

The Development of American Culture · 57

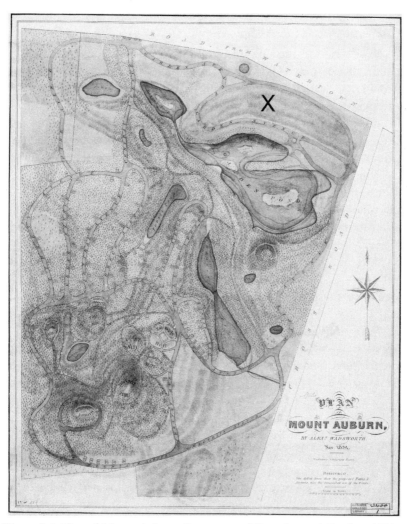

Figure 3.1 Plan of Mount Auburn Cemetery and Experimental Garden, 1831, with carriage drives and burial areas laid out. The experimental garden site is marked "X." With permission of the Harvard Map Collection, Harvard College Library.

Cemetery to its lot owners— it was receiving a percentage of all future land sales (about $2,500 annually for the next two decades). This secure source of income, and donations from local plant enthusiasts, made the MHS a permanent institution, in sharp contrast to its New York model.[16]

A Practical Man in Cambridge

Lowell's "practical men" were those directly involved in collecting, assessing, propagating, and improving varieties. A few were prosperous amateurs, such as Robert Manning (a Salem stagecoach line operator, pear tree collector, and gentleman plant seller, as well as Nathaniel Hawthorne's uncle and guardian) and Marshall P. Wilder (a Boston merchant, pomologist, and rosarian). A larger number, however, were tradesmen working in the seed and nursery businesses and in agricultural publishing. This group included John and William Kenrick, brothers who operated nurseries in Newton, and Joseph Breck, a Brighton plantsman who became editor of the *New England Farmer* in 1837. But the most important of the practical men was the Cambridge nurseryman Charles M. Hovey. Hovey recognized that participation in the seemingly disparate activities of plant improvement, horticultural society affairs, fruit tree sales, and magazine editing could in fact be mutually reinforcing, to the benefit of both himself and American horticulture. He made Boston, in spite of its substantial climatic limitations, the center for American fruit amelioration.

Born in 1810, Charles Hovey followed his older brother, Phineas, into the seed business. Working in his backyard in 1832, he began to devote particular attention to strawberries—plants that were both quick-breeding and highly malleable. Within seven years, he brought the improved "Hovey's Seedling" onto the market; the new variety (to be discussed later in this chapter) provided significant income and advertised Hovey and Company as an American pomological innovator (see Figure 3.2).[17]

During these same years, Hovey became a regular participant in the activities of the MHS. As secretary during the negotiations over control of Mount Auburn, he had the difficult task of keeping members sufficiently, but not overly, informed about the delicate discussions. During the next fifteen years, he served on committees that ran the library, organized exhibits, and cooperated with other organizations. As a consequence, he became educated as a horticulturist, became personally familiar with the society's patrician leaders, and learned routinely about new horticultural developments and, more particularly, about new varieties of fruits.

Hovey extended his influence beyond Massachusetts through work as an editor and author. In 1835 he created the *American Gardener's Magazine*, which two years later was renamed the *Magazine of Horticulture*,

The Development of American Culture · 59

Figure 3.2 Hovey's Seedling, from Charles Mason Hovey, *The Fruits of America* (Boston: C. C. Little & Jas. Brown, and Hovey & Co., 1848), facing p. 25. Reproduction courtesy of the Rutgers University Library.

Botany, and All Useful Discoveries and Improvements in Rural Affairs. Hovey's Magazine, as it was widely called, was the first vehicle for national communication among American horticulturists. A. J. Downing, a young Hudson River Valley nurseryman; Nicholas Longworth, a millionaire Cincinnati fruit lover; and Henry Ward Beecher, an Indianapolis minister and gardener, all gained national visibility through their contributions to *Hovey's Magazine* during its first decade. For thirty-four years Hovey reported annually on the "progress of horticulture," kept his readers abreast of domestic and foreign news, and provided space for often pointed discussions. Flowers, trees, and landscape design were all important elements in *Hovey's Magazine,* but fruit—especially reports and discussions of new varieties—occupied center stage.

Hovey's Nursery was an equally important institution, both locally and nationally. Purchased in 1840 with profits from Hovey's Seedling, it occupied forty acres in the center of Cambridge on the horsecar line that connected central Boston with Harvard and Mount Auburn. It was a showplace in every sense of the word: a British commentator in 1850 considered it the grandest nursery in either the United States or in England.[18] A wide promenade led from the entrance gate to a neo-Greek conservatory that held tender plants (especially camellias) and the Hovey brothers' offices. The grounds contained more than 100,000 trees of over 2,000 different varieties, planted along radiating avenues. From these grounds and from subcontractors around Boston, Hovey sent out premium stock of the newest and rarest kinds; most particularly, of course, those varieties featured in *Hovey's Magazine*.

By the early 1850s, Hovey was the leading figure in American horticulture. He sought to celebrate himself and his domain with an ambitious book project. Beginning in 1847, he issued sections of *The Fruits of America*. Modeled on fancy English publications like Lindley's *Pomological Magazine*, the book consisted of expensive color lithographs, with accompanying descriptions, of the best pears, apples, strawberries, and other species. Its frontispiece was a portrait of a young, phrenologically enhanced, fruit-headed Hovey, peering alertly into the future (see Figure 3.3). The "fruits of America" were, ambiguously, the plant parts pictured, the amelioration process that resulted in these beautiful objects, this illustrated book, the American pomological enterprise, and Hovey himself.[19]

Not surprisingly, the scale and interconnectedness of Hovey's enterprises generated challenges and questions about conflicts of interest. By the mid-1840s, Nicholas Longworth and other Cincinnati pomologists were fighting Hovey in the long-running "strawberry war" that will be discussed below. Downing broke with Hovey soon after the latter asserted that Downing's 1845 treatise, *The Fruits and Fruit Trees of America*, was long on Hudson Valley chauvinism and short on acknowledgment of its Boston sources.[20] Downing's *Horticulturist* was created as a direct competitor to *Hovey's Magazine*. Downing offered a more literary form of garden writing, and put more emphasis on "rural taste," but, like Hovey, gave greatest attention to fruit. The *Horticulturist* provided particularly extensive coverage of a scandal that enveloped Hovey in 1854: accused of manipulating an MHS prize committee in order to gain medals for his new pears and cherries, Hovey was first censured, and then, after five months of investigation and negotiation, absolved of blame.[21]

Hovey reacted to these attacks with carefully composed detachment. He prospered in business, and in 1863 became president of the

Figure 3.3 Charles Mason Hovey, 1851. Portrait facing the title page of *The Fruits of America* (New York: D. Appleton & Co., 1853). Reproduction courtesy of the General Research Division, The New York Public Library, Astor, Lenox, and Tilden Foundations.

MHS. During this critical period in American history, Hovey devoted his energy to the erection of a new and ornate headquarters for the society. On August 18, 1864, with the two American armies stalemated in trench warfare in Virginia, Hovey placed copies of the *Magazine of Horticulture* and the *Fruits of America* in the cornerstone of the new Horticultural Hall on Tremont Street, a short walk from Boston Common and the Public Garden (see Figure 3.4). Articulating the relevant principle of horticultural philosophy, he reminded his audience that their new building would eventually crumble, "for decay, though slow, is the destiny of all earthly things." The work of the society, by contrast—to bring "pleasure and happiness to the social and domestic life; to enrich and embellish our homes and country; to create a refined taste, and to open new and exhaustless sources of instruction and wealth"—would surely endure forever.[22]

Figure 3.4 Horticultural Hall, Massachusetts Horticultural Society, Boston, 1870, from *History of the Massachusetts Horticultural Society, 1828–1878* (Boston: The Society, 1880), facing p. 176. Reproduction courtesy of the Rutgers University Library.

What Were the Fruits of America?

What was the aim of the antebellum gardeners who reported their activities in the *Magazine of Horticulture*, as well as in the *Horticulturist*, *Horticultural Register*, *Western Horticultural Review*, and various horticulture-oriented agricultural publications? Their basic intention was that of their European models: in terminology that was unproblematic then, but has become archaic and strange after more than a century of semantic hegemony on the part of anthropologists, they sought to culture fruit. They worked to produce plants that would grow dependably, look enticing, taste good, yield heavily and for long periods, and, if possible, be marketed at a profit. They prized diversity in form, taste, and use. Such culture would improve not only the fruit but also the grower. Horticulturists emphasized that gardening fostered intelligence, good habits, and social mobility. At the same time, it produced peace of mind and it reinforced a man's attachment to his home, thereby contributing to the health and happiness of the individual, the family, and the community.

While culture was a widely shared human activity, the geographic conditions of the New World made American culture unique. Massachusetts plant enthusiasts were disappointed that the continent on which they lived was exceptional in its original pomological impoverishment. Prior to Europeans' arrival it lacked the ameliorated fruits that humans in Asia and Europe had produced through generations of conscious and unconscious initiatives. American horticulturists hoped to enrich the flora of their nation and thereby bring its level of culture up to that of western Europe. Success would make the United States a better place, and its residents better people—nutritionally, economically, morally, and aesthetically.

The initial key to American horticultural improvement was importation and naturalization of European fruit varieties. In a few cases, this process worked smoothly; more often, however, it involved difficulties whose nature and surmountability were unpredictable. The first problem was to keep track of varieties. The Horticultural Society of London had long worked to manage the chaos of fruit and flower names and specimens, and Americans followed English taxonomy and nomenclature as much as possible. But they faced the additional challenges of learning about the introduction of new varieties into their country and of determining when a plant growing in America was a previously unknown and genuinely naturalized variety.[23] Pomological leaders opposed local amateurs and nurserymen who wanted to give every unfamiliar specimen a new name. From a national perspective there was no reason for the Newtown Pippin suddenly to become the Albemarle Pippin when it was grown in Virginia. Additional problems were the confusing overabundance of "Superiors" and "Choices," or, even worse, names like "Sheepnose" or "Pecker." In the late 1840s, horticulturists led by MHS stalwart Marshall Wilder created the American Pomological Society to promote a nomenclatural code. Building explicitly on British standards, they proposed that names should be new, short, and in good taste, and should be originated only by a "competent person conversant with existing varieties."[24] But while Bostonians with access to the MHS library could display easy familiarity with pear names like Vessouziere and Doyenné du Comice, less-cosmopolitan Americans were resistant. In 1850, for example, Downing satirized the pomological society's initiative with an account of a convention attended by the fruits themselves. When an aristocratic French pear, the Duchesse d'Angoulême, complained that ignorant Americans mangled her name to "Dutch Dangleum," the New Jersey Red Streak apple, speaking for Downing, asserted that in the United States every foreign name needed to be "boiled down to a little pure, plain English essence."[25]

Names were less important issues than the plants themselves. A few imported varieties grew well, but many more had problems. Some froze during their first winters, while others shriveled from midsummer heat. Downing blamed "alien" English gardeners who were too rigid to abandon cultural methods developed for the cool and damp British islands, but most plantsmen understood that many European fruits (including numerous peaches, strawberries, and apricots) were simply not tough enough to endure the climatic extremes of New York and New England. More perplexing and disturbing were the imported plants that grew luxuriantly for a few years, only to droop suddenly, blighted or rotting. Were such collapses due to inferior plants, heat and drought, spring thaws and freezes, poor soils, or an unknown parasite? Discussions about these possibilities hearkened back to the fears that something about America caused European organisms to degenerate.[26]

When these complaints were raised by neophytes, American horticulturists responded with exhortations that they needed to maintain the high culture (material- and labor-intensive growing methods) to which European plants were accustomed.[27] When thinking more deeply, however, writers discussed how to make cultured American plants—to naturalize imports, select the best natives, and ameliorate ordinary specimens. The feasibility of this undertaking was evident to them from the histories of both Yankees and apples. But pinning down what naturalization and amelioration meant, and how those goals were to be achieved, turned out to be quite difficult.

The meaning of "native," and hence of "naturalization," as applied to domesticated plants, involved multiple ambiguities. Americans were not entirely certain which species predated the arrival of Europeans, nor did they know what varieties had arisen due to spontaneous interbreeding between individuals with prior American and European backgrounds during the preceding two centuries. Sometimes native meant primordially local or, more loosely, indigenous to North America. Alternatively, it could describe any plants found growing locally that were different from known European cultivars. A more relaxed practice was to consider plants naturalized if they had sprouted from seeds that matured in the United States. This situation became particularly tricky in the 1840s with the rise of nativism: organized antagonism toward (mostly Irish) immigrants. From the nativist perspective, Yankees were unquestionably native Americans, but not all foreigners were automatically naturalizable. Possibilities for ordinary usage flowed from those core convictions.[28]

The intellectual foundations for plant amelioration efforts were clearer than for thinking about natives and naturalization, but practical applications were uncertain. A handful of well-known English, Belgian, and French horticulturists—most notably Thomas Andrew Knight, Jean Baptiste Van Mons, and Antoine Poiteau—had provided sophisticated accounts of what cultured varieties were, how they changed over time, and how they could be improved.[29] The starting point for these discussions had been the widespread perception at the end of the 1700s that traditional varieties, supposedly handed down unchanged from the Renaissance, were "running out"—becoming less fruitful and more susceptible to disease. Knight explained this phenomenon by suggesting that every seedling fruit tree had a natural life of two to three hundred years, and that this aging process occurred in all that tree's tissues—including the thousands of scions that might have been taken from the original plant to form new grafted trees. Knight argued that the way to rejuvenate Old English stocks was to hybridize valuable apples with more vigorous, wilder varieties (such as the Siberian crab); more generally, he proposed to create new and improved varieties by crossing plants with complementary properties. He acknowledged that hybridization was a laborious method whose successes were few and far between, but he saw no alternative.[30]

J. B. Van Mons, a Belgian physician, chemistry professor, liberal revolutionary, and pear culturist, shared Knight's view that old varieties were in decline, but provided a different interpretation and a different set of remedies. Drawing on the ideas of French naturalists such as Buffon and Lamarck, and on his own conflicted feelings about aristocracy, he argued that cultured fruit varieties had an inherent tendency to become refined and sweet, then overrefined and feeble. He proposed to replace these degenerating old lineages with newly cultured common trees. Working for fifty years beginning in the late 1780s, Van Mons moved vigorous but ordinary pear trees from their rough roadside settings into soft-soiled gardens, and then encouraged them to put their energy into fruit production rather than into merely vegetative growth. He selected seeds from the most delicate specimens, and repeated the process over a number of generations (denying that hybridization was taking place). By the 1830s, he had produced new pear varieties that were weak growers but had soft skins and "melting" tastes.[31]

The idea that varieties ran out never became a consensus in either England or the United States. But while English botanist John Lindley, for example, merely dismissed Knight's and Van Mons's ideas and pursued his own approach to hybridization, Americans debated these issues for a generation. Knight and Van Mons seemed to offer clues regarding the most important mysteries American horticulturists faced: on the one hand, the decline in their gardens of most varieties sent from Europe, and on the other, the fact that some plants had somehow transformed themselves to fit into their new settings and—in the opinion of Americans—had become even better than their relatives remaining in Europe. The French pomologist Antoine Poiteau, who visited the United States in 1801, argued explicitly that the country was "a grand laboratory" in which the phenomena of decline and regeneration sketched by Knight and Van Mons had occurred during the two preceding centuries. John Lowell advocated for Knight, and H. A. S. Dearborn promoted Van Mons; these ideas were so common that the young Henry Ward Beecher could discuss them knowledgeably with audiences in Indiana.[32]

Few American horticulturists, however, applied either terms or theories consistently. Downing, for example, used "native" both for wild plants and for "the offspring of an old stock in the vegetable races of the other continent" transformed by "our new soil, and distinct climate." While he acknowledged that Europeans like Van Mons might produce vegetable aristocrats through enfeeblement, American improvement came either from "planting the seeds of the finest celebrated old sorts" or from crossbreeding Europeans and "natives." Ideas about culture were intellectual and commercial tools, not scientific premises. When Hovey and Downing argued about Van Mons in 1846, their real disagreement concerned the relative excellence of the climates, soils, and nursery businesses of eastern Massachusetts and the Hudson Valley.[33]

While it would be possible to explicate American horticulturists' (often only implicit) theories, doing so would not be very illuminating; looking at particular species provides a much richer picture of their beliefs. I examine the relatively simple cases of apples and pears, and then the more complicated issues involved in work with strawberries and grapes.

Jonathans and Other American Natives

Apples, everyone agreed, were model Americans whose naturalization had been effortless and whose ability to improve was remarkable. Horticulturists had to remind readers that in spite of appearances, the genus *Malus* was not a primordially New World fruit; while the crab apple was indigenous, all orchard trees had descended from European imports. At least some of the unnamed varieties brought by the Puritans had grown and maintained themselves in Massachusetts as well as, if not better than, in England. They had also been remarkably plastic and progressive. New varieties that fit into American environments had arisen, seemingly spontaneously, by the early 1700s. The three most prominent "native" apples were the Roxbury Russett, Rhode Island Greening, and, above all, the Newtown Pippin. New Jersey orchardist William Coxe described the last of these as "the finest apple of our country, and probably of the world."[34]

The identity between apples and mainstream Anglo-American settlement became fixed in the first half of the 1800s, initially through the figure of Johnny Appleseed, but then through the images of the fruits themselves. Downing, in his fruit convention satire, emphasized the apples' "honest, ruddy faces; and whether Yankee, English or German, evidently all of the Teutonic race." The Newtown Pippin described himself as "a 'native American,' and he gloried in it," as did the Jersey Red Streak and the Jonathan. Although the Lady Apple acknowledged that "my ancestors still live in France," she affirmed that she too was "a real American."[35]

This aggressively American apple culture was stippled, however, with fears of creolization. Prior to the 1820s, apple trees had been ubiquitous because of the national thirst for cider. With the simultaneous spread of temperance and the decline of traditional New England farming in the 1830s and 1840s, cider orchards were abandoned, and volunteer apple seedlings grew up in neglected fields. The puny fruits of these rogue trees made arguments about American apple degeneration immediate and real. Henry Thoreau expressed these concerns in his last essay with reverse snobbery; he celebrated the unproductive wild apple as a fellow Yankee bachelor, seemingly useless but actually superior in its tangy authenticity.[36]

The apparent alternative was uniformity. In the 1850s, western New York became the center of American apple mass production. The Baldwin became the "standard commercial variety" because its late ripening, thick skin, and hard flesh meant it could be shipped in bulk by rail and would keep through the winter. With its deep red color and supposedly bland flavor, it was the Red Delicious of its day—an urban table fruit that confirmed the impression that apples were a commodity for common people.

Bartletts and Blights

Pears *(Pyrus communis)* offered a significant variation on the apple theme. While the two fruits were botanically close and had long been grown in the same regions, pear culture was much more sophisticated. In the early 1600s French aristocrats had initiated "a veritable craze for this fruit," collecting and exchanging precious varieties. Pear trees grew more slowly than apples, and they produced fruits that had to be picked before ripening, tended in a climate-controlled fruit room, and then brought to the table during their brief moment between astringency and decay. Properly tended pears, however, rewarded discerning consumers with aromatically "rich," "buttery," and "melting" tastes. They were the perfect upper-class fruit.[37]

Growing pears was more of a challenge in the United States than in western Europe from the beginning. The country's climatic extremes and changeability caused problems for varieties that had arisen in the more even climates of England, France, and the Low Countries. In contrast to apples, only a handful of pears were described as native. The most distinctive was the Seckel, traced to a tree growing in the 1700s on the Delaware River south of Philadelphia. In 1818 David Hosack sent samples to England, touting their "exquisite" flavor and "factitious aromatic perfume"; but Seckels were small and were extremely hard unless ripened with care. The more common situation was represented by the variety that Boston merchant Enoch Bartlett found on a suburban estate he purchased in 1817 and named for himself. Promoted as an American pear, the Bartlett was eventually exposed as a Williams Bon Chrétien that had been imported from England in 1797 for the property's former owner; that name had itself been an English nurseryman's pseudo-French rechristening of a type known earlier, in Berkshire, as Stair's Pear.[38]

Both Downing and Thoreau satirized pears as foreign and pompous, but these were the characteristics that made them so interesting to enthusiasts.[39] Over a period of two decades, Robert Manning collected, largely through correspondence with Van Mons and leaders of the Horticultural Society of London, more than one thousand varieties at his orchard near Salem. Nurserymen encouraged the pear cult; Hovey, as I noted above, was displaying and marketing hundreds of named varieties by the early 1850s. Tamara Thornton has provided a discerning sketch of the degree to which Brahmin fruit enthusiasts identified with their pears, and the multiple interweavings of pear culture and elite culture.[40]

. Some degree of difficulty in growing pears was good—it stimulated experimentation and competition. The problem was that American orchardists were increasingly overwhelmed in the course of the 1800s by a blight that Europeans did not experience. In some years, often at the height of late spring growth, branches and leaves would suddenly shrivel and blacken, destroying the young fruits and sometimes an entire tree; the characteristic damage led to the descriptive term "fire blight." As with most orchard difficulties, lively arguments arose about the causes: possibilities included a borer, a beetle, poor soil, hot sun, variable winter weather, or some unknown attacker. Whatever the cause, controlling fire blight required constant attention and brutal intervention; growers needed to inspect their trees weekly and to cut off and burn all affected parts immediately. Amateurs walking the allées of Hovey's Nursery did not imagine that the trees they brought home would require repeated mutilations to survive.[41]

By the 1860s, some were questioning whether pear culture was worthwhile—the species, perhaps, was simply not naturalizable in America. The aging Hovey defended his vision for this fruit in an article dripping with sarcasm, affirming that anything valuable took effort and time. But in the 1920s leading pomologist U. P. Hedrick looked back and acknowledged that blight and climate made pear growing very difficult in most parts of the United States. Moreover, the variety most widely cultivated in the early twentieth century, the Kieffer, was neither European nor American, but a hybrid between the Bartlett and the Chinese sand pear *(Pyrus pyrifolia)*. For Hedrick, the Kieffer's mixed ancestry, gritty texture, and "potato-like flavor" made it a "pretentious cheat," but he acknowledged that it predominated because, unlike European varieties, it resisted fire blight and was cannable. Kieffers in syrup were not the rich, buttery future that Yankee pomologists had imagined three generations earlier.[42]

A Modern Amalgam

Strawberries, seemingly, were less problematic. In contrast to apples and pears, edible species of *Fragaria* grew in both the Old and New Worlds. The eastern North American strawberry, *Fragaria virginiana*, was as flavorful as its European counterpart, *Fragaria vesca,* and was larger. While orchards involved substantial commitments of both space and time, a bed of strawberries could be established on a few square feet of ground with a few days' work and would bear a crop within a year. Yet substantial continuing harvests were extremely difficult to obtain. American varieties like the Virginia Scarlet grew easily and could be propagated effortlessly through runners, but they bore sparingly and for only a short period each year. In the 1830s, when apples were available most of the year for less than five cents per pound, strawberries were a luxury fruit; a quart cost fifty cents during the brief high season in June.[43]

European developments offered great hope but introduced further complications. Strawberries had long been fixtures in aristocratic private plots and in botanical gardens. They had changed little between classical times and the end of the 1600s. During the next century, however, plant enthusiasts gathered, identified, and planted together in their gardens the world's major *Fragaria* species, including not only the European *vesca, moschata,* and *viridis,* but also the North American *virginiana,* and, most significantly, the South American *chiloensis*—a species that Andean Indians had cultivated to the point that it produced berries significantly larger than the others (though supposedly less flavorful and productive). Brought together physically for the first time in evolutionary history by humans attentive to their taxonomic similarities, *Fragaria* species spontaneously began to interbreed. In 1759 British garden writer Philip Miller described a large and flavorful berry he had obtained from the Anglo-Dutch amateur and patron of Linnaeus, George Clifford. A few years later, the young Parisian horticulturist Antoine Duchesne investigated this type—named *Fragaria ananassa*, or the Pineapple strawberry—in depth; he concluded that it was a new "race of fruit" resulting from the interbreeding, in Holland and elsewhere in Europe, between the eastern North American *virginiana* and western South American *chiloensis.* The new strawberry was quintessentially modern and cosmopolitan. While not bred intentionally, it was a product of culture—of the practices and values of the European botanical elite.[44]

In the early 1800s English horticulturists, including both Knight and commercial gardener Michael Keen, became enthusiastic improvers of what they nostalgically called the "Old Pine" strawberry. Selection and hybridization resulted in hundreds of varieties, but only a few—notably Knight's Downton and Keen's Seedling—spred widely in English gardens. Americans learned about these developments in the 1820s, and began to import the new English types. However, as with many other European fruits, they had little success. English strawberries did not reliably survive New England winters.[45]

Charles Hovey, in his first horticultural project, took up the challenge of creating an improved "American" strawberry. In 1833, when he was twenty-two, a junior partner in his brother's seed business, and a new member of the MHS, he began to crossbreed some of the dozen varieties growing in his family's Cambridge backyard. Following ideas and techniques outlined in the *Edinburgh Encyclopaedia*, he crossed varieties with complementary characters. Although he misplaced his notes on the characteristics of the different daughter plants, the project was a practical success: he selected the plant with the best hardiness, fruitfulness, and taste, named it Hovey's Seedling, and promoted it in his magazine (Figure 3.2). Hovey's Seedling caused a sensation when it was displayed at the MHS in 1838. It gave Hovey his national reputation, and its profits enabled him, in 1840, to buy the Cambridge property that would become his nursery. Horticulturists, looking back, consider this the first example of experimental fruit breeding in the United States.[46]

The problem with Hovey's Seedling was that while it produced good yields in Hovey's Nursery, where it grew near other varieties on display, it frequently disappointed those who paid five dollars per dozen plants to begin their own strawberry beds. (Cost comparison: In 2006 plant dealers routinely offered twenty-five plants of named strawberry varieties for under ten dollars.) While the plants grew and flowered, they produced few fruits; for those not associated with the MHS, Hovey's Seedling seemed one more example of the overblown claims of commercial nurserymen. Cincinnati enthusiast Nicholas Longworth gradually clarified the difficulties peculiar to strawberry culture. Around 1837 he learned from a local German American "illiterate market gardener" that strawberry plants could be either male (with developed stamens and rudimentary pistils) or female (with the reverse), and that both kinds were necessary for a truly productive patch. He saw that some of the major cultivated varieties were essentially female, or pistillate. Since the recommended practice was to pull up poor bearers in favor of the more fertile plants, he concluded that most gardeners were decimating production over time by eliminating their male plants.[47]

The separation of sexes in some strawberry races had been suggested periodically since the time of Duchesne, but growers had generally considered the issue unimportant. Longworth made it a public question in 1842 when he sent Hovey an essay declaring that Hovey's Seedling was "defective in the male organs" and required that plants of another variety, with fully developed stamens, be nearby. Initially Hovey considered this claim plausible, but then decided that the cause of Longworth's "imperfect" pistillate plants was overmanuring.[48]

For more than a decade, American pomologists debated "the great strawberry question." Issues mixed as promiscuously as the species of *Fragaria*. While Cincinnatians doubted the virility of Bostonians' plants, Bostonians questioned whether Ohio growers were capable of emulating the high culture common in Massachusetts. The millionaire Longworth's ironically deferential pose as a denizen of the "back woods" generated attacks on "Czar Nicholas." Participants could easily doubt whether people in other cities were in fact growing the varieties they described. The deeper difficulty was that while some strawberry varieties had distinct and identifiable staminate and pistillate plants, others were hermaphroditic or "perfect," some were in-between, and some, like Hovey's Seedling, existed in only one sex. Moreover, the development of sexual organs did depend to some degree on soil, manure, and other conditions of culture. Further issues were whether breeders should aim for greater sexual separation or for perfectly hermaphroditic plants, and whether the separation of sexes in strawberries was a natural condition, an improvement, or degeneration. Strawberries were so cultured that discussants could not be sure whether comparisons with plants that might or might not be wild would provide any guide.[49]

By the mid-1850s, the practical issue shifted from strawberries to people. Neither home gardeners nor commercial operations with low-paid casual labor were sufficiently disciplined to keep plants in a mixed-sex, mixed-race patch from rampant running and interbreeding. Breeders began to produce varieties that were foolproof—that did not require the repeated culling of most (but never all) males. Longworth's Prolific and Wilson's Albany were hermaphroditic varieties that would bear reliably year after year; although the berries were smaller and less flavorful than those of pistillate plants like Hovey's Seedling, they were good enough for urban markets. Wilson's Albany became the standard commercial strawberry, analogous in its ubiquity and mediocrity to the Baldwin apple. The summer price of strawberries in Cincinnati dropped to less than five cents per quart in the 1850s. It was abundance, rather than any particular high-quality type, that made "this strongly republican fruit" an international symbol of America's bounty in the decades after midcentury.[50]

The Meaning of Concord
While Chapter 1 introduced the emblematic and practical significance that wine grapes had for Anglo-American colonizers, and the fitful realization that replicating European viticulture on the eastern seaboard was an impossible task, this section examines efforts to culture what were considered native grapes.[51] Focusing on E. W. Bull's development of the Concord in the 1840s and 1850s, I emphasize the scientific and intellectual foundations from which this grape arose, and the commercial and ideological significance of its apparent success. There were, however, major ambiguities in the meaning and experience of the native as embodied in the Concord; both biologically and socially, it pointed not toward the Yankee purity that Bull and others imagined, but toward the heterogeneous mixing characteristic of twentieth-century metropolitan culture.

The history of Anglo-American efforts to establish *vinifera* in North America was a long one by the time Thomas Jefferson and Filippo Mazzei laid out their trenches on the slopes of Monticello, but, as Thomas Pinney has emphasized, it was simple in its repeated failures. From the Virginia Company in 1620 onward, prominent colonizers imported vines and vinedressers in the expectation that wine would follow, but then watched their plants fade. Growers attributed failure to inadequate or inappropriate culture (either Americans did not know how to grow *vinifera*, or immigrant vinedressers were ignorant about North American climates) or, more depressingly though unspecifically, to the idea that American environments truly were destructive to these refined European plants.

After 1800 both practical and nationalistic impulses led growers to abandon imported vines and instead to explore the possibility that American grapes might offer a solution. The initial issue was to find and culture the native vines that showed promise—that differed from the ordinary skunky labruscas that grew rampantly on roadsides. Swiss immigrant vigneron Jean Jacques Dufour was the first to move in this direction. He had planted European vines in a new settlement he established in 1799 near Lexington, Kentucky. When these failed, Dufour began to look for natives. Philadelphia nurserymen were selling a newly found local variety they called "Schuykill muscadell," or the Alexander. Dufour purchased this variety, relocated his vineyards from the bluegrass region of Kentucky to the more Rhenish topography of the southeastern corner of Indiana, and enjoyed some success as a wine producer during the next two decades.[52]

In the 1830s the Alexander gave way to the Catawba. Nicholas Longworth obtained this extremely dry and distinctively flavored variety from John Adlum, a Washington, D.C., amateur who had purchased it from a family of German Americans living in western Maryland; it was later claimed that the family had found it in the North Carolina valley from which its name came. Longworth rented hillside land along the Ohio River just east of Cincinnati to German immigrants, provided them with vines, bought their grapes, and manufactured wine. The scheme exemplified the business and civic savvy that made Longworth the Midwest's first millionaire. On the one hand, he profited from both the land and the wine; on the other, locating his vineyards on the river both impressed travelers with the beauty and promise of Cincinnati and advertised his goods. As a bonus, the vineyards improved the view from his hilltop estate, the Garden of Eden (now Eden Park). Longworth was particularly interested in selling "sparkling catawba." The product looked like champagne, but Longworth did not claim to "imitate any of the sparkling wines of Europe." He forthrightly and optimistically advertised "a pure article having the peculiar flavor of our native grape."[53]

Catawbas grew well in southwestern Ohio in the 1840s, but did not ripen quickly enough to be successful around the horticultural hub of Boston. In addition, their dryness made them useless as table grapes; protemperance horticulturists such as Henry Ward Beecher disdained fruits that were usable only in alcoholic form.[54] As a consequence, there were a number of efforts in Massachusetts to produce an earlier, hardier, and sweeter variety. By far the most important was the Concord. The history of this fruit blends smoothly the thematic strains introduced thus far.

Ephraim Bull was a literate and well-connected (he married a niece of one of Harvard's presidents) Boston craftsman who had been interested in gardening from childhood.[55] As a maker of gold leaf, he operated in a unique social space that connected modern technology, finance, and fine art with some very ancient skills. In 1836 he left Boston for the suburban village of Concord, purchasing a seventeen-acre property on the road to Lexington, where the American Revolution had begun. Bull combined management of his gold-beating shop with serious amateur gardening. He was a regular customer of New York plant importer William Prince, spending hundreds of dollars in some years on fruit trees, grapevines, conifers, and roses. He dressed well and became an active participant in the Concord Farmers Club, a group led by sometime *New England Farmer* editor Simon Brown.[56]

In Concord, fruits had philosophical import. Bull lived a short distance from Ralph Waldo Emerson, famous as the author of "Nature." In 1845 the Transcendentalist seeker Bronson Alcott, recovering from a disastrous experiment in vegetarian communitarianism at a settlement called Fruitlands, moved his family (including the adolescent Louisa May) into the house next to Bull's; like his neighbor, Alcott became a passionate gardener. A year later Henry Thoreau was contemplating beans on Walden Pond, two miles to the south. As a religious liberal with spiritualist leanings, Bull was at home.[57]

Bull was a theory-driven horticulturist who had strong views that he was glad to share. Like H. A. S. Dearborn and other Massachusetts amateurs, he was deeply interested in J.B. Van Mons's experiments refining wild plants. In 1840 he began, as had Van Mons, with a volunteer seedling found in a far corner of his garden. He later claimed that it had either been dropped there by birds or carried up from the river by local boys. He moved this unique plant to the center of his garden and tended it carefully; when it fruited in 1843 (fertilized, Bull believed, by a nearby Catawba), he planted the grapes that resulted. Following Van Mons's practice, he pulled up the seedlings that sprouted in 1844 in the belief that the more vigorous plants were more primitive; he fostered only "the more feeble vines which come up in the second and third years, and which alone bring the improved type, which is a departure from the original, requiring the hand of the horticulturist." In 1848 these vines rewarded his efforts at culture by producing grapes that were large, good-tasting, hardy, and early.[58]

Bull hoped to profit from his innovation. In 1849 he contracted with nurseryman and *Boston Cultivator* editor Samuel Cole to produce plants in quantity. When these were ready he began to share them, first with neighbors such as Thoreau and Nathaniel Hawthorne, and then, as the Concord, with the MHS.[59] Hovey contracted to distribute the Concord nationally, and launched it in the *Magazine of Horticulture* with a full-page engraving and description in February 1854, just in time to take orders at the premium price of five dollars per cutting. Hovey boosted the new variety a number of times that year, while Downing's *Horticulturist* doubted whether it was good for anything besides preserves. Horace Greeley, a weekend farmer in Chappaqua, New York, brought the Concord to general notice with a personal commendation in his *New York Tribune*.[60]

The identities of Bull, his grape, his village, and his folk merged rapidly in the mid-1850s. Bull promoted the Concord as a native grape with a decidedly Yankee character—it was early, handsome, had "good shoulders," and could be used both for the table and for wine. In an early report, this vigorous American grape was contrasted to its "too tender Syrian brothers." As a hero of American horticulture, Bull was invited to join the elite Social Circle of Concord, and, under the auspices of the nativist American (Know-Nothing) Party, he was elected to the Massachusetts legislature and then appointed to the State Board of Agriculture. As an Americanist, Bull gained a reputation as a blunt advocate for the rural natives of Massachusetts and as a critic of the urban corporate and immigrant interests that were taking over the state.[61]

The no-nonsense Yankee persona that Bull shared with his grape lost its charm as cultural sophistication rose after the Civil War. In 1869 the Concord's former partisan, Greeley, repudiated a new award for the variety because he considered the Concord a "common grape" that lacked "a high and delicate flavor"; honoring it would prevent further improvement.[62] Bull continued to develop new varieties, but was commercially unsuccessful; instead of acknowledging that profit in this area was chancy, he railed at unscrupulous nurserymen and hoarded his cultivars. And while family happiness amid economic adversity was the common theme of his sometime next-door neighbors Louisa May Alcott *(Little Women)* and Harriet Lothrop (the *Five Little Peppers* books), Bull's disappointments caused his life to collapse. His wife left him around 1870 after forty-five years of marriage, and he became isolated and increasingly eccentric. At the age of eighty-seven he fell off his roof while patching a leak and was put into an old people's home as a charity case. The crotchety epitaph of this Transcendentalist-turned-native read, "He sowed; others reaped" (see Figures 3.5 and 3.6).[63]

The Concord's identity as a native was arguably more problematic than Bull's. The search for promising native grapes meant finding varieties that had characteristics identified with wayside specimens but that were better, where "better" meant taste qualities associated with *vinifera*. The few people on the lookout for such specimens were urban

Figure 3.5 Ephraim Bull. Massachusetts botanist Walter Deane took this photograph in July 1893 and sent a copy to Liberty Hyde Bailey, copying from his journal that "Mr. B is quite old and lives in a wretched house by himself." Liberty Hyde Bailey Papers, Miscellanea, box 1, 21/2/430. Reproduction courtesy of the Division of Rare and Manuscript Collections, Carl A. Kroch Library, Cornell University, Ithaca, NY.

---

or suburban horticulturists, not backwoods farmers; they searched their surroundings, received samples from acquaintances, or contacted other better-positioned plantsmen. With two centuries of attempts to introduce and grow *vinifera* along the eastern seaboard, there were, not surprisingly, occasional spontaneous interspecies hybrids. The Alexander appeared in Philadelphia in the 1790s near an old vineyard. The Catawba's history prior to its propagation by horticultural amateur John Adlum was unclear, but morphologically it was a first-generation *labrusca-vinifera* hybrid. The Concord was a more complex cultural blend. Bull asserted that one parent was a Catawba; the other, the early sweet-tasting vine found by the wall in 1840, was probably the offspring of a *labrusca* and one of the many varieties he had planted on

Figure 3.6 In 1890, *American Garden* (p. 638), a magazine edited by Liberty Hyde Bailey, included this stereotyped "native" as an illustration for an article, "Backwood Gardens." Reproduction courtesy of the Rutgers University Library.

---

his property during the previous four years. The Concord was thus a second- or third-generation offspring of interbreeding Americans and "Syrians."

The mixed parentage of Concord, Catawba, and other American grapes made them tougher than *vinifera* and thus able to survive and bear fruit in the antebellum period. The problems that had plagued *vinifera* soon afflicted them as well, however. In the 1860s the Alexander disappeared as a distinct variety, and the Cincinnati Catawba industry collapsed. Over the next two decades botanists determined the causes of these problems. Fungi—most prominently powdery mildew *(Uncinula necator)*, downy mildew *(Plasmopara viticola)*, and black rot *(Guignardia bidwellii)*—that had coevolved with American grape species were uniformly fatal to the untested alien *vinifera*. The new hybrids carried some resistance from their American parents, but within a few decades American fungi had caught up in evolutionary sophistication and parasitized them as completely as they did *vinifera*.[64]

A new era of eastern viticulture began in the 1890s with the introduction of the copper-lime Bordeaux Mixture, the first effective sprayable fungicide, from France.[65] Development, however, continued to combine fantasies of purity with realities that were an earthier mix. Around 1870 New Jersey dentist Thomas Welch used the new technology of pasteurization to prevent grape juice from fermenting; he sold this product to fellow Methodists as a nonalcoholic sacramental wine. A generation later his son Charles saw that New York grape growers were reliably producing Concords as their standard variety, and that packaged groceries could be distributed and advertised nationally. He began to sell pasteurized grape juice as a mass market food. Advertised to older people and to children as "the national drink," Welch's Concord Grape Juice made this American grape seem fundamentally different from the European varieties used for wine.[66]

Soon after the turn of the century, however, the Concord gained another identity that would have made Bull turn in his grave. Traditionalist Jewish immigrants in New York used Concords, the most readily available and least-expensive grape variety, to make kosher wine. They added significant amounts of sugar to aid fermentation and to mask the variety's high level of tannic acid. For decades, oenophiles turned up their noses at leathery Jewish brands that threatened to make American wine synonymous with poor taste. But it is culturally fitting that Schapiro and Manischewitz vie with Welch in delineating the Concord's modern identity.[67]

The purpose of these detailed accounts of apples, pears, strawberries, and grapes has been straightforward: to convey how much culture was involved in antebellum pomology. Amateurs and nurserymen devoted a great deal of attention to the task of learning about, acquiring, propagating, and caring for new kinds of plants. They created networks that crossed class lines and extended throughout the Northeast and into the Midwest. Building on European traditions, they articulated theories of improvement and, more influentially, a web of meanings that connected particular kinds of plants, their origins, and their potentials with the future development of the American nation. Finally and most challengingly, they struggled to respond to the novel evolutionary trajectories being taken by plants—whether promiscuous strawberries, miscegenistic grapes, or rapidly adapting parasitic fungi. Plants disrupted Americans' cultural practices, and reordered their cultural categories; they generated new social relations and new national identities. All this, however, was limited to the confines of gardens. The next chapter looks outward across the nation's new boundless landscape.

CHAPTER FOUR

Fixing the Accidents of American Natural History
Tree Culture and the Problem of the Prairie

North American biotas changed dramatically in the nineteenth century as Euro-Americans moved from the eastern seaboard into the territory stretching from the Appalachians to the Rockies. Agricultural pioneers, working to establish farms linked to national and international markets, cut down forests, plowed up prairies, and exterminated animals. They replaced diverse plant populations with monocultures of wheat, corn, and cotton; they fostered their familiar complement of domesticated livestock; and they carried with them weeds, insects, and plant diseases that had earlier crossed the Atlantic. The scope of these transformations was stunning, the range of farmers' activities was complex, and the results have endured to the present.

Qualitatively, however, these changes were not new. Nineteenth-century farmers were recapitulating, albeit more quickly and on a broader canvas, the vegetational transformations their ancestors had initiated further east during the preceding two hundred years. I could extend the narratives of the preceding chapters by describing the stages by which Hessian flies chased wheat farmers across the continent, or the events that enabled codling moths to catch up with Johnny Appleseed's trees. The details might be new and revealing, but the themes would not.

What *was* new, and is the subject of this chapter, was the movement to arborize the mid-American grasslands. While farmers experienced that region as a blank slate for their familiar monocultures, tree culturists directly confronted the Midwest's biotic idiosyncrasy—the existence of vast areas populated almost entirely by grasses and forbs rather than by woody plants. Small groups of prosperous amateurs, nurserymen, and scientists took the lead in Midwestern tree culture. They interacted in state horticultural societies (most notably in Illinois, Iowa, Nebraska, and Kansas) and were linked through lectures and exchanges to national and international forestry circles. They developed governmental and communal tree planting campaigns for what they considered the common good—improvement of the regional economy and environment, and the aesthetic well-being of future generations.

Midwestern tree culturists sought to fix the accidents of American natural history. In a straightforward sense, this meant repairing damage that humans (Indians and early settlers) and destructive animals (such as bison and grasshoppers) had done to desirable organisms. More deeply, it involved reordering settings that tree-friendly naturalists believed were merely contingent consequences of historical events, some of which were recent and some of which extended back millions of years. Given tree culturists' limited understanding of species' interactions in the present and the past, and also the tensions between the goals of repairing and reordering, the futures they envisioned were divergent. On one side were opportunistic cosmopolitans—enthusiasts or nurserymen who were interested in growing any plant that might have desirable qualities. On the other were localists who wanted to restore an indigenous silva, which they believed would grow better and would also express the genius of the place. Neither tendency expressed itself fully or consistently: cosmopolitans' ties to international plant exchanges were actually quite limited, and localists were both uncertain about what was indigenous and were unable to restrain the actions of their neighbors.

This chapter examines prairie tree planting initiatives in the nineteenth century, with greatest attention to the two decades after the Civil War. It begins with the background of tree planting discussions on the seaboard. It then looks at scientific debates over the "problem of the prairie"—the reasons why the midcontinent was largely treeless. I sketch discussions regarding what kinds of trees should be planted, and resituate the two major arborizational programs of this era: the regionally significant federal Timber Culture Act, and the nationally transformative creation of Arbor Day.

The difficulties that tree culturists, along with homestead farmers, faced in the territories beyond the 100th meridian have long been discussed.[1] However, arguments about the Great Plains—involving bison, the Dustbowl, range management, and industrialized agriculture—have diverted attention from the Tallgrass Prairie region between the 88th and 95th meridians, and sometimes further west, where arborized landscapes in fact appeared during the nineteenth century. Understanding prairie tree culture is important for interpreting this environmental transformation. It also forms the background for understanding the twentieth-century efforts at prairie restoration that are discussed in Chapter 9.

A Procession of Forest Trees

European elites long had multiple interests in the maintenance and development of trees. John Evelyn's *Sylva* (1664) was formative in linking tree planting to goals that ranged from national security (maintaining the Royal Navy's "wooden walls") to the need for construction materials and fuel, the moderation of climate and water flow, and the maintenance of mental health among the gentry (through the creation of settings that were picturesque and would harbor game). French agricultural scientist Duhamel de Monceau highlighted similar issues a century later.[2]

In northeastern North America, the scope of existing forests and the small number of large estates meant that few eighteenth-century gentlemen shared these European concerns. By the middle of the nineteenth century, however, the combination of fires and accelerated stripping of landscapes to supply lumber, potash, and firewood, on the one hand, and the rural villa movement and the spread of tourism on the other, led some easterners to argue that they were facing deforestation problems similar to those that Europeans had long experienced. For artists such as Thomas Cole and Asher Durand, tree stumps symbolized the unfortunate disappearance of nature in the face of civilization. In the 1850s, pioneer environmentalist George P. Marsh saw devastation around his childhood home in Woodstock, Vermont: because Mount Tom had been stripped of the trees Marsh knew as a child, hillside springs had dried up and local streams, clogged with silt and debris, were alternately flooding or empty.[3]

Environmental historians have studied extensively the process of deforestation in North America, and the recognition—encapsulated in Marsh's *Man and Nature*—that such change had major consequences. They have cataloged the growing number of warnings about coming timber famines, and have outlined the steps through which advocates induced state and federal governments to protect and care for forests.[4]

One aspect of this history, however, has received comparatively little attention. In a reversal of the old saw, scholars have tended, in their concern about forests, to miss the trees. What mix of species did interested Americans imagine that their grandchildren would be seeing, and how did they anticipate such populations coming to be? More specifically, what did Americans do with deforested landscapes and what, if anything, did they plant? Early planting initiatives are important clues to the values embedded in tree culture; moreover, they provide a basis for understanding the later activities on the prairies—where the issue of planting was more central.

In most of the accessible parts of the Northeast, landowners did not need to do anything to transform fields into forests. Neglected farms in particular sprouted a succession of species that ultimately consisted primarily of trees. In a talk presented to Concord's agricultural society in 1860, Henry Thoreau explained that the odd mix of North American and Eurasian species—cherries, plums, apples, poplars, willows, junipers, and oaks—in the woodlots he was surveying had resulted from squirrels burying acorns, blue jays excreting cherry pits, and winds carrying seeds from trees nearby that had not been considered valuable enough to cut down.[5]

For Thoreau, this process, occurring subtly under landowners' noses but independently of their schemes, held spiritual significance that he expressed as his "faith in a seed." More prosaic tree lovers such as Arnold Arboretum director Charles S. Sargent, by contrast, questioned the value of old-field succession on the grounds that "years of struggling growth" would result in trees that were inferior for fuel and comparatively worthless for manufacturing.[6] The initial alternative, advanced by George B. Emerson in his state-sponsored 1846 report on trees in Massachusetts, was selection. Landowners could grow the trees that appeared, but should periodically cull useless species and misshapen or poorly located individuals.[7] In horticultural circles, however, interest centered increasingly on active planting. Gentleman landowners wanted both to generate economic value and to imagine legacies of varied landscapes for future generations. Much discussion revolved around the cost and labor involved in planting and managing woodlands. The more open-ended question, however, was what to plant. Responses fell into three categories: restoration, Americanism, and cosmopolitanism.

A few argued for restoration of the local "primitive forests." Massachusetts nature writer Wilson Flagg, for example, proposed that a small part of every square degree in the country be designated a "forest conservatory." These areas would serve as reservoirs of species, nurseries for animals, moderators of climate, and sanctuaries for women, naturalists, and overstressed clergymen. Flagg rejected aesthetic management of these sites, but had a particular romantic vision of their appearance and species composition. He imagined lands filled with trees and undergrowth, crossed with paths, and dotted with old-fashioned farms. He suggested that old local species could be introduced into vacant spaces and then left alone; he passed over the presence—and resistance to displacement—of naturalized trees such as willows and apples that had occupied old-fashioned farms over the preceding centuries.[8]

The second approach to tree planting, much more prominent than Flagg's, centered on the active establishment of a national silva. Massachusetts Horticultural Society leader John L. Russell argued that the many vigorous, beautiful, and useful trees found in different parts of North America should be cultivated beyond their present limited ranges. Central Appalachian species such as locusts, chestnuts, dogwoods, catalpas, and, most spectacularly, a number of magnolias, could be added to New England forests. Mountain laurels could grow on the seashore, and beach cherries far inland. The whispering pines and the hemlocks could grow in New Jersey as well as in Nova Scotia; and, in the view of A. J. Downing, the American elm was the perfect tree, which should be planted everywhere.[9]

The scope of arboricultural nationalism expanded significantly after the United States took over the Oregon Territory, California, and the desert Southwest in the late 1840s. Northeasterners added magnificent Oregon species such as the coastal redwood *(Sequoia sempervirens)*, silver fir *(Abies amabilis)*, and sugar pine *(Pinus lambertiana)* to their lists of plants that were native. Acquiring specimens, however, induced cognitive dissonance: on the one hand, the only sources for these plants were in England, where nurseries had obtained seeds two decades earlier from Horticultural Society of London collector David Douglas. On the other hand, Northeasterners discovered that while northwestern conifers apparently grew successfully in England, they often failed in the eastern United States. The most difficult situation involved the California Big Tree. While a few Americans in California had seen some living specimens, British botanist John Lindley published the first formal description of the species and gave it the name *Wellingtonia gigantea*. Americans were livid that what was now their great tree would forever memorialize a British general, and pushed instead for *Washingtonia californica*. French taxonomist Joseph Decaisne mooted the dispute by declaring that the Big Tree was in the same genus as the already-named redwood, and hence should be named *Sequoia gigantea*.[10]

A. J. Downing conveyed the high stakes involved in silvicultural nationalism when he proposed, in his 1851 plan for landscaping the National Mall in Washington, that the Washington Monument should be surrounded by a grove of "American trees, of large growth."[11] The alternative site was Mount Vernon: the 1860 federal agriculture report, being typeset as Southerners fired on Fort Sumter, presented a pathetic vision in which, a thousand years hence, "a forest of gigantic trees, like those of Maine, Mississippi, and California," were

> mingling their branches hundreds of feet above the tomb of Washington, and say[ing]: "These trees were planted by our ancestors, and have been raised to their present height by the Creator, while yonder marble monument has crumbled to the earth. Here are the links that connect the Present with the Past, for the Park is the *living* memorial of Our Country's Father, preserved to us since the early days of this great Republic by the patriotism of our mothers."[12]

For the next two decades William Saunders, the U.S. Department of Agriculture's superintendent of gardens and grounds, tried to grow sequoias and other western conifers on the National Mall in Washington, but was unsuccessful. Visitors to Mount Vernon and Washington in the year 2861 would not in fact be able to experience the symbolically compelling national groves imagined by their ancestors—one absorbing the substance of the father of their country, and the other, having grown year by year closer in height and form to the ancient obelisk, comprising a living memorial to that era of conquest.[13]

Whatever their hopes for sequoias, arboricultural nationalists disdained European and Asian trees. J. L. Russell argued that foreign plants were popular largely because Americans maintained a childish willingness to "defer to extranational opinions and modes of life." It was demeaning, he declaimed, for Nature to be thus interpreted "to us; to us, who are Nature's sons."[14] Columnar Lombardy poplars, planted widely during the neoclassical first decade of the century (but ragged and dying by the 1850s), were the prime exemplars of Old World trees that were inappropriate for America and inferior to natives. The more pressing problem, however, was the ailanthus, which had been introduced in the late 1700s from China as a fast-growing ornamental. In 1852 A. J. Downing warned his readers that this "Tartar" "smells like the plague" and should not be planted (see Chapter 7 for more on Downing's views).[15]

The third approach to planting—the cosmopolitan—was gaining energy just as Downing was damning the ailanthus because a new cohort of Asian shrubs, vines, and trees was becoming commercially available. Britain's Royal Botanic Garden and its nursery dealers were promoting Himalayan rhododendrons that were much more spectacular than the American species that had interested John Bartram and the Duke of Argyll a century earlier. American collectors associated with Admiral Matthew Perry's "opening" of Japan in 1853 sent both pressed plants and live specimens back to Asa Gray at the Harvard Botanic Garden. A Japanese cedar (*Cryptomeria japonica*) was planted in New York's new Central Park in 1860 to honor the first Japanese diplomatic mission to the United States, and New York nurseries soon began to receive cultivated Japanese seeds and plants—especially from the American consul, James Hogg, a plantsman and former Central Park commissioner.[16] By the late 1860s Japanese conifers, as well as Japanese honeysuckle (*Lonicera japonica*) and Japanese creeper (*Parthenocissus tricuspidata,* which gradually gained the common name of Boston ivy), were all spreading through and beyond gardens in the Northeast.[17]

What were the virtues and defects of the rapidly lengthening list of available woody plants? In the 1870s, Massachusetts horticulturists—including the aged George Emerson, Massachusetts Horticultural Society leader George Manning Jr., and Asa Gray's arboreal protégé, Charles S. Sargent—discussed these questions before the state's Board of Agriculture. While Brahmin amateur Leverett Saltonstall pushed the argument that Old World trees were always inferior to American ones, the consensus among the experts was that some foreign species had proven themselves desirable additions to American forests. This perspective propagated itself later in the decade through New York tree advocate Franklin Hough's federally published *Report on Forestry.* Wealthy Bostonian Joseph Fay demonstrated that imported seedlings including oaks, elms, beeches, Scotch pines *(Pinus sylvestris),* and European larches *(Larix decidua),* could flourish on the seemingly barren lands he owned on western Cape Cod near Woods Hole.[18] Sargent developed his plans for the Arnold Arboretum within this context. He proposed to grow all trees that could survive in Massachusetts; the goal was to enable landowners and nurserymen to compare form, foliage, and vitality in a large number of species from many parts of the world, so that they could make planting choices wisely.[19]

The new wave of imports had consequences that extended beyond practical horticulture to theories of biogeography and ideas about the boundaries and significance of North America. In the late 1840s Harvard botanist Asa Gray was lecturing his students on the biogeographic unity of temperate North America, which extended from Maine to Oregon.[20] A decade later, however, having examined some of the first shipments of Japanese plants, he questioned the value of sharply distinguishing American and Asian trees. Gray compared the floras of eastern North America, Pacific North America, Europe, and eastern Asia. He concluded that eastern American plants were more closely related to those of Japan and China than they were to those of Europe; and that the Pacific North American species were the outliers. He explained this situation as the consequence of the physiographic and evolutionary history of the Northern Hemisphere. More than once between the Miocene and the present, plants of eastern Asia and eastern North America had intermingled across the Bering land bridge and then separated, leaving geographically disjunct populations that then diverged in form. Glaciations had produced a relatively depauperate flora in Europe, but one that still participated in some circumpolar exchanges. By contrast, California had long been separated from the rest of the North American continent by mountains and oceanic influences. Many of that state's plants were unique or related to those of areas further south. They were, by a Massachusetts standard, less American than those of Japan.[21]

For Gray, vegetation history was more important than contemporary geography. He initially presented this argument within the context of the debates over Darwinism, but in 1872 he drew out its national implications. His presidential address to the American Association for the Advancement of Science (AAAS) was a scientific sermon, appropriately, on *Sequoia gigantea* (see Figure 4.1). Gray dismissed enthusiasm about the sequoias' size, and disdained the practice of naming trees for celebrities such as General Sherman. He emphasized, first, that sequoias were not particularly American. Originally circumpolar in distribution, they were equally related to the southeastern North American bald cypress *(Taxodium distichum)* and to the southern Chinese water pine *(Glyptostrobus pensilis)*. Gray's more important point, however, was that sequoias were evolutionary losers. They had disappeared nearly everywhere, and were not replacing themselves in their last refuges on

Figure 4.1 The second edition of George Emerson's *Trees and Shrubs Growing Naturally in the Forests of Massachusetts* expressed silvicultural nationalism and, more specifically, sequoia enthusiasm, by facing its title page with an image of the Big Tree, misnamed a pine. With proper care the species could survive in Boston. Reproduction courtesy of the Rutgers University Library.

---

the Sierra slopes. Nature, Gray preached, was not an ocean with merely tidal fluctuations, but a great river that moved, if imperceptibly, in one direction. Migration, replacement, and extinction were integral parts of this divinely ordained process. Americans could choose to preserve the Big Trees, but they should not interpret them as exemplars of their nation's natural greatness.[22]

The Problem of the Prairie

The AAAS had originally planned to hold the annual meeting at which Gray presided in San Francisco. However, the unwillingness of the

Union Pacific Railroad to provide passes in bulk for men of science resulted in a last-minute relocation to the terminus for free westerly train travel—Dubuque, Iowa.[23] Talking about sequoias in a Mississippi River city surrounded by plowed prairie was superficially incongruous. But Gray's larger subject of trees, and his argument that history mattered in natural history, were just right for that place and time. Forestry discussions in the Northeast had revolved around secondary issues: the kinds of trees to plant and the amount of labor to employ. No one doubted that, in the absence of farming and lumbering, forests would eventually reoccupy seaboard lands. Beyond the 88th meridian, by contrast—in Illinois, Iowa, Nebraska, and Kansas—tree enthusiasts confronted major challenges, whose resolution depended on correct interpretation of the region's vegetational history.

While Midwestern tree culturists believed that woody plants were essential for the economic, aesthetic, and environmental development of their region, they were unsure whether forests could be established. Why was this vast area so bereft of trees? Was the prairie primeval, or was treelessness an accident—a historical contingency that would disappear with the development of tree culture? These questions were collectively understood in the nineteenth century as "the problem of the prairie."[24]

Naturalist Caleb Atwater had raised the issue of the origins of American prairie in the initial volume of the first regularly published scientific periodical in the United States. Focusing on the treeless tracts in the central and western parts of his state of Ohio, he suggested that this flat landscape had once been covered by a much larger Lake Erie. The formation of Niagara Falls, perhaps through a giant earthquake, had partially drained the lake, leaving sands and bogs. Trees were unable to take hold in such soil, and so the result was bleak, grass-dominated flat barrens and meadows.[25]

Within a few months, however, Missouri lawyer R. W. Wells presented an alternative account. He argued that Atwater's geology was inconsistent with topographic facts (were Niagara Falls blocked, Lake Erie would find another outlet only a few feet higher). He emphasized, however, that such "speculation" about ancient natural history was unnecessary because the cause of prairies was directly visible. He asserted that they "were occasioned by the *combustion of vegetables*" initiated largely by Indians trying to make hunting and travel easier. He had personally seen large tracts of woodland destroyed by Indian-set fires. Longtime residents of St. Louis informed him that forests had expanded significantly in recent decades with the retreat of the Indians westward.[26]

This polarization of opinion surfaced repeatedly in the course of the nineteenth century. Naturalists interested primarily in geology and meteorology, discussing at various times lands in Illinois, Kansas, and the High Plains, provided physical explanations for treelessness. Leo Lesquereux emphasized flat topography and lack of drainage, Josiah Whitney focused on the fineness of the soils, James D. Dana and John S. Newberry emphasized insufficient soil moisture, and J. G. Cooper pointed to lack of rain and soil porosity. All their explanations pointed toward the need for settlers to approach occupation of this land more cautiously than they had further east.[27]

Joel Allen, Lorin Blodget, George P. Marsh, and George Sternberg, among other naturalists interested in plants or people, periodically responded to these explanations for the cause of prairies with the refrain that grasslands existed because Indians had burned the trees. They emphasized the various ways that the Indians had used prairie fires—to attack enemies, to eliminate dry growth that impeded travel, and to foster new grass that would attract bison. They noted that fires were easier to start than to stop, and hence grasslands had spread with abandon. The implication of this history was that with settlement, plowing, and the end of reckless burning, trees could be established on the prairie (see Figure 4.2).[28]

Resolution of this debate depended on giving up the idea that the American prairie was a geographically unitary phenomenon. In 1878 Asa Gray suggested that while Indian burning accounted for much of the prairie province, it did not explain the treelessness of the entire expanse. Charles S. Sargent elaborated on this perspective in a government report a few years later. He proposed that grasslands west of the 97th meridian existed primarily because of aridity. The area between the 97th and 95th meridians involved multiple factors. But prairies east of 95° W (Kansas City) were "accidental"—they had existed because of the Indians' use and misuse of fire. Federal geologist John Wesley Powell acknowledged this division in his 1879 *Report on the Lands of the Arid Region of the United States;* in focusing on territory west of the 100th meridian, he tacitly abandoned the application of aridity arguments to lands further east. This consensus acknowledged what people on the ground experienced: that much of the prairie region could support the familiar eastern cultural landscape of farms and trees.[29]

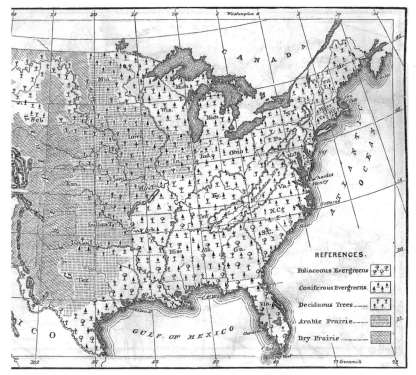

Figure 4.2 In 1857 the Patent Office Agricultural Division's *Report* included a map, "Forest and Prairie Lands of the United States," which tried to please all parties. It clearly demarcated prairie from nonprairie, and distinguished the modern tallgrass and short-grass regions as "arable" and "dry" prairies. At the same time, however, it expressed the perspective of tree planters by marking nearly all the arable prairie as forest—putting the western tree boundary at about 97°W, with extensions along the rivers. Reproduction courtesy of the Rutgers University Library.

Prairie Tree Culture

Buckeyes and Hoosiers planted trees on their prairies throughout the nineteenth century with little self-consciousness. Beyond the 88th meridian, however—at the beginning of the Grand Prairie, in Champaign County, Illinois (fortuitously, the site of the Prairie State's new land grant university)—tree culture was a more serious matter. The first horticultural experiment at the University of Illinois was the planting in

1869 of thirteen acres of trees, comprising twenty-four species.[30] At least some Illinois tree men accepted the geologists' argument that "the very existence of these prairie lands shows an *original* unfitness for tree growth," and they recommended that farmers plant pioneer species such as white elm and cottonwood to "make the soil fit for other growth" in succeeding decades. But others advanced the view that treelessness had resulted from Indian fires and animal activity, and they dismissed the idea that bogginess, thinness, or dryness of soil would make growth of most trees difficult. It was only necessary to plant species that could tolerate local conditions. In the 1870s, horticultural society meetings in Illinois, as well as in Iowa, Kansas, and Nebraska, featured reports about trees with clean trunks more than a foot across that had been planted only a few decades earlier. Railroads such as the Kansas Pacific planted demonstration plots along their lines to show travelers that trees could flourish.[31]

The problem faced by Midwestern tree culturists was to get debt-burdened farmers to set aside land for trees and to maintain their plantings to maturity. Arguments, directed at both individual and communal interests, ranged from the utilitarian through the aesthetic to the environmental. Advocates offered the well-honed argument that trees would ultimately pay in the form of marketable wood. Their broader point was that farmers could and should contribute to regional prosperity by providing raw materials for industries. The most sweeping claims concerned the timber famine that would come once all the forests had been cut.[32] Farmers could readily agree with these concerns, but—diffident and hard-pressed for both money and time—were unwilling to do work that would not generate a rapid return. As a consequence, tree culturists pushed state legislatures in Iowa, Kansas, and Nebraska to offer property tax abatements for land devoted to trees. Nebraska senator Phineas Hitchcock dramatically expanded this approach in 1873 by gaining congressional passage of the Timber Culture Act. This law enabled settlers to claim 160 acres of public land by planting forty acres (soon reduced to ten) with timber trees. By 1891, 40 million acres had been preempted under this law, primarily in Kansas, Nebraska, Dakota, and Minnesota; large amounts of business went to nurserymen and to contract tree planters.[33]

These economic considerations coexisted comfortably with aesthetic or civilizational concerns. The greatest Midwestern innovation in tree culture was the creation of Arbor Day. Democratic politician, newspaper editor, amateur pomologist, and nursery owner J. Sterling Morton induced the Nebraska Board of Agriculture to declare an official annual tree planting day in 1872. Republican governor, newspaper editor, amateur pomologist, and nursery owner Robert W. Furnas embraced the idea enthusiastically. Nominal rewards were provided to county societies that promoted tree planting and to individuals who started the most seedlings, but the event was essentially a statement about community. Morton's focus was on trees, not timber; he consciously chose the individualistic "Arbor Day" over the forest-centric "Sylvan Day." Tapping into springtime urges that went back to the Druids and resonated with modern liberal Christianity, Arbor Day enthusiasts rapidly spread the holiday throughout the country; because scheduling depended on a particular state's climate, the celebrations had a distinctly local quality. By the 1890s, a tree planting ritual formed the centerpiece of a springtime day of pageantry in most American elementary schools. The official themes of Arbor Day were nature, growth, and concern for the future. But children absorbed more straightforward messages: that tree planting was necessary and that any tree anywhere was better than none. As they carried this sentiment, with varying intensity, into adulthood, prairie tree culture became universal.[34]

The broadest arguments for tree culture were environmental—that treelessness had deleterious consequences for water supply, soil stability, and wildlife. Such assertions, associated with the names of Alexander von Humboldt, George P. Marsh, and contemporary European forestry professionals, were well known in Midwestern horticultural circles. In the early 1860s, Illinois tree culturists urged communal planting of windbreaks in order to protect grain from being flattened by storm gusts.[35] During the next decade, the state horticultural society's *Transactions* detailed the more complex claims made by naturalists. Because forests retained moisture longer than "nude" prairies, tree planting would control flooding, maintain stream flow during dry periods, and prevent the drying up of springs. Woodlands moderated local temperatures in both summer and winter, and lessened the violence of storms. Trees would lower the incidence of malaria by drying up boggy areas.[36]

The bolder claim—that forests would increase moisture on a regional scale and hence that tree planting would lead to more rainfall over large areas—received sympathetic consideration, but was considered unproven.[37] This belief, memorably excavated by Henry Nash Smith in the environmental history classic *Virgin Land*, has stood at the center of scholarly interpretations of nineteenth-century prairie tree culture. In the 1880s, it was a significant element in the disagreements between federal forester Bernhard Fernow and federal geographer Henry Gannett over the region west of the 100th meridian, and it was an enthusiasm of some Great Plains naturalists, notably the Nebraskan Samuel Aughey.[38] But most Midwestern tree culturists gave the argument little heed. In 1873, for example, the Northern Illinois Horticultural Society passed a resolution rejecting the prominent Philadelphia botanist and horticulturist Thomas Meehan's broadside attack on the idea that forests influenced climate. Their emphasis, however, was on Meehan's denial that deforestation altered stream flow and dried up springs, not on his assertion that forests had no effect on total rainfall. When a speaker at the Iowa State Horticultural Society raised the climate issue in 1879, the discussion that followed was almost completely skeptical; the concluding comment was that "this talk is largely visionary" and was unnecessary because the commonsense environmental arguments for tree planting were sufficient.[39]

What to Plant

Midwestern ideas about tree planting were much more activist than in the East. As an Illinois writer noted in 1857, "these prairies seem as if purposely left by the Creator for man to impress upon them his own ideas of convenience and beauty."[40] Their strategies, however— restoration, Americanism, and cosmopolitanism—were similar to those discussed in New England.

I have found only one individual who advocated the restoration of local tree associations. A. N. Godfrey's slogan was "Kansas trees for Kansas forests," but he was neither consistent (he made an exception for the chestnut) nor very knowledgeable (he characterized the ailanthus and catalpa as equally alien).[41] Americanism—in this context, the belief that all temperate-climate trees found east of the Rockies were native to the Midwest—was more widespread. American trees were familiar and were considered more likely to grow and less susceptible to disease than species from further away.[42] This view was shaken, however, by the widely valued black locust's increasing susceptibility to destruction by borers, and by the inability of growers to get paradigmatic Eastern trees such as the chestnut to flourish.[43]

The predominant attitude in prairie tree circles was a market-mediated cosmopolitanism. Scottish-born Waukegan nurseryman Robert Douglas articulated this perspective in impromptu remarks at the Illinois State Horticultural Society in 1871. Responding to a warning about "relying on foreign trees," Douglas argued that "we can't do without foreigners in any department of industry." He pointed out that improved grapes, grains, domesticated animals, and fruits were all foreigners and, to "renewed laughter," he ridiculed the native crab apple for being so sour "the hogs won't eat it." "Look at your men!" he went on, affirming that "white men," even Irish and Chinese, were improvements on the "true native American." Just as deer, elk, and buffalo had given way to cattle, the Illinois Indians had "retired before the better breed." From this perspective, which Douglas knew his audience shared, he drew the conclusion that "prejudice against foreign *forest trees*" made no sense. The European larch and Norway spruce were more adaptable to a wider range of conditions than any native tree, and Scotch and Austrian pines resisted prairie winds better than the native white pine. University of Illinois founder Jonathan Turner reinforced this viewpoint with the argument that the most useful plants were "the results of universal intelligence and civilization of the whole race," and their geographic origins were lost in myth. As a historian of Mormonism, Turner reminded his audience that some Americans believed "this is the oldest continent on the globe, once densely populated, but run down, to rest; . . . who, then, can tell us what is originally American and what is not?"[44]

Douglas's highlighting of the European larch was not a random choice. Arguing that it was inexpensive, fast growing, tolerant of different soils and climates, pest free, rot resistant, and easily worked, he promoted it so heavily for mass prairie planting that "Mr. Douglas's favorite tree" became a running joke at the Illinois horticultural meetings. Americanist Willard Flagg countered Douglas with a public wager that the "American white oak" was a stronger wood.[45] Douglas, however, was not parochial in his enthusiasms. His widely shared viewpoint was that any location, even "the bleakest knoll that can be found on any prairie in our State," could support some tree species. The only way to determine what would flourish in the different conditions of soil, moisture, wind, and temperature was to plant.[46]

Midwestern horticultural societies put considerable effort into preparing lists of desirable species, with notes on habitats, culture methods, and uses. Conference speakers described the properties of recently identified species, debunked earlier fads, and compared soil requirements, hardiness, pest resistance, and growth rates of different kinds. Major issues included the susceptibility of the black locust to borers, the toughness of the western catalpa, and particular comparisons—between the white pine and the Scotch pine, sugar maple and Norway maple, and the black spruce and Norway spruce. In preparing tree lists, committees sought to balance their interest in novelty and diversity with farmers' concerns for economy and reliability.[47] As early as 1867, however, more than one hundred species were included in the Illinois recommended list. Most were eastern North American, but a substantial fraction were European or east Asian, and a few were from west of the Rockies.[48]

In the 1870s and 1880s, tree culturists projected this cosmopolitan silva westward. The United States Interior Department, concerned about keeping homesteaders from counting orchards, hedges, or ornamentals toward Timber Culture Act claims, was initially restrictive in the species it allowed to count toward fulfillment of a claim. However, with the support of national forestry leaders prairie horticulturists soon pressured the government to include such favored tree culture types as Osage orange (*Maclura pomifera*), catalpa (*Catalpa speciosa*), and ailanthus. An 1883 report by former Nebraska governor Robert Furnas recommended twenty-two genera ranging from the indigenous cottonwood to such exotics as Russian mulberry (*Morus alba* var. tatarica) and Austrian pine (*Pinus nigra*). Now that the buffalo, Indians, and prairie fires were gone, tree culturists could reasonably imagine that the landscape from Lake Michigan westward would soon be dotted with woodland.[49]

Tree planting initiatives along the 100th meridian stalled in the 1890s following repeal of the Timber Culture Act.[50] In the next decade, however, the Agriculture Department created the Nebraska National Forest, planting 200,000 acres of the Sandhill Region in the northwestern part of the state with jack pine, ponderosa pine, juniper, Norway spruce, and other species.[51] State and federal agencies provided free tree seedlings to farmers through a variety of programs into the 1920s. And during the Depression, the federal Shelterbelt Project planted 18,600 miles of 130-foot-wide strips of woodland along the 100th meridian line from the Canadian border into Texas. These 200 million trees included the mix of types common from the 1870s, plus a number of Asian species, such as Siberian pea shrub *(Caragana arborescens)*, Chinese elm *(Ulmus parvifolia)*, salt cedar *(Tamarix ramosissima)*, and Russian olive *(Elaeagnus angustifolia)*, which had been introduced in recent decades because they resisted wind, drought, or cold.[52]

By the end of the nineteenth century, tree culturists had created something new. From eastern Illinois to eastern Nebraska and Kansas, and in patches further west, trees grew in places that had been grass-covered for centuries or more. More significantly, many of these woody plant assemblages were absolutely novel in their composition. In contrast to earlier oak-hickory or cottonwood groves, the new landscapes combined prior residents, geographic varieties transported from the East and the South, species from other parts of North America, and trees imported from the rest of the world.

The hope that these woodlands would be profitable to their owners was largely illusory. By the late nineteenth century, small local lumbering operations could seldom compete against the standardized products that large timber companies could deliver cheaply via railroad. Farmers eliminated many plantings when crop prices rose or properties were consolidated. But trees continued to provide windbreaks and erosion control. Both in towns and beyond, at least some Midwesterners saw beauty or comfort in landscapes that contained features other than corn or pasture.

The trees that had been planted or allowed to grow propagated themselves when they could. Junipers kept down by fire and grazing shot up and soon spread with the help of birds. Catalpa, native only to the southern tip of Illinois in the early nineteenth century, spread generally after being planted in the northern part of the state. Osage orange, found only in parts of Arkansas, Oklahoma, and Texas around 1800, was growing throughout the Midwest a century later after widespread use as a hedge plant. Ailanthus, Siberian pea shrub, Siberian elm, and white mulberry were all listed as naturalized in the 1986 *Guide to the Vascular Flora of Illinois*. On the other hand, while Robert Douglas's larch had "escaped from cultivation" in a few northeastern Illinois counties (i.e., near Waukegan), the millions of individuals he had marketed more widely had disappeared.[53]

Most of the unbuilt Midwest continued to be plowed and planted annually into the twentieth century and beyond. A small percentage consisted of managed, and generally temporary, pasture. What changed was the residual land that was not used intensively; it was much more likely to be dominated by trees or, less flatteringly, by "brush," than by the grasses and forbs that early settlers had emphasized. "True prairie" became a memory, maintained largely in the names of towns and counties.

Recognition of this change formed the basis for the gradual reemergence of the problem of the prairie in the twentieth century. The questions still were why particular assemblages of plants grew or did not grow on unplowed terrains in the Midwest, and how Americans could influence the distribution of species. But the goal shifted 180 degrees. Rather than trying to determine whether prairies could be transformed into woodlands, the groups involved were seeking to preserve or even to create prairies. This necessarily involved eliminating trees. That development is examined in Chapter 9.

CHAPTER FIVE

Immigrant Aid
Naturalizing Plants in the Nineteenth Century

In late 1897, Agriculture Secretary James Wilson agreed to create an office in his department for "systematic seed and plant introduction." Assessing national needs and regional capabilities, government botanists were to search the world methodically for species and varieties that could be grown beneficially within the territories of the United States, either immediately or through programs of selection and hybridization.[1]

Prior to this time—as the name of the office implied—federal plant introduction activities had been unsystematic. They were not, however, insignificant. During the nineteenth century, the national government was involved in the introduction of thousands of different kinds of plants from around the world, and it promoted the circulation of plant material around North America. Less visible, but also important, were efforts by government scientists to respond encouragingly to the tendencies of plants to migrate and evolve on their own. This chapter examines these initiatives and responses. It then sketches how introductions were systematized at the end of the century, and to what ends.

Federally sponsored plant introduction initiatives in the decades before the Civil War included land grants for promoters of new crops, instructions to naval officers and consuls, publication of descriptions and cultural guides, free distribution of both specially imported and commercially available plants and seeds, seed exchanges, and greenhouse propagation. Plants of interest included mulberries, tea, legumes, grains, garden vegetables, temperate and tropical fruits, and forage and fiber plants. Few of these programs lasted for long, however. Why were they so diverse and yet so halting?

Half of the problem was the rudimentariness of the relevant infrastructure. Americans did not have botanic gardens, horticultural societies, nursery industries, and scientific communities comparable to those that had developed over the previous century in Britain and France. The other half of the problem, however, was the range of Americans' interests in, and therefore disagreements over, plant introduction issues. European activities primarily involved distant and dependent colonies, and therefore could be pursued by the elites centered around the national botanic gardens without interference. American introductions, by contrast, were into the continental United States itself, potentially affecting the national future and the livelihoods of varied groups of people. Individuals and organizations could move plants as they wished, but had limited capabilities. Government actions, which were more important and more visible, became foci for aspirations and divisions involving class, slavery, and bureaucratic power. These political issues were evident in the very multiplicity of initiatives. The Civil War dampened these disputes, but also put the United States in the anomalous status of a mature nation that was a horticultural neocolony.

Responses to moving plants—rather than initiatives—can best be studied by focusing on a single group, the *Poaceae*. Grasses—more specifically, pasture (non-cereal) grasses—are remarkable for forming a major element of agricultural civilization while behaving more as opportunistic symbionts than as cultivated domesticates. They moved around, interbred, and evolved in humanized settings for millennia, but received only cursory attention from humans. In the nineteenth century, American botanists sought to clarify the muddle of grass taxonomy, geography, and evolution. Unassuming, like their objects of study, these government scientists realized that mastery of detail could enable them to transform both vegetation and society.

Systematic plant introduction depended on comprehension of the deeper foundations of the peculiarities of North America that had been evident to John Mitchell and other early American naturalists. It entailed ongoing cooperation within an elite-guided context between governmental and private actors, and between national and local interests. Its proponents imaged a future United States that was cosmopolitan in outlook, global in reach, and imperial in perspective.

Unsystematic Plant Introduction

The Reasonableness of the Silk Craze

During the first three decades of the nineteenth century, Americans modeled their public plant introduction projects on English colonizing and mercantile efforts of the previous century. One strategy was to provide land for foreigners who proposed to create communities that would produce high-culture products, notably wine. In 1802, for example, Congress granted 2,500 acres in southeastern Indiana to the community of French-Swiss vignerons led by Jean-Jacques Dufour (mentioned in Chapter 3). In 1817 a group of Napoleonic émigrés who promised to introduce the culture of the vine and the olive received a larger grant in Alabama. The usual problems with grapes, olives, and frontier living meant that these settlements generated no significant innovations. Congressional dissatisfaction with this approach displayed itself in the rejection of the 1822 petition of the American Coffee Land Association—a group of French exiles from San Domingue living in Philadelphia—to gain title to Key Largo in southern Florida.[2]

A second option was to encourage government-affiliated travelers to send back promising plant material from abroad. The most notable of a number of initiatives in the 1810s and 1820s was the circular distributed to diplomats in 1827 by Richard Rush, John Quincy Adams's treasury secretary. The New York Horticultural Society (as mentioned in Chapter 3) saw this request as an opening for a permanent public-private partnership. They were unsuccessful, both because of the ordinary difficulties that private organizations faced in obtaining public money, and because of the presumed local character of a botanical garden in New York. In the absence of such a partnership, however, Rush's circular was fruitless. A request from one federal department to another with no funding attached and a timeline extending into the indefinite future was not a priority; and, as Chapter 8 shows, the one diplomat who did respond substantively experienced severe problems due to the lack of an official garden to receive his plants.[3]

During these decades of discussion, a generally unremarked elephant—cotton—gradually filled the room. Upland cotton *(Gossypium hirsutum)*, a beautiful yellow-flowered relative of the hibiscus, grew in the New World tropics. Beginning in the 1790s, its culture spread across the South, and by the mid-1820s, cotton fiber had become the most valuable export of the United States. From the perspective of plant introduction, cotton offered mixed messages. It represented an overwhelming demonstration of the fact that a newly introduced species could have revolutionary agricultural potential. But how the cotton revolution happened and how it might be replicated were unclear. The predominant understanding was that *Gossypium hirsutum* had been available in the South long before the 1790s, and may actually have been indigenous; it was also believed that development of the cotton industry had resulted from new technology—Eli Whitney's invention in 1793 of a machine to separate fibers from the species' furry (hence *hirsutum*) seeds.[4] A more complex history that emphasized horticulture was also advanced, however. In the 1830s Georgia planter Thomas Spalding traced the cotton boom back to 1785, when loyalist exiles sent samples of the smooth-seeded *Gossypium barbadense* from the Caribbean island of Anguilla to relatives who had remained behind on Georgia's Sea Islands. Anguilla cotton was a languidly bearing tropical perennial. However, through selection and possibly through hybridization, American planters soon isolated a variety that flowered and fruited on the accelerated schedule of a temperate-zone annual when planted in the unique high-latitude, long-growing-season environment of the Sea Islands. Spalding claimed that the combination of the rapidly mechanizing English textile industry's demand for Sea Island cotton and the inability of planters to grow it on the Georgia mainland formed the essential stimulus for Whitney's invention of machinery to remove seeds from the cruder but less-finicky *hirsutum*.[5]

Could the cotton revolution be repeated? The "silk craze" that extended from the mid-1820s to the early 1840s becomes less crazy, while its collapse is more understandable, when viewed from this perspective.[6] Chinese silk culture, based on the silkworm moth *(Bombyx mori)* and its preferred food, the white mulberry *(Morus alba),* had been replicated successfully in Italy and France as long ago as the late Middle Ages. Georgia colonists in the 1740s and Pennsylvanians around 1770 had successfully but laboriously produced silk from worms that were fed leaves from both imported trees and the North American red mulberry *(Morus rubra).* By 1825, when southern cotton was supplying both foreign and native mills, influential Americans imagined that silk culture could soon be equally important. Silk had the additional value of offering a new source of income to northeastern farmers, who were struggling to compete in commodity production with settlers further west. Potential silk culturists needed information, plants, start-up subsidies, and organization. The hope was that, as with cotton, demand would stimulate labor-saving inventions and thus generate productivity and profitability.

The national government took the lead on the informational side. Prodded by Albany aristocrat and congressman Stephen van Rensselaer and his distinguished French American neighbor, E. C. Genet (in the 1790s the notorious French revolutionary diplomat Citizen Genêt), Congress produced a series of reports on silk, culminating in the subsidized translation and distribution in 1828 of *A Treatise on the Culture of Silk in Germany* by Bavarian aristocrat Joseph von Hazzi. The main source of plant material came from New York City. The American Institute, a private technological improvement society, created a silk committee, led by the French American physician and naturalist Felix Pacalis, which imported and distributed mulberry seed from France.[7] Enthusiasm grew after 1829 when a shrubby variety from China (promoted as a new species, *Morus multicaulis*—the "many-stemmed mulberry") that was adapted to the needs of leaf pickers became available, probably through the nursery of the well-connected Belgian American plant importer Andre Parmentier. Both private organizations and state governments, notably in Connecticut and Massachusetts, provided subsidies and guarantees so that farm families would learn silk culture and produce the thread that would support the rest of the industry.[8]

The relations among plant introduction, naturalization, labor, innovation, and industrial growth were, however, much more complex with silk than with cotton. Cotton was an annual crop with one bulk local processing step (ginning), so planters received rapid feedback from failure and adapted quickly. Silk production, by contrast, could begin only when mulberry trees were sufficiently mature, a half decade after sprouting; culture involved both weeks of tending a generation of caterpillars and the delicate work of unraveling, or reeling, the cocoons. In the 1830s silk enthusiasts focused on the first and most familiar stage of the process, the propagation of *Morus multicaulis*. In 1837 demand for cuttings outstripped supply, leading to a price bubble and widespread speculation in nursery stock. A series of problems then emerged. Some growers were disillusioned when the price of *multicaulis* nursery stock dropped precipitously in 1839. High mortality among young trees during that winter raised questions about the viability of a variety that, confessedly, had been imported from south China via the Philippines. On the processing side there was no invention comparable to the cotton gin, and farm families had difficulties inserting the intense periods of labor tending caterpillars, and the tedious work of reeling, into the rhythms of rural life. Southern slaves had few reasons to take the care necessary for successful silk culture. Finally, the financial incentives, both directly from states and indirectly through a federal tariff, were not large enough. Silk mills continued to rely on Chinese and Italian imports, and by the middle of the 1840s domestic silk production had faded.[9]

The Public Side of the Patent Office

At the end of the Jacksonian period, federal officials for the first time took a proactive rather than opportunistic approach to plant introduction efforts. This initiative arose within the Patent Office. On first consideration, it seems anomalous that an organization created to secure private monopolies over profits from technological innovations would collect and distribute self-replicating natural objects *gratis*. Part of the explanation is that the task went to the Patent Office by default, since it was the only part of the federal government that focused on problems of knowledge. The more significant reason, however, is that important Jacksonian leaders believed that plant introductions could reconcile their support for a patent system with their larger commitment to equal opportunity for all white men.

The Patent Office took shape under the leadership of Henry Ellsworth. A graduate of Yale and the son of Oliver Ellsworth, the second U.S. chief justice, he rejected Federalism and began his political career as a Democrat in the 1820s by attacking bank "monopolies." He experienced the continental expanse in 1832 when he traveled to Indian Territory (with Washington Irving) to resolve land control problems that arose after Jackson forced the southeastern tribes to relocate there. His reward for this work as a fixer was appointment in 1835 as superintendent of the Patent Office—at that time, a small appendage of the State Department. Within a year, however, Ellsworth had acquired the status and salary of commissioner, and the Patent Office had gained organizational autonomy, funds to hire expert staff, and authorization for a monumental fireproof headquarters.[10]

The concept of the Patent Office was problematic for Jacksonians like Ellsworth. They were glad to honor individualism and entrepreneurship, but had ostentatiously rejected government protection of monopoly in their campaign against the Bank of the United States. Ellsworth thus emphasized that patent holders, in return for their limited-period privilege, had to make their ideas public so that any individual could use them to make his own further improvements. The office facilitated this diffusion of knowledge by publishing the patents in its annual reports and by creating a museum of patent models in its new public building. For Ellsworth, the creation of an agricultural division was an additional way that the Patent Office could mitigate monopoly and serve the people. The division's primary task was to collect and publish agricultural statistics so that dispersed farmers were not victimized by metropolitan speculators with inside information. The division could also become, to use modern but not totally anachronistic terminology, a national germplasm exchange.

According to Ellsworth, the stimulus for this program came from agricultural inventors, who often brought remarkable local varieties of crop seeds along with them on their visits to the Patent Office.[11] They knew that seeds, as natural objects, could not be patented, and that, while costs of propagation were minimal, there was little chance of enduring profit and thus little incentive in marketing particular varieties. Local sharing of seeds and plants was a traditional form of mutual aid; Ellsworth in effect envisioned the Patent Office as the clearinghouse for a national friendly community of seed sharers. Ellsworth's examples were not ornamental rarities or unfamiliar exotics, but garden vegetables, and especially local races of maize. He suggested that sending local corn varieties from Maine or Mexico to the upper Midwest could extend the geographic range of that plant's culture substantially. No one variety had notable value; but the free and public distribution of different sorts of plants around the nation, with local testing and further replication and selection, would improve productive capabilities for both individuals and the nation.[12]

The problem was that this scheme was free and public only in ideal conception. Since Ellsworth had no specific appropriation for seed exchange, he organized the project off budget. Patent Office clerks sorted seeds in the midst of other work, and the packages were mailed free through the offices of friendly congressmen—with preference, of course, for their constituents and friends. This link between seeds and votes induced Congress to fund the Patent Office's agricultural program beginning in 1839, but created new logistical demands—not only to mobilize more clerks, but also to acquire enough seed that would be free, new, and valuable.[13] Ellsworth, who possessed the high-mindedness, efficiency, and savvy to continue in office through four presidents (Jackson, Van Buren, Harrison, and Tyler), balanced these demands without being trapped in accusations of waste and corruption. But when he left the Patent Office in 1845, his program declined.

Tea on the Mall and Boston Lettuce

The issues surrounding unsystematic federal plant introduction came to a head in the 1850s. The seizure of northern Mexico in the previous decade meant that the United States controlled territories that were environmentally quite different from the humid East. Government leaders thus became interested in trying out organisms that had successfully been domesticated in Old World arid lands. (In 1856, for example, the army introduced camels into Texas as transport vehicles.) On a more explicitly political level, Democratic presidents Franklin Pierce (1853–1857) and James Buchanan (1857–1861) sought ways to dampen antagonism about slavery in the western territories. Emphasizing that different regions of the country had the potential to grow different new and valuable crops, but might then require different labor systems, had the potential to transform ideological and moral arguments into more temperate discussions about the natural and national economies. Democrats also wanted to show citizens—especially ordinary people in the Midwest, and the increasingly disaffected Southerners—that the general government could better their lives while keeping taxes low and staying out of local issues. In pursuit of these goals, the Patent Office made major efforts, beginning in 1853, to collect, import, publicize, and distribute whatever potentially useful exotic plants were available. Northeasterners attacked the programs as wasteful and vulgar, and as threats to the high level of New England culture.

As a former army officer (he had edged out Robert E. Lee for highest marks in the West Point class of 1829) and a federal judge in Iowa Territory, Charles Mason combined nationalism, the belief that the sphere of the general government was circumscribed, and an interest in agricultural innovation. Appointed commissioner of patents in 1853, he immediately launched initiatives to find and distribute new plants. His agricultural assistant, D. J. Browne, traveled to Europe and purchased large quantities of many kinds of seeds, primarily from French commercial dealers. Farmers and gardeners around the country were soon receiving packets of vegetables, flowers, herbs, grains, forage plants, and grasses from the Patent Office, both through local organizations and directly from their congressmen.[14]

A much-expanded Agricultural Division annual report described and promoted these initiatives. It announced that the Patent Office was collecting wheat varieties from around the world to renew the vigor of American stock, and that it was importing crop plants from exotic places such as Paraguay, Abyssinia, Cashmere, and New Zealand. It particularly emphasized eastern Asian plants, including the opium poppy, "the soja bean of Japan . . . used for making the celebrated 'Soy sauce,'" and syrup-producing temperate-zone Chinese sorghum. The division's reports framed these programs climatologically, emphasizing the similarities between North America and East Asia, and suggesting how far north corn could be grown and how far west cotton (and implicitly slavery) could flourish.[15]

Joseph Holt, Mason's successor as patent commissioner, dramatically advertised the Democratic Party's vision of the nation's horticultural foundations and the importance of government plant introductions for its future. His first annual report in 1857 emphasized that cotton was the basis for the economies of both South and North. It then presented D. J. Browne's long prospectus for a plan to introduce tea culture throughout the South and along the Pacific coast. Browne predicted that Americans, with their improved machinery, efficient transportation, well-nourished slaves, and "the aid of a few Chinamen, at first," would be able to outcompete "enfeebled and poorly-fed Asiatics" (see Figures 5.1 and 5.2). A second report elaborated that tea culture would fit smoothly into the slave economy because the work could be performed by "the weakly and young, who are unfitted for the culture of cotton and tobacco, and the heavier duties of the plantation."[16] Holt hired the celebrated British plant collector Robert Fortune, who had earlier transplanted tea culture from China to India, to send thousands of seeds and seedlings to Washington. A "Government Experimental and Propagating Garden" was rapidly constructed on the National Mall, just south of the present site of the National Gallery of Art, to receive this material (see Figure 5.3). In 1859, 32,000 tea bushes were growing under glass at the foot of the Capitol. With women nationwide brewing a "grateful beverage" picked and processed by American slaves, the meaning of tea would alter dramatically from that established in Boston in the context of the Revolution.[17]

The Patent Office Agricultural Division's global reach, scattershot distribution techniques, search for dramatic innovations, and sympathy for Southern interests made sense inside Washington. Northeastern agricultural journalists and seedsmen, however, were critical of an alien bureaucracy that used public funds to compete against their own carefully nurtured enterprises. *American Agriculturist* (which distributed free seeds to subscribers from its Manhattan offices), for example, reported extensively on the "agricultural humbug at Washington," with

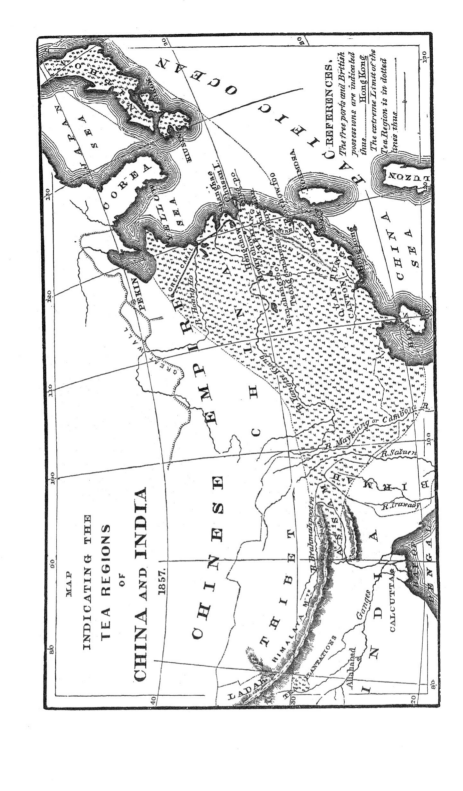

Figures 5.1, 5.2 These two images fronted the Patent Office's agricultural report for 1858. They emphasized the geographic similarities between southeastern Asia and the American South and West, and the extent to which American cotton regions were adaptable to the tea culture. Reproduction courtesy of the Rutgers University Library.

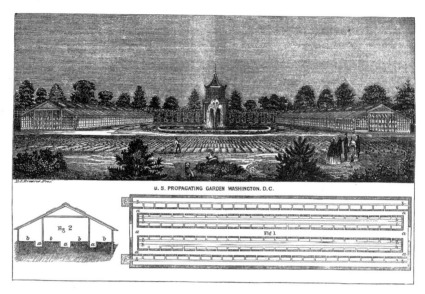

Figure 5.3 U.S. Propagating Garden, National Mall, Washington, D.C., with glass houses built to receive the tea plants sent from China by Robert Fortune, the English special agent hired by the Patent Office. The complex was located just south of the present main building of the National Gallery of Art. *Report of the Commissioner of Patents: Agriculture,* 1859. Reproduction courtesy of the Rutgers University Library.

---

strictures on mislabeled seeds, the distribution of standard commercial varieties as new, Browne's "pleasure tours" in Europe, political favoritism in contracting, and poor distribution of reports. The Philadelphian Thomas Meehan's *Gardener's Monthly* complained that the Agricultural Division was "a perfect Augean stable of corruption and shameless ignorance."[18]

The deeper criticism came, appropriately, from the Massachusetts Horticultural Society. Asked to report on the Patent Office program, longtime honorary horticulture professor John L. Russell laid out the foundations for New England horticultural sectionalism. He sketched the development of the human condition from savagery through pastoralism to the "barbarous or agricultural," and finally to the highest stage, the horticulturist—"the most ingenious and elegant of occupations pertaining to the cultivation of the earth." Russell emphasized that all plants, animals, and settings were fundamentally "local and peculiar." Introduced weeds had lost their "transatlantic character" after centuries in the New World, and the important pests (notably fungi) were "exclusively *American*," and even limited by region. Successful introductions of new plants depended on "home observations by our own home naturalists or experimentalists," and on the study of "the particular contingencies of local conditions."[19]

From Russell's perspective, the Patent Office's indiscriminate distribution of "weeds and useless foreign trash" to anybody with a garden would pull "the advanced culture of Massachusetts" back down to the level of barbarism prevalent in the southern and western parts of the United States. Distributing artichokes and endive (which, Russell reported, intelligent Yankee farmers confused with *"thistles and dandelions"*) was pointless and wasteful when lettuce was so much better. Massachusetts culturists had long ago decided that most of the forage plants broadcast by the government were noxious weeds—notably sweet clover *(Melilotus officinalis),* spurry *(Spergula arvensis),* and chufas *(Cyperus esculentus).* Russell was particularly distressed that Washington officials would send out "the coarsest and meanest sorts" of plants to replace refined local types, and that they would suggest that the Japan pea *(Lespedeza striata)* or Chinese broom corn *(Sorghum bicolor)* could replace "the more juicy and sweeter stem of the Indian corn" that had long filled New England fields. He patronizingly acknowledged that "there may be some portions of the United States, perhaps, where these and similar plants may possibly be useful," but Massachusetts needed no such assistance from Washington.[20]

Russell called on local organizations to provide home observations: to report continuously on what plants were growing in their neighborhoods, and which ones were valuable. During the next few years Massachusetts Horticultural Society members pursued these goals. The Committee on Flowers, for example, led by wealthy amateur Edward Rand Jr., advocated the "culture of our native plants." They highlighted wild morning glory (*Convolvulus sepium,* now *Calystegia*) and bindweed *(Convolvulus arvensis),* and suggested that ordinary plants such as *Aquilegia canadensis* could be greatly improved through "high cultivation and rich manuring," and through hybridization.[21] They asked Asa Gray to report on the native plants that the Harvard Botanic Garden had brought into cultivation, and he obliged with a list of species drawn mostly from Texas, New Mexico, and Arizona.[22] Charles Sprague published an overview of the introduced plants that had become "intolerable pests to the agriculturist." These included such "troublesome usurpers of the soil" as buttercups, watercress, and "nearly all the *Clovers.*"[23]

Relying on home observations, however, was problematic. Few places in the United States could emulate Massachusetts's combination of 240 years of Anglo-American settlement, an insignificant agricultural economy, and a large number of naturalists. Even there, information was partial. Views on the agricultural value of clover varied widely, the nativity of bindweed was unclear, and, at the moment when Gray submitted his list of far southwestern native plants, he was preparing his major contribution to science (discussed in Chapter 4) showing how the eastern North American flora was more closely related to that of eastern Asia than to the plants of the American Southwest. The attachment of New Englanders to the local and native was, in any case, superficial. Russell's central argument for the high level of culture in Massachusetts was the fact that plant importers enabled the state's gardens to blossom "with every hardy exotic nearly cotemporaneously with her sister gardens of Great Britain."[24]

From Nation to Colony

Private importers became an increasingly prominent alternative to federal plant introduction between 1850 and 1875. The combination of British and French colonial ventures, steamships, Wardian cases, and printed catalogs transformed the international market for plants. Companies with deep ties to imperial botanical gardens—most notably Veitch and Loddiges in Britain and Vilmorin in France—dramatically expanded the number of plant types they sold to northeastern American nursery companies such as Peter Henderson, Parsons & Co., and Ellwanger and Barry. These firms sent seeds, cuttings, and bare-root plants around the country at specially subsidized postal rates.[25]

By embracing commercial imports, Americans confirmed their positions as subordinate participants in an international plant market centered on northwestern Europe and bifurcated into mass and class segments. Catalogs featured standard fruit and ornamental trees, vegetable varieties, and bulbs, adapted primarily to France and England but usable in parts of North America with varying degrees of care. Ornamental novelties were available to wealthy enthusiasts directly from England through boutique plant lists.

Edward Rand Jr. exemplified the change at the high end of the market. Between the late 1850s and the late 1860s, he shifted from promoting native flowers to advocating for a group of shrubs that were rapidly losing their American character. As noted in Chapter 1, rhododendrons, azaleas, and kalmias had gained the name "American plants" in the 1700s, when English fanciers like Peter Collinson and the Duke of Argyll planted Appalachian specimens on their estates. After 1850, however, British interest turned completely to the showier rhododendron species first brought from the Himalayas by the Royal Botanic Garden's Joseph Hooker. Elegiacally placing the established name in quotation marks, Rand published *The Rhododendron and "American Plants,"* a guide to the new culture, in 1871. He helpfully listed the English nurseries from which the best specimens of Asian "American plants" could be obtained.[26]

Missing from the commercial system was the Patent Office's ability to send odd things to marginal places. Also, there was no longer—granting the vagueness of the concept—a North American and national perspective on introductions. The predictable consequence was that new things arrived haphazardly. Both hard winter wheat varieties and the Russian thistle *(Salsola kali)* were carried into the Plains region in the 1870s by Russian Mennonite immigrants. Kudzu *(Pueraria montana)* began its North American career because Japanese gardeners at the Philadelphia World's Fair in 1876 needed a fast-growing and characteristically Asian vine to decorate their pavilion.[27]

Federal agricultural bureaucrats were in a difficult position. The Patent Office *Annual Report* written in the midst of the secession crisis commented wistfully that thousands of little tea plants had been sent out during the previous year, but that "little more can now be said of them than that, from Maryland to Louisiana and Texas, they are known to have taken root and prospered."[28] Plantation owners' inability to develop tea culture while dealing with invasion and emancipation was less consequential, however, than the fact the Patent Office's hothouse was replaced by a military hospital. By 1870 the new and autonomous Agriculture Department had secured headquarters at the western end of the National Mall and had laid out an impressive display garden with greenhouses (see Figure 5.4). But it did not regain the Patent Office's prewar mass propagating capability.[29] Unable to import, maintain, or distribute significant numbers of plants, the department was an inconsequential force for finding and distributing new species and varieties by comparison with commercial nurserymen. On the other hand, "Congressional seed distribution"—the Agriculture Department's mailing of millions of free packets of standard varieties of seeds, purchased from commercial sources, under the names of legislators—became one of the characteristic boondoggles of the Gilded Age.

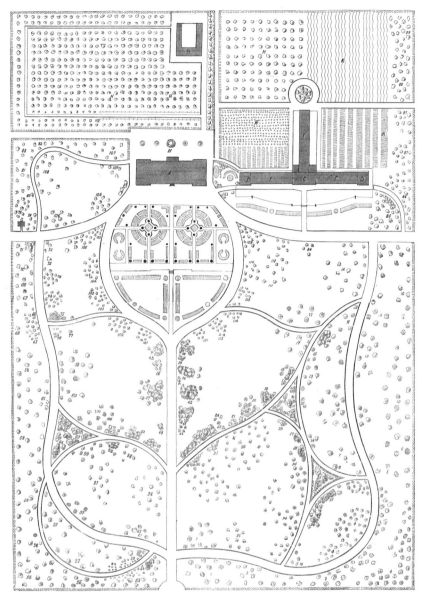

Figure 5.4 From 1865 to around 1900, the Agriculture Department controlled the part of the National Mall bounded by modern Twelfth and Fourteenth streets, and Independence and Constitution avenues. Each year, hundreds of thousands of visitors to Washington could see a large number of either potentially useful or merely curious plants on the grounds and greenhouses supervised by horticulturist William Saunders, a founder of the National Grange and the designer of Gettysburg Cemetery. The department had minimal propagating capability, however. Plan of grounds from U.S. Department of Agriculture, *Report*, 1870. Reproduction courtesy of the Rutgers University Library.

Grasses: The Culture of Public Plants

Assisted plant migrations were important, not only economically and politically, but also environmentally. To convey this dimension, I shift from people and policies to one strategically chosen plant group. Pasture grasses were fundamental components of western European agricultural civilization. Yet, by comparison with livestock, fruit trees, or cereal grains, these plants were only lightly domesticated. In permanent pastures, individual grass species appeared and disappeared as seeds were transported, climate fluctuated, and grazing pressures changed. Few botanists paid close attention to the identities of these unspectacular-looking plants; the propensity of some genera (notably *Poa* and *Elymus*) to hybridize spontaneously and evolve rapidly made classification particularly difficult.[30] In contrast to the private garden treasure of, for example, a named pear clone replicated through grafting, grasses—in their necessity, anonymity, and mobility—were uniquely "public" plants.

In the nineteenth century, agricultural promoters sought to increase the number and diversity of grasses in the United States. Difficulties arose at both ends of the process: searches for new and promising species were haphazard at best; and a new grass could be as much a cause of dissension as of general interest. Ultimately, these plants became the particular concern of the public's agricultural scientific agent, the U.S. Department of Agriculture (USDA). "The sorting and classification of hay," as a bureaucratic rival in the 1890s dismissively termed it, became a major Agriculture Department endeavor.[31] By the end of the nineteenth century, operating from presumptions that were both practical and evolutionary, USDA botanists had identified and characterized thousands of grass species in North America and elsewhere, and had moved hundreds of these both intra- and intercontinentally. This scrambling of the ranges of grasses provided the model for more ambitious vegetational rearrangement plans that the department undertook in the early twentieth century.

Turf Management

As noted in Chapter 2, inadequate pasture grass was one of the significant issues that Anglo-American colonials faced. Eastern North American grasses grew for only a few months in the year—mostly in late summer, when other food was readily available. They did not stand up to grazing, and they made poor hay. The poverty of American grasses was considered one of the causes of the degeneration of Creole cattle and horses.[32]

The introduction of Old World pasture grasses, both intentionally and adventitiously, mitigated this problem in the Northeast by the early nineteenth century. Species such as Kentucky bluegrass *(Poa pratensis)*, tall fescue *(Festuca arundinacea)*, and orchard grass *(Dactylis glomerata)* spread so quickly and widely that nearly all farmers, and some botanists, believed that they were indigenous to North America.[33] Settled communities accepted their situation with the commonplace wisdom that "grass is grass." With the exception of commercially available seeds used to lay down a one-time pasture in a crop rotation (generally a mixture of timothy [*Phleum pratense*], redtop [*Agrostis vulgaris*], and red clover [*Trifolium pratense*]), particular species and their differences drew little attention.[34]

The situation south of Kentucky was more difficult. While cool-season grasses could grow through the southern winters, and thus offer the possibility of year-round grazing, they seldom lasted because they were either choked out by rampantly growing summertime plants or killed by heat and drought. The low state of southern grass culture was epitomized in farmers' reliance on crabgrass *(Digitaria sanguinalis)*. With the rise of cotton monoculture, many wealthier Southerners became indifferent toward grasses; they were willing to pay cash for Kentucky-bred horses and for the feed necessary to maintain them. But promoters of mixed agriculture believed that poor grass was the major barrier to the region's progress. Just before the Civil War, Georgia agricultural writer and minister C. W. Howard used the Patent Office report to deny that depressed land prices in the South were caused either by the institution of slavery (it was humane and efficient) or by the lack of intelligence of slaves (while inferior to Englishmen, they were at least as smart as Belgians). The problem was that the output from land devoted solely to artificially fertilized cotton could not match that from farms where grass supported livestock, which then provided power, meat, and milk, and also the manure that would make rotated cash crops cost-free.[35]

Improvers, both north and south, argued that increasing the species diversity of pastures would result in greater productivity—especially by adding persistent types with distinctive seasonal or nutritional qualities. This was easier said than done, however; difficulties included both lack of knowledge about the properties of grasses and the lack of a market in the seeds of any but the most common species.

American botanists believed that the first and most promising path for grass improvement was to culture local species. With varying degrees of providentialism, they anticipated that among the dozens of kinds of grasses growing obscurely in any given region, some would have characteristics making them suitable for pasture or hay. Pennsylvania botanist and minister Henry Muhlenberg advanced this vision in 1817 with the first catalog of North American grasses. A generation later John Torrey's state-funded flora of New York included descriptions of more than a hundred grasses, with efforts to specify nativities and to provide notes on habitats.[36]

Few if any previously unknown eastern American grasses entered agriculture in the first half of the nineteenth century, however. By the 1840s, while continuing to bow rhetorically to the potential of "indigenous" grasses, improvers began to call for "a free introduction of foreign varieties."[37] The problems of finding, identifying, transporting, assessing, and distributing these grasses were, however, even more difficult than with local material. Introductions resulted from individual interest or happenstance, and distribution was informal and uneven. The most notable example of this kind of initiative, in retrospect, was John Hugh Means's introduction of *Sorghum halapense* into South Carolina in the 1830s. The southern-born son of a Bostonian, Means toured the Near East as a young man. On returning to his plantation, he sowed seed of a lush warm-weather perennial grass he had noticed during his travels. Means freely shared seed of this grass with visitors and correspondents. One Alabama recipient, William Johnson, was particularly enthusiastic: his promotional efforts soon made "Johnson's grass" an integral part of the southern agricultural landscape.[38]

"Willard's bromus" displayed some additional difficulties involved in grass introduction. In 1850, Massachusetts farmer Benjamin Willard reported that he had found "the richest feed that grows," which he had propagated "from a single spire" in a batch of imported seed. In the face of both biological and customary obstacles, Willard sought to profit from this discovery: in marketing his special variety, he required purchasers to pledge solemnly that they would not compete with him by reselling seed they harvested. Commentators were offended, not only by what they considered Willard's "monstrous prices," but by his presumptuous idea that restrictions could be placed on grass distribution "other than those which the Creator has put upon it."[39]

These problems were similar in kind, if not degree, to those ordinarily experienced by horticulturists. Grass introduction, however, raised a larger issue. Anyone who planted a grass that was not already prevalent could affect not only himself but the community around him. Desirable field grasses germinated readily, crowded out other plants, grew rapidly at odd seasons, and withstood grazing, drought, heat, cold, and insects. They were plants that would survive and spread on their own indefinitely. They thus lived in tension with the other side of mixed farming—the creation of corn, wheat, or other annual monocultures. Farmers had learned to keep most of their semidomesticated grasses under control, but a new species with different properties could interfere with established cultural practices; or, as culture changed, a useful grass could become a noxious weed.

Issues of grass weediness ranged from the straightforward to the exceedingly complex. Many vigorously spreading introduced grasses were so widespread by the middle of the nineteenth century that, while excoriated, they were considered inevitable. In newly settled areas, however, there were scattered disputes over the wisdom of planting the persistent but only seasonally valuable Kentucky bluegrass. The popularity of Willard's bromus declined when Massachusetts state agricultural bureaucrat and grass specialist Charles Flint denounced it as merely a variety of *Bromus secalinus,* the wheat field weed long known as chess or cheat; while Massachusetts farmers did not grow much wheat, they did not want to cultivate something condemned as noxious.

The most prominent New England grass controversy involved *Elymus repens,* commonly known as creeping wheat, Chandler grass, Durfee grass, Fin's grass, Quick, Quack, Quake, Quitch, Twitch, Witch, Squitch, Couch, or Scutch. Classed in the nineteenth century as *Triticum repens,* a rhizomatous congener of wheat, it was long sold commercially and valued for its ability to green up early (hence quickgrass), fill a field, survive drought, and provide sweet hay. As farming became more mechanized and crop rotations more sophisticated, however, the species became problematic. Eliminating it from fields being switched from pasture to grain required repeated plowing, disking, and harrowing; the more difficult problem was that the plant, generally called "quackgrass" after the Civil War, would reenter a cleared field from neighboring pastures, either by creeping or as bird-carried seeds. The moderate judgment was that quackgrass was "too much of a good thing"; the stronger position, taken by Michigan agricultural botanist W. J. Beal, was that it was an "execrable weed" classed, as an object of his personal hatred, in the same category as slavery and alcohol.[40]

Johnson's grass was the most controversial introduction, however, because its particular properties made it a player in Southern political disputes in the decades after the Civil War. Advocates of northern-style diversified agriculture considered it (and its less-obtrusive partner, Bermuda grass [*Cynodon dactylon*]) a "great boon." Johnson grass grew more than six feet tall in bottomlands and formed rhizomes that enabled it to come back immediately after cutting. Farmers found that in favorable locations they could harvest it three or four times a year without affecting future productivity. The real significance of Johnson grass, however, was that it pushed even the unwilling to accept it. The seeds were readily dispersed by birds, so that it was soon sprouting, not only in fields where it was planted, but also on any well-watered, cleared ground nearby—notably cotton fields. Its rhizomes, like those of quackgrass, made elimination through the usual control methods for annual weeds impossible: cultivation only spread the roots and sped up the grass's takeover of a field. By the 1880s, it was flourishing from the Atlantic to California. Federal agricultural scientists advised Southern farmers that their best course of action when faced with a field of Johnson grass was to adapt themselves to it.[41]

Cotton planters, not surprisingly, considered Johnson grass "an unmitigated evil." The immediate problem was that it "infested" their laboriously prepared fields; the broader issue was that it gave a competitive boost to mixed farming over cotton because the more acreage in a locale that was devoted to the alternative form of agriculture, the stronger the infrastructure of supply, marketing, and transportation would become. Those arrangements threatened the labor relations, and thus the racial hierarchy, that planters had laboriously reestablished around cotton monoculture in the aftermath of Reconstruction. In 1895 supporters of cotton culture in Texas outlawed the planting of Johnson grass, bringing police power to bear on those who favored it; they also pushed the USDA to find ways to eliminate the pest. This tension continued until the 1910s, when the spread of a more powerful pest, the boll weevil, made protection of cotton monoculture from a weed moot.[42]

George Vasey and the Value of Sorting Hay

In the 1890s, the area between the headquarters and the public museum of the USDA on the National Mall was planted as a grass garden; it included a prominent square of Johnson grass (see Figure 5.5). The garden and its plantings indicated both the degree of interest and the particular outlook of the organization. In 1872 the Illinois plant collector and

Figure 5.5 The Agriculture Department featured Johnson grass in the garden planted adjacent to its Washington headquarters. Photograph from Bureau of Plant Industry *Bulletin* 11, 1902, frontispiece. Reproduction courtesy of the Rutgers University Library.

science magazine editor George Vasey (Figure 5.6) became the botanist in the still tiny, nine-year-old federal department. While his predecessor, the explorer C. C. Parry, had commented offhandedly that "the largest number of the species" of grasses on the Great Plains "could be dispensed with," Vasey made the increase and diffusion of grasses the main focus of his division for the next two decades.[43] He understood that grasses were a particularly appropriate subject for this public agency. They filled unoccupied and unvalued American landscapes—both explicitly on the western "public lands" transitioning from bison habitat to cattle range, and more broadly across the continent. In contrast to fruits and grains, and even trees, there was little private-sector presence. Finally, Vasey recognized that grass work required very particular kinds of expertise that encompassed both North America and beyond.

Vasey advanced the combination of basic survey data collection and practical problem solving that was characteristic of federal science in

Figure 5.6 George Vasey, chief botanist of the U.S. Department of Agriculture, 1870–1893, and advocate of new grasses. *Botanical Gazette* 18 (1893): plate 18. Reproduction courtesy of the Rutgers University Library.

the Gilded Age; this mix was highly integrated when applied to grasses. He believed, first of all, in the necessity and utility of systematic botany—of sorting hay. By the 1890s, he had amassed a major grass herbarium containing more than 15,000 sheets, he published a long series of notices about new species in botanical journals, and he sought to make this knowledge widely available by describing and illustrating "American grasses," both native and naturalized. His increasingly elaborate books culminated in a posthumous two-volume government printed quarto, *Illustrations of American Grasses*. Vasey's herbarium and illustrations formed the foundation for Albert S. Hitchcock and Agnes Chase's comprehensive *Manual of the Grasses of the United States*. Finally, Vasey argued that educating farmers about insignificant plants contributed to their intellectual uplift.[44]

Vasey's more practical work began with a report, prepared in 1878 in collaboration with the department's chemist, on the habits, cultures, and nutritional values of about thirty widely grown grasses. He then adopted a strategy common among federal scientific bureaucrats:

in 1881 he sent out a circular letter with open-ended questions about grasses to hundreds of people he hoped would be interested. Respondents provided some information, demonstrated considerable gaps in knowledge, and expressed the desire for better varieties. Vasey used these statements to argue that the American public needed and demanded investigation and guidance about grasses. His initial program was to "select from our wild or native species some kinds which will be adapted to cultivation in those portions of the country which are not yet provided with suitable kinds." A few years later, he expanded his reach to include inquiries "as to any additional benefits we may obtain from the introduction of foreign grasses and forage plants."[45]

Since Agriculture Department seed distribution was controlled by a separate division more interested in cabbages and congressmen than in grasses, Vasey organized his efforts through field stations in the two most grass-challenged regions of the country—the South and the Great Plains. S. M. Tracy, a professor at the Mississippi Agricultural College, used federal funding to obtain and plant seeds from around the world. Hoping to find cool-season grasses that would not die from drought or crowding in summers, he distributed nearly four hundred species among cooperating state experiment stations in Louisiana, North Carolina, Georgia, and Florida. His recommended species came from places as geographically diverse as the Great Plains and Rockies (*Agropyrum glaucum* and *Oryzopsis membranacea*), Japan *(Agropyrum japonicum)*, and Australia *(Andropogon erianthoides)*.[46]

Vasey's own interests centered on the Great Plains. In 1886 he questioned the viability of dry farming in the arid region of western Nebraska, western Kansas, and eastern Colorado. He argued that the land should be used for grazing, but noted that the local buffalo grasses and grama grasses were not very productive. He proposed creation of a special station at Fort Wallace in western Kansas, recently abandoned with the end of the Indian wars, for testing both local and introduced species. This proposal was ignored until 1888, when widely publicized blizzards, droughts, and overgrazing wiped out both farmers and ranchers. A year later, Congress funded a federal grass experiment station in the southwestern corner of Kansas near Garden City.[47] The aim of the station, Vasey explained, would be to answer the "vital question, affecting the interests of thousands of settlers," namely, what could be grown in the arid region. The USDA's agent would experiment with "grasses and forage plants, both native and foreign—with any kinds, in fact, which give promise of utility and adaptation to the climatic conditions of the arid plains, and furnish a substitute for the scanty pasturage now existing." An 1891 report listed both local and introduced grasses that had thrived, and others that were particularly rugged. It suggested that smooth brome *(Bromus inermis)* might anchor the grass future of the Plains.[48]

Academic scientists situated Great Plains grass introduction within a range of geographic frameworks. University of Nebraska naturalist Samuel Aughey emphasized that grasses were not demographically stable—that the distribution of species on particular prairies had changed repeatedly over periods of only a few years. Harvard geologist Nathaniel Southgate Shaler argued that the geological recency of aridity on the Plains meant that the diversity of drought-adapted grasses was limited; but since grasses were "more easily feralized than any other of our domesticated plants," he was optimistic that species from long-dry regions—Australia, South Africa, Patagonia, and Siberia—would grow well there. Charles Bessey, Aughey's successor at Nebraska, expressed amazement and pleasure in 1889 that at least twenty-two "useful" foreign grasses had already become naturalized in the state. W. J. Beal framed the policy and status issues clearly in his comprehensive *Grasses of North America* (1887). Complaining that the national government had sent travelers "at great expense" to explore the Arctic and to observe astronomical events abroad, he argued that public money would be better used to send "competent persons around the earth in search of better grasses."[49]

Such an initiative was easier to imagine than to realize, however. Vasey's attention up to his death in early 1893 was on his catalogs and field stations. A few months later, the second Cleveland administration closed the Garden City experiment station, and it did not hire a grass scientist to replace Vasey for more than a year.[50] The deeper problem, however, was that locating useful new grasses was, to adapt the old adage, like finding a particular straw in a haystack. Where in the world might desirable grasses be? How would a collector know that a particular plant, among the thousands of unidentified non-American species, subspecies, and hybrids, was new and possibly worthwhile? How could introductions be tested for viability and utility in multiple regions of the United States? Creating the intellectual, bureaucratic, and gardening infrastructure to make such activities successful required political will, individual initiative, and institutional continuity. Once a search engine was developed, however, there was no reason to limit it to grasses.

Introducing the Fruits of Empire

In the 1890s, Americans ended their reliance on Europeans for new plants. At a time when British botanical and horticultural leaders were questioning the value of the elaborate system of plant introduction gardens they had established around the world over the previous century, their American counterparts became active and systematic searchers. I examine why this change occurred, how the new American system worked, and the consequences that followed.

The initial stimulus for systematic plant introduction was the intensification in the late 1880s of the long-standing interest of Americans in plants and gardens of the temperate Far East. This occurred because American plant scientists were, for the first time, spending extended periods of time in Japan. Hired by the modernizing Meiji government to introduce Western methods of agricultural education, they gained sympathy for at least some aspects of Japanese culture. More concretely, they acquired access to the countryside and directly experienced what had previously been abstractions: Japanese summers were just as muggy as American ones, the climatic differences between Hokkaido and Kyushu were similar to those distinguishing Massachusetts and Georgia, and cultivated and wild plants in Japan were both wonderfully diverse and similar to those of the eastern United States.

The prominent naturalist E. S. Morse, who taught zoology in Tokyo in the 1870s, reported on both foods and gardens in *Japanese Homes and Their Surroundings* (1885). In 1890 "the largest audience ever assembled" at the Massachusetts Horticultural Society heard William Brooks, who had recently returned from a mission to establish agricultural education on Hokkaido, report on that region's plants.[51] C. C. Georgeson taught for three years at the Imperial College of Agriculture and then, as professor of horticulture at Kansas State Agricultural College, published a long series of articles on Japanese "economic" plants in Liberty Hyde Bailey's *American Garden*.[52] Charles S. Sargent, who was then editing Asa Gray's collected scientific papers, grasped the practical implications of his former mentor's claims regarding the close affinities between the New England and Japanese floras. In late 1892 he traveled to Japan to see its forests, trees, and shrubs; his reports to *Garden and Forest* were compiled into an illustrated book.[53]

Books were not plants, however, and individual horticulturists—even those as wealthy and well-connected as Sargent—were unsure how to find, assess, and introduce new things. The USDA provided no leadership.

The primary goal of the second Cleveland administration was to shrink the size of government in response to the Depression that began in 1893. Agriculture secretary J. Sterling Morton, the Nebraska nursery owner who had invented Arbor Day, made strenuous efforts to eliminate the seed distribution program because it competed unfairly with seedsmen and nurserymen. Nurserymen, however, were selective in their interests and limited in their capabilities. Luther Burbank introduced striking cultivars like the blood-red Satsuma plum, and importers delivered Chinese chestnut trees *(Castanea mollissima)* to the New York Botanic Garden. But dealers did not reach beyond either commercial Japanese sources or established American markets.[54]

The election of 1896 produced major change. Republicans looked outward to the world from a deeply nationalistic perspective. They sought new resources, new markets, and new territory that would make the nation more self-sufficient. The Agriculture Department generally, and plant introduction specifically, became integral to this policy. James Wilson, secretary of agriculture from 1897 to 1913, advanced a long-term program to increase the diversity of economic plants in the United States and its new territories.

Wilson had a strong commitment to useful agricultural research. He came to the United States from Scotland at age sixteen in 1852, reached Iowa four years later, and prospered greatly as a farmer, stockbreeder, journalist, congressman, and state railroad commissioner. When new federal money for agricultural experimentation became available to the states in 1889 through the Hatch Act, he worked with Rev. Henry Wallace (father of the Harding-Coolidge agriculture secretary Henry C. Wallace and grandfather of Franklin Roosevelt's agriculture secretary and vice president, Henry A. Wallace), the influential editor of *Farm and Dairy* (later *Wallaces' Farmer*), to make sure the program at Iowa State College would be "directly for the benefit of farmers," and only "incidentally" for students. To implement this plan, Wallace engineered Wilson's appointment as director of the experiment station as well as the college's professor of agriculture.[55]

Iowa agricultural scientists believed in the value of plant introduction but knew the pitfalls involved. In the early 1880s, horticulture professor Joseph L. Budd had coordinated a State Department distribution of apple varieties from Russia. The effort had no discernable effect because the nurseries that received the plants discarded everything that did not show obviously superior market qualities and soon lost track of the names of the varieties they kept. Having a real impact,

Budd understood, would require long-term maintenance of a large, carefully labeled varietal collection, and an extended program of crossbreeding to blend the virtues of Russian and "native" apple trees.[56] A more paradoxical development along the same lines was botany professor Louis Pammel's collaboration with Wilson himself on the Russian thistle. This weed had overrun the upper Midwest in the 1880s. While most efforts were directed at suppressing the species (see Chapter 6), Pammel and Wilson drew out the optimistic lesson that the success of one Russian steppe plant on the North American prairie indicated that other more valuable types might succeed as well.[57]

Both party policy and personal experience thus led Wilson to promote plant exploration at the USDA in ways that departed from prior scattershot acquisition and broadcast distribution. Soon after he became secretary of agriculture, he hired Budd's former student Niels Hansen to visit Siberia and Turkestan as a "special agent" to collect plants adapted to dry summers and cold winters. Hansen's introductions included downy brome *(Bromus tectorum),* now called cheatgrass, and crested wheatgrass *(Agropyron cristatum).*[58] A year later Wilson created the Section of Systematic Seed and Plant Introduction, which would manage distribution of the railcar loads of seeds and stock that Hansen was shipping to Washington and would organize a permanent program to find, test, and promote new plants.

The central figure in systematic plant introduction (SPI) for the next quarter century was David Fairchild (see Figure 5.7). The son of an abolitionist minister and teacher who became president of Kansas State Agricultural College (later Kansas State College, now Kansas State University), Fairchild began his career at the USDA in 1889 as a mycologist in Beverly T. Galloway's Division of Vegetable Pathology; in 1894 he wrote the USDA's first assessment of the uses of Bordeaux mixture as a fungicide. Fairchild dreamed about travel to the tropical Far East, but he more prosaically decided in 1893 to take a leave of absence from the USDA to do graduate work in Germany on fungi. On his transatlantic passage, however, Fairchild fortuitously met the wealthy Virginia-born bachelor Barbour Lathrop. The forty-five-year-old Lathrop, who had diffuse interests in agriculture and well-formed young men, gave Fairchild $1,000 to study termite gardens in Java on completion of his German work. Three years later, Lathrop dramatically appeared on Fairchild's doorstep in Buitenzorg (now Bogor) and spirited him away on a tour of the Pacific with vague plans to search for new plants to grow in Hawaii.[59]

Figure 5.7 David G. Fairchild, 1893. Reproduction courtesy of the Fairchild Tropical Gardens Research Center, Miami, FL.

Contact with Lathrop transformed Fairchild from a shy provincial egalitarian puritan into a self-confident cosmopolitan who defended racial hierarchies and loved sensuous fruits. Returning to Washington in late 1897, he was the logical choice to lead SPI. His position was shaky for some years because Secretary Wilson disapproved of his companion and of his lack of bureaucratic regularity (he took a second leave of absence in 1899–1900 to tour South America with Lathrop). But in 1905 he settled down and solidified his position in Washington's scientific and social circles through marriage to Alexander Graham Bell's daughter, Marian. Galloway, his former mentor, was now chief of the Bureau of Plant Industry, created in 1900 to consolidate all the USDA's plant science initiatives. Fairchild built a freewheeling operation within that framework.

SPI's most visible activity was collecting. While the initial approach, exemplified by Hansen, was *omnium gatherum*, SPI emphasized focused missions to established foreign agricultural centers. Kansas State–educated cerealist Mark Carleton visited southern Russia to collect hard winter wheats. The elderly Seaman Knapp, Wilson's predecessor as agriculture professor at Iowa State, traveled to Japan to find short-grained rice varieties for Louisiana. Over the next decade, collectors gathered thousands of specimens in different parts of the world.[60]

In particular circumstances Fairchild supported heroic exploration. The most promising locale for temperate-zone introductions, but the hardest to succeed in, was China. An effective collector there needed to be a competent botanist, a practical nurseryman, a skilled cross-cultural negotiator, and a traveler willing to endure primitive conditions for extended periods. In 1906 Fairchild hired the immigrant Dutch gardener Frank Meyer to fill this role. Meyer spent most of the next twelve years collecting in China and southern Russia, and sending potentially useful seeds and plants to the United States. His activities (like those of his Arnold Arboretum competitor, Edward ["Chinese"] Wilson, who focused on more ornamental species) were featured in newspapers and enhanced the public image of SPI; Meyer's exploits in the mysterious East ended, fittingly, when he fell, jumped, or was pushed in the dark from a riverboat on the upper Yangtze in 1918.[61]

Work inside the United States was more important, however, than junkets or expeditions. The major lesson from previous government plant introductions was that broadcast distribution did not work. New plants needed caretakers who would keep track of them and prevent loss of identity, and then would insert them into the American agricultural system. In the case of actual or potential bulk commodities, SPI functioned as part of the larger Bureau of Plant Industry. Carleton spent a few months in Russia, and then devoted more than a decade to persuading farmers to grow, millers to process, and the public to consume *Triticum durum,* marketed as "macaroni wheat."[62] Knapp, a Methodist minister who had run schools for girls and the blind before prospering from swine breeding, returned from Kyoto to educate Louisiana Delta planters about rice culture.[63] Bureau of Plant Industry agrostologists promoted Hansen's downy brome and crested wheatgrass across the West. The soybean *(Glycine max)* succeeded only after SPI introduced thousands of varieties, USDA botanists came to understand that the species consisted of a number of latitudinally distinct types, and bulletins outlined milling technology, explained uses for oil and cake, and provided recipes. American soybean acreage rose from 50,000 in 1907 to 1.8 million in 1924, 11.4 million in 1941, and 55.7 million in 1973, when it began to surpass the great European staple, wheat, as the nation's third largest crop (after corn and hay). It was one of the major transformations of American agriculture.[64]

Fairchild's personal interest was in the more unusual exotics not taken over by the industrial divisions of the Bureau of Plant Industry. But he had neither the staff nor the facilities to shepherd small-value types such as avocados, bamboos, mangos, and papayas toward general use. Like his father, Henry Wallace, and Seaman Knapp, he drew on the hybrid heritage of pastoral outreach and agricultural education to advance his cause. Fairchild organized and maintained a national circuit of horticultural congregations, consisting of propagating gardens with local patrons, and of dispersed "cooperators" who included land grant college professors, commercial orchardists and nurserymen, and serious amateur gardeners.[65] Cooperators received new plants and returned testimony about their qualities; all were kept abreast of new introductions through published inventories and a mimeographed bulletin aptly titled *Plant Immigrants*. Fairchild preached the virtues of new things, such as the mango, chayote, dasheen, and udo; he urged Americans to overcome their gustatory puritanism and to indulge in richer and more varied tastes.[66]

Fairchild's broadest vision concerned the ways that exotic plants could reconfigure national sensibilities. In 1905 he and his wife imported Japanese cherry trees and a Japanese gardener to the rustic property they had purchased beyond Chevy Chase, Maryland. While the gardener soon died of tuberculosis, the trees flourished, and four years later, Fairchild sought to introduce them into the nation's most public space. Working with travel writer Eliza Scidmore, he proposed that a Japanese "field of cherries" be planted around the Tidal Basin, the extension of the National Mall then under construction south of the Washington Monument. The idea was endorsed by Mrs. Taft and the president, and in late 1909, Fairchild was coordinating a large shipment of trees that Tokyo officials obligingly donated to their "sister city" of Washington, D.C.[67]

This action took Fairchild into unexpectedly deep waters, with swirling currents of ethnic politics, imperialist plotting, diplomatic posturing, bureaucratic rivalry, and scientific competition. On January 28, 1910, Fairchild's cherry trees were burned on the Washington Monument grounds. While a different collection of specimens were successfully planted two years later, by the end of the 1910s, the government had adopted restrictions on the importation of nursery stock that systematically changed plant introduction. The instigator of both of these upheavals was entomologist Charles Marlatt—Fairchild's childhood neighbor and best man at his wedding. Chapter 6 explores the impulses that culminated in these events, and it sketches the transition from horticultural cosmopolitanism to agricultural nativism between 1910 and 1930.

CHAPTER SIX

Mixed Borders
A Political History of Plant Quarantine

In the early 1910s, the monthly bulletins of the California State Commission of Horticulture's Quarantine Division prominently displayed a bull's-eye seal declaring the agency's mission (see Figure 6.1). In the center was a drawing of the Mediterranean fruit fly, *Ceratitis capitata*, with its distinctive mottled wing and thorax pattern carefully delineated. Surrounding the insect was the Latin motto, *Finis Rationem Excusat*—in English, "the End Justifies the Means."

Why were these scientific bureaucrats so melodramatically determined? They were a small group who had few tools to counter threats that seemed biologically relentless. In addition, they saw themselves surrounded by people who were at best indifferent about their mission. The deeper issue, however, was that the systems they defended and relied on were so unstable. On the one hand, they were seeking to keep landscapes comprised of human newcomers and their exotic plants free from the pests that had coevolved with those plants, or that had already succeeded in similar settings around the world. On the other hand, the borders they manned had a multiply mixed character. U.S. political units were not congruent with North American biogeographic provinces. Authority over boundaries was politically divided. And while borders were established to control movement, American leaders worked hard to facilitate the transport of many kinds of living organisms.

This chapter tells the history of American efforts to interdict pests.[1] Crop-destroying insects were the most important issues. But people also worried about weeds, and the fungi, bacteria, and viruses that were

Figure 6.1 The official seal of the Quarantine Division, California State Commission of Horticulture, from its *Bulletin*, 1914. Reproduction courtesy of the Rutgers University Library.

classed together in the late nineteenth century as plant diseases. I begin with the background of pest activities in America and attitudes about their control. I then examine the efforts of localities and states in the nineteenth century to selectively arrest the increasing horticultural and economic homogenization of the continental nation. California was the most aggressive proponent of this kind of local control. In the 1890s, scientific bureaucrats sought to combine diverse state interdiction efforts into an effective and equitable national policy, and they concerned themselves increasingly with international and intercontinental issues. They were stymied, however, by the proliferation of pests, the weakness of American governments, and the dramatic changes in American boundaries that occurred in 1898. Federal entomologist Charles Marlatt resolved these issues. In the 1910s, he developed the legal and bureaucratic capability to regulate the movements of species across both international and interstate boundaries, and to attack pest populations locally. This new federal power, however, limited both the freedom of horticulturists to plant what they wished and the ability of Americans to move freely within their country.

This chapter is written as a traditional political history narrative oriented toward its end point: Marlatt's development of what came to be the United States Plant Quarantine and Control Administration. Such a form gives a first level of coherence and significance to events that individually seem trivial and disparate; by linking actions on different fronts—ranging from scientific to diplomatic—it illuminates how issues sometimes shifted suddenly and without (at the time) explanation. Finally, it emphasizes the relations among issues and personalities. In the early twentieth century, pest control and horticulture were small worlds with outsized people. Marlatt, Leland Howard, David Fairchild, and J. Horace McFarland are not household names today, but their passions formed the environments in which Americans now live.

Legislating against Pests

As noted in Chapter 2, George Morgan named his insect the Hessian fly as a comment on its behavior and as a way to build nationalism; the British were the ones who cared most about where the species actually came from. Prior to the 1870s, Americans worried about a large number of pests, but they distinguished them more on the basis of their novelty and severity than by their origin. Agricultural writers knew that many of the weeds that plagued their fields had come from the Old World, but some of the worst problems—including poison ivy *(Toxicodendron radicans)* in the east and locoweed *(Oxytropis campestris)* further west—were clearly indigenous. The situation with insects was similar. Farmers were unhappy that the plum curculio *(Conotrachelus nenuphar)* and the apple-boring codling moth *(Cydia pomonella)* had crossed the Atlantic. But American arthropods, including various ticks, the chinch bug *(Blissus leucopterus)*, Rocky Mountain locust *(Melanoplus spretus)*, and Colorado potato beetle *(Leptinotarsa decemlineata)* were larger and more immediate challenges to the spread of Anglo-Americans and their exotic cultivars across the continent.[2]

Foreign pests became more prominent toward the end of the nineteenth century, partly because most American organisms that could feed on crops had by then made their appearance, and partly because global steamship travel introduced a new set of foreign species. Scientific commentators noted the growing predominance of Old World crop insects and weeds in North American fields. Their interpretations of these phenomena blended the long-standing impressions about the inferiority of New World creation (discussed in Chapter 1) with newer Darwinian ideas about competition and progress. In 1870 entomologist Charles Valentine Riley, who had migrated from England to Illinois as a young man, reported with some pride that Old World insects were much more aggressive than their New World counterparts. The only consolation he offered was that American pests were evolutionarily superior to Australian insects, which were spreading nowhere and declining fast. Asa Gray similarly reported on the "pertinacity" of European weeds in America, and placed their spread within a Darwinian biogeographic framework. In 1892 Liberty Hyde Bailey argued that the spread of Old World pests was one stage in the progressive transformation of balanced but unutilized natural areas into managed crop settings: while European pests followed European culture, they would be controlled through higher culture.[3]

At times, the power of pests—indigenous or not—could seem overwhelming. Hessian flies were marauding armies. Midwestern farmers believed that the clouds of grasshoppers that appeared intermittently in the 1850s, 1860s, and 1870s were as devastating as the locust plague that had preceded the biblical Exodus. Entomologists such as Riley mixed Jeremiah and General Sherman in chronicling how the potato beetle moved slowly eastward from Colorado until "the advance guards of the vast army pushed to the extreme eastern limit of New York," where they blanketed Coney Island beaches; their crushed bodies made railroad tracks so slimy that trains were immobilized. Riley predicted that the insect would soon "be safely borne over to the land of 'murphies' " (Ireland) and then beyond to continental Europe.[4]

In some cases, pests—primarily plant diseases—ended particular forms of culture. New Englanders had largely abandoned wheat in the 1700s because of stem rust. The enthusiasm for pears faded from most parts of the Northeast after the Civil War due to the difficulty of controlling fire blight. On the other hand, many devastating agricultural plagues—notably the Rocky Mountain locust—disappeared, like human epidemic diseases, both suddenly and spontaneously.

The most difficult situations were those where pests became intermittently irritating components of domesticated landscapes. Humans then had to choose among responses. One possibility was acceptance and adaptation: apples with codling moth damage were unsalable as high-end table fruits, but few cider buyers objected (if they knew) that their drink included larvae and their feces. The second alternative was individual control. Farmers routinely kept down field weeds through a mix of plowing, mowing, harrowing, and crop rotation. Orchardists managed insects and blights with "clean culture"—picking up dropped fruit, removing and burning infested branches, maximizing light and air, and controlling soil conditions. Washes, oils, and systemic poisons became increasingly important in the last third of the century.[5]

Agricultural leaders intermittently argued that community coordination in pest control was preferable to seemingly endless individual labor. Attacking particular weeds or insects throughout a town, county, or even state could lower their numbers significantly. To be effective, however, such projects required full participation. If even a few farmers declined to cooperate—whether from ignorance, indifference, skepticism, or disinclination to make life easier for their competitors—a communal campaign would be futile.

Beyond the Reach of the Sheriff

As early as 1726, Connecticut leaders had concluded that the presence of barberry plants increased the wheat rust problem, and they empowered town meetings to mandate eradication.[6] In 1791 Vermont required property owners to cut down what locals called the "Canada thistle" *(Cirsium arvense),* an Old World plant that had established itself along the St. Lawrence River and was expanding its range southward along Lake Champlain and the Connecticut River. Those who refused could be fined thirty shillings plus costs.[7] In 1812 New York authorized towns to apply regulations that governed noxious wild animals to the thistle, and a decade later, Connecticut required farmers to control both the thistle and the wild carrot or Queen Anne's lace *(Daucus carota).*[8] Both the number of states with antipest laws and the number of targeted species expanded in the 1870s and 1880s: four prairie states required farmers to destroy grasshopper nests, Michigan legislated against peach yellows, and the Dakotas mandated coordinated local campaigns against cocklebur, Canada thistle, and the new Russian thistle.[9]

The efficacy of these laws was limited. People might complain privately about their neighbors' pests, but they were loath to worsen relations further by reporting weeds to local government. Prominent people, moreover, questioned whether coercive pest elimination programs worked. University of Vermont president Daniel C. Sanders at the beginning of the century, and Cornell horticulturist Liberty Hyde Bailey at its end, each ridiculed the idea that legislation could change nature; Bailey concluded an address to the 1894 meeting of the American Association of the Advancement of Science with the message that "weeds are beyond the reach of the sheriff." He applied ecological thinking to the problem of the Russian thistle, arguing that prairie farmers had created a "vacancy in nature" by stripping the vegetation off the land and planting a wheat monoculture. The only way to keep down weeds was to complete the transition from grassland to farmland; this required not laws but intelligence, knowledge, labor, and time.[10]

Protecting California

The one part of the United States where sheriffs routinely fought pests was California. It differed from the rest of the country both geographically and economically. The warm and dry climate of the Central Valley was more like that of southern Europe than the East. The California botanical province was separated from eastern ecosystems by plains, mountains, and deserts; and, as Asa Gray had understood, its evolutionary history differed from that of the rest of the United States. A number of major pests plaguing the East did not exist in California during its first decades as a state. As a consequence, Old World fruits—including *Vitis vinifera*—thrived in the region around San Francisco Bay. By the 1870s, Californians were building up horticultural enterprises much larger than those in the East. They initially concentrated on preservables such as wine, raisins, and prunes, but they imagined using the railroads to compete directly with eastern growers of fresh fruits.[11]

The people who dominated this field were aware of the uniqueness of their setting and familiar with the use of law, bureaucracy, and technology to control pests, insure quality, and keep up prices. In 1880 Sonoma Valley wine producer Arpad Haraszthy, son of the Hungarian nobleman who had introduced the Zinfandel grape from his homeland, learned about the havoc that the American root louse, *Phylloxera vastatrix*, was wreaking in European vineyards. Concerned that this aphid had also reached California, he persuaded the state to create a viticultural board that would have the powers of a "Board of Health" over "diseases of grapevines and vine pests." Orchardists, who were concerned about the codling moth, red spider, and a new scale insect found a few years earlier near San Jose, soon expanded the board's scope to cover all commercial fruits.[12]

The central activity of the California State Board of Horticulture was a local vineyard and orchard pest control regime. County inspectors had the power to enter private property, search for pests, and require remediation. Growers pruned and washed vines and trees, sprayed them with oil soap, arsenic, and copper salts, and enclosed them in special tents for fumigation with cyanide. Officials directed the work on the property of those who did not comply, and the cost was automatically added to property tax bills. Growers financially unable to maintain sanitary practices went out of business.[13]

The Gypsy Moth in Massachusetts

In the East, official pest control was a suburban rather than an agricultural issue. E.L. Trouvelot, an amateur lepidopterist working as a scientific artist at Harvard's Museum of Comparative Zoology, is notorious in entomological circles for importing eggs of the gypsy moth *(Lymantria dispar)* to the Boston suburb of Medford in 1868, and hatching them in his backyard under poorly secured netting. The insects spread into nearby vacant lots but initially were kept down by local boys who burned the vegetation frequently for fun. In 1889, however, the caterpillar population of Medford exploded. Within a few weeks, shade trees and backyard orchards were completely defoliated, and pedestrians endured a rain of feces and squirming larvae. Within a few years the gypsy moth spread into woodlands recently acquired by the Massachusetts Metropolitan Park Commission.[14]

In 1890 the state legislature created the Gypsy Moth Commission, which proposed to eradicate the pest completely. Naturalist Edward Forbush, advised by University of Massachusetts entomologist Charles H. Fernald (who, as a Union sailor during the Civil War, had experience in blockades), planned to push the insect back upon its center (see Figure 6.2). Large numbers of seasonal laborers were given uniforms, authorized to enter private property, and trained to hunt for gypsy moth eggs in the highest tree branches (see Figure 6.3). For more than a decade, they sprayed poisons in the woodlands (in 1892 a chemist working for the commission developed lead arsenate, the standard insecticide for the next half century), clear-cut infested areas, and employed the "cyclone burner," the first modern flamethrower, to destroy egg masses hidden in rock crevices (see Figure 6.4). Commission reports featured maps showing the yearly advances and retreats of the caterpillar from its bases around Medford. Their aim was to show the country that Yankee technology and determination could produce total victory over an undesirable alien insect.[15]

The San Jose Scale: Panic and Response

One solution to the problem of pests was to prevent their entry into an area. In the nineteenth century, the borders that mattered were within the United States. Entomologists and horticulturists were unable, however, to devise a workable and comprehensive response to the propensity of pests to cross state lines.

As Anglo-Americans spread across the continent, they repeatedly experienced the sequence of settlement outlined by Liberty Hyde Bailey. Land was cleared of its vegetation and exotic species were planted. The combination of fertile soil and lack of crop pests resulted in high levels of productivity and quality for periods of years. Gradually, however, weeds followed settlers along canal paths and railroads. Insects and plant diseases arrived with travelers and supplies, and on nursery stock sent through the mail.

In theory, the movement of at least some pests could be controlled, but in practice, Americans in the nineteenth century realized that interdiction involved great difficulties. Scientists or officials in any one place

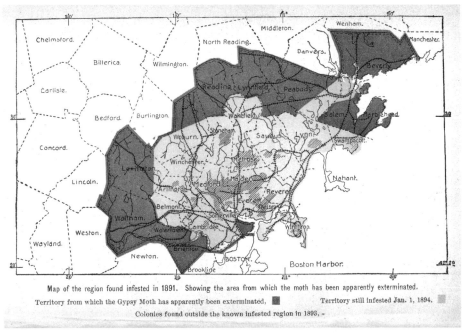

Figure 6.2 In early 1894, the Massachusetts Gypsy Moth Commission published this map displaying their progress in pushing the enemy back upon its center, with anticipated eradication. Photograph from *Agriculture of Massachusetts*, 1893, facing p. 285. Reproduction courtesy of the Rutgers University Library.

were often unaware of the existence, or the motile capabilities, of a new organism living elsewhere. Even if a species was known and generally loathed, a small number of enthusiasts (as in the case of Johnson grass) could spread it widely. Finally, there were issues that reached to the foundations of the American political system. How could the movements of weeds, insects, or fungi be controlled across untaxed and normally unpoliced borders? How were local interests and the powers of individual states to be balanced against national concerns and the authority of the federal government?

Stated concretely, the issue was whether wormy apples and mildewed grapevine cuttings were analogous to poisonous chemicals and Irish laborers, or rather were more like prostitutes and potential carriers of cholera. In the early nineteenth century, the Supreme Court balanced the new Federalist principle that states could not interfere with the movement of people or merchandise across state lines against an

Figure 6.3 Each year in the 1890s, Massachusetts Gypsy Moth Commission employees climbed notable trees, such as this elm, and scraped off egg masses by hand. Photograph from Edward H. Forbush and Charles H. Fernald, *The Gypsy Moth* (1896), facing p. 246. Reproduction courtesy of the Rutgers University Library.

Figure 6.4 In 1899 gypsy moths were found in Georgetown in northeastern Massachusetts, well beyond the Gypsy Moth Commission's defensive perimeter. Commission employees used flamethrowers in an effort to wipe out the insects. The state legislature eliminated the commission a few months later. Photograph from *Agriculture of Massachusetts*, 1899, facing p. 368. Reproduction courtesy of the Rutgers University Library.

---

acknowledgment that local governments retained police powers, including the authority to declare quarantines, in the face of epidemic diseases. In the decades after the Civil War, however, the federal judiciary's interest in facilitating national commerce substantially limited state and local quarantine power. In 1878 the Supreme Court declared that a Missouri law prohibiting railroads from shipping cattle through the state for nine months each year—designed to protect local livestock from Texas cattle fever (and incidentally to support the price of Missouri cattle)—was illegal interference in interstate commerce. A ban on all livestock for most of every year because of a general suspicion of disease in a few individuals was considered too broad by the court. A state needed either to reasonably limit the duration of a quarantine or to offer an inspection procedure whereby potentially dangerous individuals were separated from those who were not.[16]

California resisted this movement toward national commercial homogenization. In the early 1880s, state officials imagined a permanent "horticultural quarantine" to keep common eastern pests from entering the state on nursery stock or fruits. They were unable to stop small shipments of plant materials sent through the mail, and they had neither the manpower to inspect large shipments nor the technical capability to find problematic insects or spores. However, as New Jersey entomologist John B. Smith commented acidly, they used their authority to inspect and disinfect bulk plant shipments "in such a way that when the stock is thoroughly treated, it is of not much use to set it out; so, practically, there is not much inducement for a Californian orchardist to buy his stock outside of his own State." This was good enough.[17]

Eastern nurserymen were deeply antagonistic toward California because of the asymmetries the state introduced into the national commercial system. Not only were Californians like Luther Burbank able to advertise and ship their special "creations" freely around the country, while easterners were hampered in the Golden State; the more galling consequence was that if California inspectors entered one of their state's nurseries and condemned its trees as diseased, nothing except "qualms of conscience," as U.S. Department of Agriculture (USDA) entomologist Leland Howard noted hopefully, could stop the owner from packing up the stock immediately and selling it to dealers in the East.[18] From the perspective of the Californians, the solution to this difficulty was for other states to emulate their practices (as Oregon and Washington did in the early 1890s). But the smaller scale of the orchards, along with the multiplicity of governments and the much denser commercial and ecological networks, made such advice unrealistic east of the Rockies.[19]

California exceptionalism shifted from irritation to crisis in August 1893. A hobby orchardist in Charlottesville, Virginia, sent some pears to USDA vegetable pathologist Beverly Galloway in Washington, with a query about the spots that mottled their surfaces. Unable to recognize the problem, Galloway carried a pear down the hall to the Entomology Division. Leland Howard reported, "I nearly jumped out of my seat" on seeing eastern fruit covered with "that scourge of western orchards," the tiny insect that Californians were calling the San José scale *(Quadraspidiotus perniciosus)*. Howard soon found scale infestations scattered throughout the East, and traced them to a New Jersey nursery that had received a shipment of plum trees from California about five years earlier. Eastern orchardists were unable to control this insect. As their trees (especially peaches) died in the midst of the major economic depression following the Panic of 1893, panic about the future of fruit growing east of the Rockies spread rapidly.[20]

What could be done about the scale, or about other pests that might be carried across state lines? Democrats, in control of the White House in 1893, had no interest in giving new powers to the federal government. Two weeks after the first scaly pears surfaced at USDA headquarters in Washington, Assistant Secretary of Agriculture Edwin Willits explained to the World's Horticultural Congress in Chicago that the USDA's activities were limited to research and education. Pest control had to grow from the development of "wholesome public sentiment" and state inspection systems. Only when these grass roots were established could the general government consider coordination of interstate relations and action at the nation's borders.[21]

Federal and state scientists tried to work within this restrictive framework. Howard, who replaced C. V. Riley as head of the USDA Division of Entomology in 1894, provided orchardists with basic information on the San Jose scale, discussed how far it might spread, and advised on treatment techniques. USDA staffers produced bulletins compiling state laws dealing with the control of insects, weeds, and plant diseases.[22] State entomologists, while discussing among themselves whether they could accurately determine whether nurseries were scale-free, lobbied for orchard inspection laws. Only fifteen states had passed such laws by early 1896, however, and their provisions varied significantly. Nurserymen involved in interstate business struggled to keep track of the different and often conflicting rules.[23]

The problems that resulted from leaving pest control to the states were displayed further in the contemporaneous failure of Texas to prevent the spread of the boll weevil *(Anthonomus grandis)*. This species had coevolved in Central America with upland cotton, but, like the Hessian fly in relation to wheat, had been left behind when its host plant spread across the United States. In contrast to the Hessian fly, the boll weevil did not have to cross an ocean; in the early 1890s, it flew over the Rio Grande and found large fields filled with its preferred food. USDA entomologist C. H. Tyler Townsend, who made the first scientific study of the "Mexican cotton-boll weevil" in 1894, recommended that extermination in the area around Corpus Christi, combined with establishment of a fifty-mile-wide cotton-free zone along the international border, might keep the insect out of the United States. But Assistant Secretary of Agriculture Charles Dabney was unable to persuade Texas leaders to adopt and implement this coercive policy.[24]

By early 1896, a significant number of eastern entomologists and orchardists sought a national response to the problems posed by the San

Jose scale and other pests. They fully understood the political dimensions of their movement. Ohio entomologist Francis Webster used the old language of Radical Reconstruction in calling for "United States laws, enforced by United States authority and by United States officers, who . . . will faithfully perform their duty anywhere and everywhere, when called upon to do so." Later that year, the Ohio State Horticultural Society invited experiment station scientists and fruit growers to come to Washington the day after the Cleveland presidency ended for the National Convention for the Suppression of Insect Pests and Plant Diseases by Legislation. James Wilson, the new Republican agriculture secretary, welcomed them, and the USDA published their proceedings.[25]

Convention leaders struggled to shape a consensus plan of action. Realizing that they could not agree on a uniform inspection law because California was too different from the eastern states, they adopted only a set of guidelines in that area. They focused instead on the federal level, drafting a "Bill Relating to Interstate and International Legislation against Insect and Fungous Pests." This proposal gave the Agriculture Department the power to regulate interstate commerce in nursery stock (but not fruit), and also to screen both plants and fruits coming into the country. Entomologists believed that a bill supported by the organizations represented at the convention, and by the USDA, would easily get through Congress within the year.[26]

They were wrong. Nurserymen, the group most affected by the proposed law, had not been included in the convention. *National Nurseryman* criticized the "Washington bill" for mandating inspection of trees but not fruits, and for allowing states to open crates of carefully packed nursery stock during transit. The American Association of Nurserymen drafted its own bill for a truly national pest inspection system, in which plant materials would be examined and certified at the point of origin, and could then travel without interference.[27] This bill, and not that of the convention, was reported favorably from the House Agriculture Committee on February 16, 1898.[28] Movement stalled, however, first because two weeks earlier Germany had banned American fresh fruit and nursery stock due to the danger posed by the San Jose scale (leading to calls for war), and more immediately because on the day of the committee report the USS *Maine* exploded in Havana harbor. The American invasions of Cuba and the Philippines put the issue of pest invasions within the United States on hold.[29]

The failure of the National Convention produced dissension among entomologists. John B. Smith used the broad forum of *American*

Agriculturist to criticize his peers for "losing their heads" over the San Jose scale. The insect, he asserted, was "not a particle more destructive than many of our native species or than those to which we have become used." He doubted, in any case, whether inspection of foreign plant and fruit shipments would keep out new pests.[30] At the next annual meeting of the Association of Economic Entomologists, Herbert Osborne of Ohio State University criticized the German-born Smith with the argument that American entomologists had a duty to "stand together in final recommendation to the public." Disagreements about the feasibility of quarantines or extermination campaigns should be discussed only "among ourselves."[31]

The most vigorous criticism of the previous years' campaigns came from Leland Howard's assistant, Charles L. Marlatt (see Figure 6.5). The son of an abolitionist minister who had settled in Manhattan, Kansas, in 1856 to farm, promote education, and fight slavery, Marlatt had learned entomology and horticulture at Kansas Agricultural College, and then joined the USDA Division of Entomology in 1889. A witness to the

Figure 6.5 Charles Marlatt, 1920s. Photograph from the National Archives (007, H, 27 Marlatt-774).

Rocky Mountain locust plagues and an expert on the periodical cicada, he understood both reasonable and unreasonable panics over insect outbreaks.[32]

Elected president of the Association of Economic Entomologists in 1899, Marlatt decided to use the San Jose scale as a text for heretical "sermonizing." He began by declaring that the panics, laws, and extermination programs of the previous six years had done little to stop the spread of that insect. Migrations of insects were fundamental "world movements," and efforts to resist them could be ridiculous. He proposed that "either we must build a Chinese wall and live entirely apart from the rest of mankind, or make up our minds ultimately to be the common possessors of the evils as well as the benefits of all the world." He proposed a policy of "laisser-faire" regarding those pest movements that were "beyond practicable reach by human agencies." While acknowledging that introduced insects severely damaged previously unexposed plants, he reminded his listeners that the Darwinian "vital stimulus" experienced by organisms when they arrived in a new setting was the reason "our forefathers" had become a "race of stalwarts" who were able to take over America. In insects, at least, this vigor was temporary; the histories of pests such as the horn fly *(Haematobia irritans)* and the potato beetle showed that new species could be controlled within a few years. Entomologists should focus on the severe temporary disruptions that sometimes occurred in particular places. "Local control," Marlatt concluded, was "the chief, if not almost sole, legitimate field of effort in applied entomology."[33]

Marlatt's address generated intense discussion. Some objections were fact-based—that quarantines and extermination campaigns did work. The main criticism, however, was that "general distribution of the paper would lead thousands of farmers to fold their arms and await a harmonious and perfect balance of nature, ignoring and ridiculing the suggestions of economic workers." Marlatt's unyielding answer was that responsibility for pest control should be in the hands of those who were directly affected and who could act.[34]

The practical meaning of laissez-faire soon became evident in Massachusetts. In 1899 gypsy moths were found well beyond the perimeter established and defended by the Gypsy Moth Commission. This failure triggered a legislative inquiry into the decade-long eradication campaign. Critics attacked the commission's extravagance, destructiveness, alarmism, and lobbying. Advised by prominent noneconomic entomologists, the legislature concluded that wiping out the insect was not feasible, that the gypsy moth was no worse than other familiar pests, and that the better policy was local control. The commission was disbanded and its duties transferred to a committee of the state's Board of Agriculture.[35]

In spite of Marlatt's message, Leland Howard and other entomologists continued to push for national regulation of interstate commerce that could spread pests. An experienced Washington hand and a genial raconteur, Howard patiently maneuvered entomologists, state officials (organized as the American Association of Horticultural Inspectors), horticulturists, and nurserymen toward agreement. But with Pacific Coast officials demanding radical action and the authority to develop their own rules, and eastern nurserymen pushing for a minimally intrusive uniform procedure, consensus was elusive. In 1906 the stakeholder organizations formed an official joint committee to craft mutually acceptable legislation. But the membership of the American Association of Nurseryman could itself not reach a uniform position, and in 1908 declined to cooperate further.[36]

Japanese Cherry Trees and a Chinese Wall

The person who broke through this impasse was Marlatt. He abandoned his advocacy of laissez-faire and instead took charge of the campaign to implement an activist national policy against pest movement. In place of Howard's efforts to harmonize interests, Marlatt adopted tactics of maneuver and confrontation. He also shifted the primary focus from the borders between states to the external boundaries of the nation. These new initiatives resulted in conflict at the highest levels of government, but were ultimately successful: in August 1912, Congress passed the Plant Quarantine Act.

Marlatt's change in perspective was personal. His life as an ordinary scientific bureaucrat was transformed in 1901 when, at age thirty-eight, he married a wealthy and socially prominent Bostonian, Florence Brown. Taking a leave of absence from government service, he embarked on a half-year tour of the Far East that would combine the upper-class custom of an extended wedding trip with opportunities to meet Japanese agricultural scientists and find the home territory, and possibly some importable predators, of the San Jose scale. In China the Marlatts saw masses of displaced and distressed people, and more generally the ruin left by the native Boxers' attacks on foreign missionaries. Most immediately distressing, Florence Marlatt was attacked by intestinal parasites and died soon after her return to America. What began as a romantic and scientific tour ended as a set of encounters with Yellow Perils.[37]

Marlatt mourned and returned to work, studying the dramatic resurgence of the gypsy moth in Massachusetts following the end of the state's extermination campaign, and trying to rename the San Jose scale the "Chinese scale," since he had shown that it existed in China before reaching San Jose. A few years later he remarried, to the socially prominent Philadelphian Helen McKey-Smith. And he became a Washington player. In 1903, as acting chief of the Division of Entomology, he developed the plan to transform the division into a more structured bureau organized around crop types rather than insect taxa. As owner of a mansion not far from the White House, he socialized with scientific, bureaucratic, and congressional leaders, and gained the capability to make things happen.[38]

In late 1908 Marlatt wrote a new bill under the title, "Inspection of Nursery Stock at Ports of Entry of the United States." This legislation was framed as a response to an international emergency—the discovery of brown-tail moth *(Euproctis chrysorrhoea)* eggs on imported French nursery stock. Regulation of interstate shipments of stock from areas with particular pest infestations was a subsidiary provision of the bill. Marlatt's fellow Kansan, C. S. Scott, chairman of the House Agriculture Committee, quietly introduced this bill in the January 1909 lame duck session. It was approved by the House and had passed the parallel Senate committee before the nurserymen's legislative committee learned about it and was able to block it.[39] Marlatt met with this group in hopes of making a deal: he conceded to their objections regarding inspection procedures in exchange for their agreement that the bill could go forward in the next Congress. The nurserymen's convention that summer, however, declined to endorse what their committee had arranged. They proposed a different bill, and persuaded the Association of Horticultural Inspectors to support it.[40] Marlatt was frustrated that these groups did not recognize the urgent need for action. The extended discussions, cancelled deals, and repeatedly revised proposals were recapitulating the irresolution of the preceding decade.

Hardwood Politics

How could he break through what were, to him, irresponsible attitudes? He found an answer a few days after the December 1909 meeting of the Association of Horticultural Inspectors when David Fairchild's collection of ornamental cherry trees arrived from Japan at the USDA's Washington greenhouses. The Bureau of Entomology had the authority to inspect all USDA plant introductions for pests. Marlatt's staffers reported that they found crown gall, root gall, two kinds of scale, a potentially new species of borer, and "six other dangerous insects." As acting chief of the bureau, Marlatt recommended that the entire shipment be condemned.[41]

Destroying the Japanese cherry trees would be an object lesson concerning the danger of imported pests and the courage of the USDA. The recommendation, however, disrupted both international relations and Marlatt's personal relationships. The trees were integral to a complex political web that had developed during the previous decade. In 1905 labor leaders in San Francisco began to express dissatisfaction with the growing numbers of Japanese immigrants who were outcompeting Irish American natives for jobs and property. A new Japanese and Korean Exclusion League urged Congress to prohibit immigration from both those lands and, more immediately, pressured the city to limit Japanese, Koreans, and Chinese to a single segregated elementary school. For the Japanese, this was a double insult: declaring them inferior to other Americans and lumping them with, in their minds, the inferior Korean race. In 1907 rioters attacked Japanese in San Francisco, and there was talk of war in newspapers on both sides of the Pacific.[42] President Theodore Roosevelt admired the Japanese but considered them alien. He pressured them to accept the euphemistically named "Gentlemen's Agreement," whereby the United States rescinded official discriminatory policies but Japan restricted emigration to the United States and accepted U.S. curbs on the movement of Japanese between Hawaii and the American mainland. Roosevelt then secretly agreed to support Japanese hegemony over Korea and part of China.[43]

Fairchild's cherry tree concept was floated soon after the conclusion of these discussions. William Howard Taft, the new president, approved the proposal because it nicely expressed the ideal of American-Japanese friendship based on equality and separation: a field of Japanese flowering cherries in Washington was preferable to a settlement of Japanese working families in California. Japanese officials embraced the idea publicly as a gesture that officially affirmed that Tokyo and Washington were equal "sister cities." But it meant more: as a correspondent to the *New York Times* explained, it was equivalent to the United States placing in a foreign capital "some symbol of nature that would embody all that Plymouth Rock, all that the Declaration of Independence, all that the Emancipation Proclamation means, of liberty, patriotism, union." The

Japanese emperor was planting "the symbol of the very soul of the manhood of Japan" in the capital of the United States.[44]

Marlatt's condemnation of these trees was problematic enough as a subordinate's disruption of delicate high-level diplomacy. The action had harsher overtones, however, because the entomologists' emphasis that these beautiful trees hid insidious borers and other damaging insects resonated with widespread Yellow Peril arguments, notably the claims during the recent Russo-Japanese War that Japanese politeness masked treachery and atrocities against Caucasians.[45]

The personal dimension of Marlatt's destruction of Fairchild's trees was not so consequential but it cut deeper. The two men, both sons of abolitionist ministers and educators, had grown up together on the Kansas prairie during the grasshopper years; Fairchild was an undergraduate at Kansas Agricultural College when Marlatt was an assistant in the Department of Entomology and Horticulture. They followed parallel paths of upward mobility through work at the USDA and marriage into prominent families. Their differences were ultimately more important, however. The entomologist Marlatt had apprenticed in Kansas and had been subservient at the USDA for five years to C.V. Riley. He had done fieldwork in Texas cotton fields and Virginia orchards; his first marriage had ended tragically due to circumstances he had created; his position as a Washington insider had come through hard work and hard dealing. Fairchild, by contrast, was a golden boy. As the son of Kansas Agricultural College's president and a nephew of the prominent mycologist Byron Halsted, he had an easy path to the USDA, where he became a protégé of the rising Beverly Galloway. He traveled the world with Barbour Lathrop for years without mishap, and then effortlessly penetrated the highest Washington circles by marrying Marian Bell. His work was gardening, and his home, called "In the Woods," was far from official Washington; yet, with the Japanese cherry tree proposal, he quickly gained direct access to the president and the diplomatic corps. Marlatt wanted to convey to Fairchild that life was not always a bowl of cherries.

The recommendation that the cherry trees be destroyed went from Marlatt to Secretary Wilson to the president. Taft was unwilling to countermand his government's experts, and on January 28, the trees were taken from storage on the Washington Monument grounds, arranged in conical piles, and burned (see Figure 6.6). The action was presented as an apolitical scientific necessity. But the *New York Times* noted skeptically that "we have been importing ornamental plants from

Figure 6.6 Burning of the first shipment of Japanese cherry trees on the Washington Monument grounds, 28 January 1910. Reproduction courtesy of the Fairchild Tropical Gardens Research Center, Miami, FL.

Japan for years, and by the shipload, and it is remarkable that this particular invoice should have contained any new infections." Even if it did, the editorialist did not understand why the government provided public notice that the trees were infested and were being destroyed; if the only issue were those particular insects, a tactfully arranged "accident of the obviously unavoidable sort" would have been preferable.[46]

The Japanese declined to take offense at this example of American boorishness. Over the next two years, however, they rejected American efforts to maintain the "open door" in Manchuria, formally annexed Korea, and pushed the United States to take the lead in terminating the discriminatory trade treaties that Western powers had imposed on their country a half century earlier. At the end of what diplomatic historian Walter LeFeber summarized as a "string of horrors for U.S. hopes in Asia," they sent to Washington a new shipment of trees that scientists at the Imperial Quarantine Service, the Imperial Horticultural Station, and the Imperial University had selected from an area that was free of scale insects, raised in ground that was free of nematodes, sprayed with fungicides and insecticides, and fumigated twice before packing. These trees were accepted and were planted in 1912 around the Tidal Basin, along the Potomac, and on the White House grounds "as a living symbol of friendship between the Japanese and American peoples."[47]

The Plant Quarantine Act

Marlatt, meanwhile, returned to his legislative battles with stronger opinions and a reputation as an official to be feared. At a House hearing in April 1910, he reintroduced his Chinese Wall metaphor of a decade earlier; but having seen the Great Wall and having committed himself to protecting America against pests, he drew precisely the opposite conclusion. He declared that the United States did not need nursery stock imports, and that "if this country should set up a Chinese wall against such importations we could take care of our own needs," growing seedlings at home "as they were grown in the days of our fathers."[48] He interrupted the testimony of William Pitkin, president of the American Association of Nurserymen, to accuse him of mischaracterizations. Pitkin then explained that his primary opposition to the USDA's proposed bill came from fear of putting his industry into a regulatory "noose" controlled by a single official, "especially if one end . . . is going to be held by our friend Mr. Marlatt."[49]

The situation became tenser later in the year, when Hawaiian entomologist E. M. Ehrhorn, formerly California's horticultural quarantine officer, found Mediterranean fruit flies in trees around Honolulu.[50] This cosmopolitan pest had first been identified scientifically in the 1810s and 1820s in "the East Indies," Mauritius, and the Azores; it had the common name of the "orange fly" in London markets in the 1840s, the "Bermuda peach maggot" at the USDA in 1890, and the "West Australian fruit fly" in Perth, but Sydney entomologist W. W. Froggatt called it the "Mediterranean fruit fly" to distinguish it from another in Queensland. Californians followed that lead. State horticultural officials after 1900 routinely strove to interdict the flies by cooperating with federal customs inspectors to find and destroy the small amounts of fresh fruit that crossed the Pacific in ships' stores.[51]

Discovery of the species on Oahu posed a significant problem for this arrangement. The American annexation of Hawaii had disrupted the long-standing presumption that the continent's limits were the nation's borders. After 1898, ships from Honolulu, like trains from Arizona, were engaged in interstate commerce. A century of legal precedents had established that states could not routinely search domestic travelers and their personal effects.[52] How then could California officials confiscate souvenir mangos from Hawaii? Marlatt sought to deal with this problem in early 1911. He brought to Congress a bill entitled "Quarantine against Importation of Diseased Nursery Stock," which seemed almost identical to that introduced the previous year. Slight changes in wording, however, empowered the secretary of agriculture to declare both foreign and domestic quarantines on both nursery stock and picked fruit. The bill thus would give entomologists the authority to stop interstate commerce in citrus, apples, or even wheat, if deemed necessary; this would be sufficient to keep Hawaiian Mediterranean fruit flies out of California orchards.[53]

When Congress did not act, the Californians pushed ahead unilaterally. In 1911 the state established a new quarantine division within the State Commission of Horticulture, strengthened the horticultural quarantine law, and "abrogated" their previous deference to federal authority.[54] Adopting the aggressive motto that introduced this chapter, the Quarantine Division prohibited commercial shipments of nearly all fruits from Hawaii, and sent agents to assist the territory's entomologists in fruit fly eradication. They circumvented their inability to force travelers to submit to search by persuading the steamship companies running between Honolulu and San Francisco to require a waiver of rights as a condition for ticket purchase. Horticultural quarantine officials then built a dockside "corral" where they inspected all incoming passengers.[55] The Bureau of Entomology cooperated with California by sending them lists of all plants entering the country destined for California; they intercepted not only private shipments, but also "30 tons of Asiatic plants" that David Fairchild's heroic explorer, Frank Meyer, had shipped from the frontiers of China to the USDA's experimental garden in Chico, near the northern end of the Central Valley. California officials fumigated and cut the tops off of these "badly infested" specimens.[56]

Entomologists continued to push for comprehensive federal legislation. In April 1911, they published an exposé of bad-faith tactics that the small group of "importing nurserymen" who controlled the American Association of Nurserymen had employed during the previous three years.[57] Marlatt publicized his arguments, with copious close-up photographs of pests, in the widely read *National Geographic Magazine*.[58] The nurserymen's association responded by setting three conditions before they would support Marlatt's bill: regional quarantines should not be declared for pests that were already established in the United States, like the gypsy moth and brown-tail moth; state inspectors, rather than federal agents, should enforce rules locally; and federal power should be held by an interagency board, not a single official.[59] Because Marlatt was concerned about getting both foreign and federal domestic quarantine power—specifically against the Mediterranean fruit fly—he accepted these limits, and the stakeholders affirmed agreement at the meeting of the Association of Economic Entomologists in December 1911. The Plant Quarantine Act, adopted in August 1912, included specific language concerning the Mediterranean fruit fly. Marlatt immediately left for Hawaii and soon declared a federal quarantine blocking shipment of fruits from that territory to the rest of the United States.[60]

Limits on Horticultural Freedom

From 1912 to 1928, Marlatt pressed incrementally toward the goal he had enunciated in his 1910 congressional testimony: building a regulatory Great Wall to prevent foreign pests from establishing themselves in the United States. Most of these "alien plant enemies" were herbivores or parasites that had coevolved with plants domesticated in the Old World or the New World Tropics. The aim of plant quarantine was to keep them from emulating the Hessian fly or boll weevil by catching up with hosts that were now planted widely in temperate North America. A smaller group of problem organisms were generalists or opportunists. These were much harder to specify beforehand, either because they had just discovered American plants (e.g., the European corn borer [*Ostrinia nubilalis*] and pine blister rust [*Cronartium ribicola*]), they were not major problems in their home ranges (the gypsy moth and Japanese beetle [*Popillia japonica* Newman]), or they were simply unknown prior to their appearance in the United States (San Jose scale and chestnut blight [*Cryphonectria parasitica* (Murrill) Barr]). For the purposes of the Plant Quarantine Act, pests could be either insects or plant diseases (weeds were not covered). But as the names of both the act and the board indicated, living plants were the main objects of scrutiny—not lumber, packing materials, ships' ballast, or other mobile environments for small organisms. The law's primary impact was thus on nurserymen, plant enthusiasts, and botanic gardens.

The exclusion of foreign pests depended first on establishing the policies that the United States should be as self-sufficient as possible in plants, and that agricultural commodities were more important than horticultural novelties. Second, it involved the development of the legal and technical tools that would enable government to limit the "horticultural freedom" of individual Americans—their ability to transport and propagate plants without government restriction. Customs regulations would prevent the entry of problematic plant material from abroad. A horticultural "Ellis Island" was built outside Washington to inspect, detain, and sanitize desirable but suspect specimens. Federal and state officials cooperated to regulate interstate commerce in plants. Most ambitiously, Marlatt acquired police powers for the USDA, so that its uniformed agents could stop and search individuals, and seize their property, to prevent the spread of pests from one state to another.

Quarantine 37

The Plant Quarantine Act established only the sketchiest bureaucratic foundation for this work. Its creation, the Federal Horticultural Board (FHB), consisted of five already employed USDA staffers; it had a budget of $25,000 and the power to recommend quarantines to the secretary of agriculture. Relations between bureaucrats and the political leadership were very uncertain in 1913 with the end of sixteen years of Republican rule. Marlatt used his expertise and efficiency, his social connections (he golfed with Franklin Roosevelt, among others), and his savvy (the new Democratic agriculture secretary, former University of Texas president D. F. Houston, strongly supported the FHB's campaign against the pink bollworm of cotton [*Pectinophora gossypiella*]) to control the board and build it up. By 1920 its annual appropriation had grown to nearly $750,000. The board produced a continual flow of quarantine notices and other regulatory announcements. These were distributed over the sole signature of "C.L. Marlatt, Chairman of Board."[61]

Initially, the FHB declared quarantines on specific pests, hosts, and locales (e.g., fruit from Hawaii to deal with the medfly, and potatoes from Newfoundland to keep out potato wart).[62] Between 1916 and 1918, however, its authority broadened dramatically. Forest pathologists had been shaken by the destructive and unstoppable chestnut blight, a disease that was completely unknown until a decade after it arrived, probably on Chinese chestnut trees imported to the New York area (possibly for the collection of the New York Botanical Garden) in the 1890s.[63] In the mid-1910s their new fear was pine blister rust. Northeastern state foresters, riding the crest of conservationism after 1900, had advocated mass replanting of the great native timber tree, the eastern white pine. Because American nurseries did not have the capability (in particular, low-paid skilled labor) to produce conifer seedlings in bulk, a number of government agencies purchased tens of millions of high-quality American pines from German, French, and Dutch companies.

These shipments introduced the well-known pathogen of Old World five-needle pines, *Cronartium ribicola*. The New York State Experiment Station stopped an infestation on its grounds in 1906 by burning millions of young trees, but the rust established itself within a few years via some other shipment. The FHB was too late when it blocked imports of five-needle pines in 1912. In 1916 Massachusetts foresters organized an interstate (and Canadian) conference on tree pathogens; the resulting American Plant Pest Committee pushed for a more proactive plant exclusionary policy.[64]

The issue entered general discussion through the annual meeting of the American Forestry Association held in Washington in January 1917. It was a fraught moment in the relations between the United States and the rest of the world. A new immigration law, just passed over President Wilson's veto, excluded all non-Japanese Asians, and sought to discriminate more effectively among Europeans with tests for intelligence and literacy. Wilson's recent effort to end the world war diplomatically had collapsed after German publicists argued that general adoption of Wilson's principle of nationality would mean the return of Texas to Mexico (and German foreign minister Arthur Zimmerman proposed alliance with Mexico on essentially those terms).

Addressing the Forestry Association the same day that Zimmerman was secretly offering to help Mexico restore its original borders, Charles Marlatt set the problem of imported pests within its biogeographic perspective. Emphasizing that the settlers of North America had originally possessed an enormous advantage due to the freedom of the continent's "virgin lands" from the plant enemies that had developed in the Old World "through centuries of cultivation of special crops," he outlined how that beneficial asymmetry was being eroded by the continuing arrival of pests. Damage from imported insects totaled more than $500 million annually. It could be limited, however, by excluding "from the American continent" foreign species such as the Mediterranean fruit fly and the (Asian) pink bollworm of cotton, the latter of which had established itself in Mexico. For Marlatt, the primary goal was "conservation of the big commercial crops . . . and our enormous natural forests, which are and must always remain our chief productions." New field plants, fruits, and "novelties and curiosities" for gardens and parks were of secondary significance.[65]

Marlatt concluded his address with a call for "very much restricting the further entry of all foreign plants and plant products." The American Forestry Association immediately pushed the idea forward with a resolution favoring "the principle of absolute national quarantine on plants, trees and nursery stock." When a proposed amendment to the Plant Quarantine Act went nowhere amidst the rush of war activity, Marlatt acted administratively. Quarantine 37, declared in November 1918, stated that because injurious insects and plant diseases existed "in Europe, Asia, Africa, Mexico, Central and South America, and other foreign countries," entry of foreign nursery stock, seeds, and bulbs would (with specified exceptions) be prohibited. Forestry Association president Charles. L. Pack, comparing Quarantine 37 to the literacy test, described it as the end of the "open door to plant immigrants," and hoped that "the treasonable activities of these enemy aliens will be curbed"[66] (see Figure 6.7).

Figure 6.7 This cartoon, which accompanied American Forestry Association president Charles Lathrop Pack's "Excluding Enemy Aliens with Appetites De Luxe," presented an image of foreign pests that blended fantastic monsters, insect "pilgrim fathers," and hoards of Asians waiting on boats. *American Forestry* 25 (1919): 1053. Reproduction courtesy of the Rutgers University Library.

By this time, Marlatt had established the government's predominance over commercial interests. He announced the new bureaucratic direction in early 1918 in the FHB's Service and Regulatory Announcements, which were routinely distributed to the trade press. The board emphasized the extent of the foreign plant pest problem with carefully marshaled facts, and pointed out that while the elimination of imports would cause disruption, it would also create opportunities for domestic producers. Meetings in Washington enabled stakeholders to present particular concerns in a context that affirmed the board's authority. Groups who demonstrated special needs (in this case, narcissus bulb dealers) received concessions, while public critics (such as the importers Henry Dreer and John Scheepers) were isolated through exposure of their self-interested motives. The consequence was that while nurserymen and florists complained loudly about Quarantine 37, they did not mobilize effective opposition[67] (see Figure 6.8).

What we may expect if the health of the people as well as plants is given in charge of the Federal Horticultural Board

Figure 6.8 National Nurseryman 27 (1919): 292, alluded to controversial U.S. Public Health Service anti-syphilis campaigns in imagining federally mandated ornamented kissing barriers. Reproduction courtesy of the General Research Division, The New York Public Library, Astor, Lenox, and Tilden Foundations.

J. Horace McFarland and the Amateur Institutions

The people who did not fit within Marlatt's calculus were the noncommercial horticulturists. Prior to 1919, they had supported the FHB as a help to American gardens. But they recognized that Quarantine 37 would fundamentally change their culture, which had been based, for at least a century, on the ability of scientists and fanciers to share novelties freely and for free. The quarantine rules included provisions for experimental introductions. But requirements for permits, inspections, and the posting of bond, all designed around commercial shipments, severed the shoestring arrangements that linked enthusiasts around the world. Horticulturists wanted their special status to be acknowledged and their interest in free interchange accommodated.

David Fairchild, as the chief government advocate for "novelties and curiosities," was early and forthright in his opposition to the new policy. Asked to speak at the January 1917 forestry convention on the ways that Americans could grow previously imported trees, he presented instead a plea for horticultural openness and interdependence.[68] He argued that although any individual foreign variety might have minimal value, the total economic worth of plant imports was substantial—and would only be known in the future, as a result of continued importation. He appealed to American principles of justice and charity, arguing that "it would be eminently unfair to assume that because we do not know that these little apple seedlings from the old world or from Japan are as clean and free from disease as any which we can produce in America, they represent undesirable immigrants and should be excluded from the country." He pointed to the inextricable commingling of evil and good: chestnut blight had led Americans to discover both new varieties of Chinese chestnuts, and also oriental pears resistant to blight. Finally, he directly evoked Marlatt's arguments regarding world movements and walls:

> We can say to ourselves, "let us be independent of foreign plant production. Let us protect our own by building a wall of quarantine regulations and keep out all the diseases which our agricultural crops are heir to and have the great advantage over the rest of the world." But the whole trend of the world is toward greater intercourse, more frequent exchange of commodities, less isolation, and a greater mixture of the plants and plant products over the face of the globe.[69]

Loss of some cherished species was unavoidable; the best that could be done was to foster research, ingenuity, and attention to individual diseases.

As a bureaucrat responsible primarily for his own turf, Fairchild declined to challenge Quarantine 37 further after it was approved by Agriculture Secretary Houston.[70] Private horticultural leaders did not have these constraints. The Massachusetts Horticultural Society was the initial center for opposition to what Boston patrician W. H. Wyman denounced as the "autocratic rulings of the Federal Horticultural Board." In early 1919, the society invited Beverly Galloway, who described himself as the USDA's "shock absorber" on plant introduction and quarantine policy, to explain the new rules. Audience members attacked the quarantine and Galloway's "weak and pitiful" defense of it, and they proposed resolutions regarding inspection procedures, greenhouse plants, and the inclusion of a horticulturist on the Federal Horticultural Board. Galloway responded that "this quarantine would go into effect on June 1 and stay there forever, no matter if you pass forty resolutions," and that the orchids and similar plants in which they were most interested did not amount to a "bagatelle." When the "spirited and at times acrimonious discussion" continued, Galloway walked out of the meeting.[71]

Later that year, Arnold Arboretum plant explorer Ernest Wilson publicly attacked Quarantine 37, and arboretum director Charles S. Sargent organized opposition. The horticultural societies of Massachusetts, New York, and Pennsylvania convened a national conference of botanic gardens, horticultural organizations, flower societies, and garden clubs at the American Museum of Natural History in New York in June 1920 to call for revision of Quarantine 37. After hearing messages from Sargent and Massachusetts Horticultural Society president Albert Burrage, advice from a lawyer, and testimonials from amateurs and nurserymen, the meeting created and funded a Committee on Horticultural Quarantine to work for change in the Federal Horticultural Board's rules.[72]

The head of this lobby group was the entrepreneur and reformer J. Horace McFarland (see Figure 6.9). McFarland's Mount Pleasant Press was the leading producer of the illustrated books, magazines, and catalogs central to American horticulture in the early twentieth century. This work, and his own garden interests, linked McFarland to the national horticultural leadership: closely associated with Liberty Hyde Bailey for thirty years, he had recently organized the American Joint Committee on Horticultural Nomenclature, composed of USDA botanist Frederick Coville, landscape gardener Frederick Law Olmsted Jr., and nurseryman Harlan Kelsey. Their *Standardized Plant Names* realized the decades-old hope that the various groups interested in horticulture

Figure 6.9 J. Horace McFarland, around 1920. Photograph reproduced from the J. Horace McFarland Papers, courtesy of the Pennsylvania Historical and Museum Commission, Pennsylvania State Archives, Harrisburg, PA.

could agree on terminology that would enable them to communicate reliably about varieties. McFarland's *American Rose Annual,* and its associated American Rose Society, showed that amateurs, commercial growers, and scientists could cooperate productively within an organization run primarily by fanciers; the society's central mission was to create new "American" roses.[73] McFarland's additional qualification was as a public activist: as founder and longtime president of the American Civic Association, he had participated in Theodore Roosevelt's formative White House Conservation Conference in 1908, and he had later campaigned to keep power companies from drying up Niagara Falls, to prevent the flooding of Yosemite's Hetch Hetchy Valley, and to create the National Park Service. McFarland's conviction was that gentlemanly outsiders could succeed best in the rough-and-tumble of Washington politics by unapologetically being themselves—by publicly displaying reason, rectitude, and concern for both the common good and the finer things.[74]

McFarland's committee entered the quarantine debate with a typical civic reform manifesto. In self-consciously genteel language, they asserted that "horticulture . . . is an important agent of civilization," building love of home, stable citizenship, and a sense of beauty. The United States, as a "new country," still relied on the "inherited knowledge and labor" of skilled horticulturists in Europe and Asia for "many beautiful, rare, and valuable plants." Their ideal was a "sane and efficient quarantine" that would give citizens willing to provide safeguards the ability to import the plants they "reasonably require." They called on "those who are opposed to a Chinese Wall plant-policy for America" to support the committee's moderate mission: to "search for facts" to present to "the Federal authorities at Washington."[75]

Through two decades of tough negotiations with provincial legislators and small businessmen, Marlatt had learned the value of condescension and stigmatization of critics. He applied these techniques to the genteel botanists and plant culturists on the Committee on Horticultural Quarantine. He noted in a FHB service announcement that a few unidentified "enemies" of the Plant Quarantine Act were spreading "propaganda" that misrepresented the facts and misled amateurs. He therefore explained that many kinds of plants entered the country freely under Quarantine 37; moreover, he emphasized, individuals could import "any plant for which a necessity can be shown," provided they obtained a special permit, posted bond, and routed the material through Washington so that it could be inspected. Difficulties in getting unusual plants through the inspection system in good condition could be resolved if more funding were available.[76]

McFarland's experience had been that entrenched bureaucrats could be defeated if their political supervisors could be persuaded that big-picture issues outweighed technical concerns. He prepared the ground with the name of a rose. In June 1921, McFarland's American Rose Society held its annual meeting in Washington, visited the private rose garden of Office of Systematic Plant Introduction (SPI) plant breeder Walter Van Fleet, and agreed to aid the USDA in distributing Van Fleet's roses to the American people. In addition, the society proposed that Van Fleet's most admired new variety—a cross between the American prairie rose *(Rosa setigera)* and the SPI-imported Chinese memorial rose *(Rosa wichuraiana)*—be named for Agriculture Secretary Henry C. Wallace's daughter. *Rosa* "Mary Wallace" would show the secretary that free horticultural introductions beautified American homes while at the same time highlighting the beneficence of the USDA (and the Wallace family).⁷⁷ With that foundation laid, McFarland sent the secretary a detailed brief challenging the legality and the value of Quarantine 37 and calling for reform.⁷⁸

Months of maneuvering resulted in a new stakeholders' meeting in May 1922. The gathering was chaired by Marlatt but overseen by Wallace and officially witnessed by a panel of three independent observers reporting to the secretary.⁷⁹ McFarland objected from the floor to Marlatt's proposed division of the meeting into caucuses of state officials, nurserymen, florists, amateurs, and importers; this taxonomy ignored "the great educational gardens of the country" and the "amateur institutions." He then made a long presentation on the importance of plant improvement, the need for individual creativity, and the negative consequences of allowing bureaucrats to exclude whatever foreign plants they did not consider "necessary." He also asserted that Quarantine 37 illegally overreached the goals of the Plant Quarantine Act because it focused on the exclusion of plants rather than of insects and diseases. McFarland's opposition was reinforced by Massachusetts Horticultural Society president A. C. Burrage, by florists' organizations, and by plant pathologists representing the British, Dutch, and Belgian governments.⁸⁰ Marlatt deflected these pressures. The FHB's postconference press release reported blandly that the meeting had gone "a long way toward dispelling some prejudices and mistaken ideas"; a page-long list of the attendees did not include McFarland and his committee. Marlatt provided follow-up guidance to the USDA's legal counsel and to Wallace, who ultimately declined to countermand his subordinate.⁸¹

In the aftermath of this defeat McFarland reported briefly to his supporters on his inability to overcome what he considered "scientific hysteria." Marlatt responded with an attack on McFarland's "violent propaganda of misrepresentation."⁸² Then, in response to an article by the young Indiana University entomologist Alfred Kinsey endorsing federal pest policy, McFarland addressed the issues from a broader standpoint. The fundamental question, he declared, was "What is America, anyhow?" He contrasted Kinsey's "Utopian America" of "native plants and people" with the reality of a "composite people" who lived in forty-eight states that differed widely in climate and vegetational capabilities, but who passed each other freely, with their different license plates, on roads that were under one flag. He supported efforts to "scrutinize plants for bugs and bothers," but hoped that America "ought to continue to be cosmopolitan in plant relation."⁸³

Search and Seizure

Each year in the second half of the 1920s saw an enhancement of the FHB's influence. In 1925 Marlatt passed smoothly through another round of criticism following his extension of Quarantine 37 to include narcissus and other spring bulbs.[84] A year later the Supreme Court was neutralized. The court had finally extended *Gibbons v. Ogden* to plant quarantine in a case involving Washington State's ban on out-of-state hay to keep out the alfalfa weevil. Chief Justice William Howard Taft's opinion declared that the federal government's entry into this area in 1912 superseded regulations that individual states imposed. The FHB supported strong state laws like those of Washington and California, however, and it relied on state officials and state money for enforcement. Within two months of the decision, Marlatt had induced Congress to legalize concurrent state authority over plant commerce across interstate borders.[85]

In 1927 Marlatt was before Congress requesting that USDA agents have the power to search and seize. The problem was that motorists were carrying sweet corn infested with the European corn borer across state lines contrary to FHB rules. Federal agricultural agents blocked much of this pest movement by stopping cars and requesting that contraband ears be surrendered, but they had no authority to act if a driver refused to open his or her trunk. (Americans were sensitized to search issues due to Prohibition, which banned interstate transport of alcoholic beverages.) While Marlatt saw no difference between body-searching Mexican laborers crossing the Rio Grande at El Paso for mangos, and inspecting autos for corn at state lines, members of the House Agriculture Committee drew a sharp distinction between international and interstate borders, and were concerned that authorizing federal agents to search domestic travelers was a major extension of government power. Congressman Franklin Fort (R, NJ) remarked that "this law is very clearly unconstitutional." The bill moved forward, however, following assurances from USDA lawyers that seizure of contraband would not entail arrest, and that agents would not feel motorists' bodies for fruits.[86]

In 1928 Marlatt reached the bureaucratic promised land. He persuaded the secretary of agriculture to promote the FHB to the status of a bureau, with himself as chief, and Congress acquiesced by approving an appropriation for a new Plant Quarantine and Control Administration (PQCA). (Marlatt also succeeded Howard that year as chief of the

Bureau of Entomology.) The PQCA drew staff and legal authority from existing bureaus; as its name indicated, its responsibilities extended beyond foreign and domestic quarantine to encompass efforts to control pests. Marlatt's divisions were directed against the different pests for which the FHB had obtained appropriations: the pink bollworm, date scale, gypsy moth, European corn borer, Japanese beetle, and Mexican fruit worm. With a budgetary blank check derived from an 1800 percent increase (to more than $5 million) in the pink bollworm control appropriation in 1928, the PQCA was well equipped to handle new threats.[87]

CHAPTER SEVEN

Gardening American Landscapes
From Hyde Park to Curtis Prairie

While the previous chapters have examined the relations between horticulturists and particular plant and pest types, here I investigate the history of more holistic concerns about the composition of humanized vegetation. In the nineteenth and early twentieth centuries, these issues fell within the rubric of landscape gardening. The central problem, as framed by Andrew Jackson Downing in 1841, was to adapt European landscape gardening to North America. Most immediately, this involved learning how to plant American landscapes, notably private estates and public parks. That goal passed smoothly, however, into the project of making gardened landscapes, in some meaningful way, American. Landscape historians have studied these questions largely by examining ideas about design—shaping of land; construction of water features, paths, and architectural elements; and arrangement of masses of vegetation. I suggest that the horticultural tasks of siting specimens, deciding which species should be encouraged or suppressed, and transplanting, whether from nearby woodlands or from eastern Asia, raised issues that were equally important and more difficult. Questions centered largely around the meaning of what can be called "American naturalism." The general issue in naturalistic landscape gardening was the tension between imitation and improvement. The geographically more specific problems concerned the paradoxes involved in planting natives, and the assumption that plant species could become as naturalized to America as Anglo-Americans believed themselves to be.[1]

This chapter examines plant choice in the formative contexts identified by landscape historians. I focus first on the development of Hudson

River Valley estates such as Hyde Park, and on valley gardener A.J. Downing's changing views on naturalization and on natives. I sketch how Calvert Vaux, Frederick Law Olmsted, and Ignaz Pilat planted New York's Central Park in the years around 1860, and how Olmsted and his associates fought with Charles S. Sargent over the native and the naturalized in Boston around 1890. The proliferation of styles after 1890 undercut the naturalness of American naturalism in the Northeast, but new naturalistic efforts arose in the Midwest in the early twentieth century. Design historians have focused on Chicagoan Jens Jensen; by emphasizing plant choice, I shift attention to forester and environmentalist Aldo Leopold. In the 1930s, Leopold and his colleagues at the University of Wisconsin Arboretum created the garden that would come to be called Curtis Prairie; it was an exemplary modern naturalistic American landscape.

The Gentrification of the Hudson River Valley

Landscape gardening became a movement in the United States in the Hudson River Valley during the second quarter of the nineteenth century. Manhattanites with money and leisure sought to escape the summertime smells and diseases of the city. A long strip of riverfront landscape became accessible via steamboat in the 1810s and 1820s. While most vacationers lodged in hotels—some of which were convenient to scenic destinations such as Kaaterskill Falls—a few of the most prosperous inherited or purchased properties along the river. Proprietors of these Hudson Valley estates faced choices about the kinds of landscape they wanted to experience each year.[2]

Estate gardening practices had long-established antecedents in England, and a series of designers and writers during the previous century had articulated theories for this type of rural improvement. As Downing noted, however, European landscape gardening required substantial adaptation to work in North America. English country homes were scattered over the countryside and generally located in the midst of income-producing farmland. Proprietors developed isolated private microcosms over periods of decades: earth moving, the creation of water features, and informal plantings improved nondescript views and distinguished pleasure grounds from pasture and cropland.[3] The situation in New York was different. Prime Hudson Valley estate properties stretched linearly for miles along the river. Few owners expected that their lands would produce large profits; the appeal of the sites derived

Gardening American Landscapes · 167

from their accessibility and visibility. Owners could reach their summer homes in a few hours and could then enjoy panoramic views of the sublime nature celebrated by Thomas Cole and other New York City painters. They knew that their properties could be seen by thousands passing on the river each day. Interest, therefore, was in ways to display taste quickly. Acquisition of a topographically dramatic and well-vegetated site was helpful; beyond that, estate owners sought to mark their land as genteel. In the first decades of the century, the most prominent method was to line straight drives with fast-growing visually distinctive trees—notably Lombardy poplars (see Figure 7.1). These markers of republican taste quickly distinguished patrician estates and proud towns from plebeian working farms.[4]

David Hosack, the physician-botanist who had established the Elgin Botanic Garden and spearheaded the creation of the short-lived New York Horticultural Society, showed valley proprietors how to use knowledge, imagination, and money to produce more tasteful gardened

Figure 7.1 Scenic view of the town of Hudson, New York, around 1820. The straight rows of trees crisscrossing the landscape are maturing Lombardy poplars. Plate 13, from William Guy Wall, *The Hudson River Portfolio* [New York: Henry I. Megarey, 1825]. Typ 870.25.874 PF, Department of Printing and Graphic Arts, Houghton Library, Harvard College Library.

landscapes. In 1828 he purchased Hyde Park, a seven-hundred-acre property north of Poughkeepsie with a ridgetop villa overlooking the Hudson. During the next eight years he spent perhaps $100,000 of his and his third wife's money to make it a showplace. Recognizing that such a project required a European professional, Hosack hired Belgian horticulturist André Parmentier, who had recently settled in Brooklyn, to lay out the ridge as a *Jardin Anglais*—the more elaborate French interpretation of the landscaping style popular in England in the previous decades.[5] Parmentier featured a broad lawn sloping down from the house to the river. This gave Hosack an unobstructed view toward the Shawangunk Mountains to the west, and, conversely, enabled river travelers to see the house and garden clearly and to gain "a vast idea of the proprietor." Otherwise, Parmentier explained, "the taste and expense are, in a great measure, thrown away." He emulated the Marquis de Girardin's design for the tomb of Jean Jacques Rousseau by creating a grove (rather than rows) of Lombardy poplars; he also planted magnolias, an Appalachian species not found along the Hudson, but long popular on English and French estates. A rustic bridge spanned a creek enhanced as a cascade, and paths led to classical pavilions. A hothouse was stocked with citrus trees and with tropical flowers such as orchids and the bird of paradise *(Streltzia reginae)*.[6]

Hosack developed Parmentier's plans in the years following the Belgian's death in 1829, but himself died in 1835. His children sold the property to John Jacob Astor, who gave it to a daughter who seldom lived there. Leadership in Hudson Valley landscape gardening then passed to nurseryman Andrew Jackson Downing.[7] Born in Newburgh on the west side of the river in 1815, Downing learned basic concepts about landscape as an adolescent from the English-American painter Raphael Hoyle and the Austrian consul in New York, Baron Alois von Lederer, both of whom were his neighbors in the summer. In 1832 he joined the family business, which supplied fruit trees and ornamentals to the river estates. Downing also began writing on rural embellishment, and particularly emphasized the "highly improved" Hyde Park. In Parmentier's spirit, Downing advertised both his trees and his taste by transforming part of the family nursery, visible from the river, into a landscape garden with a faux-sandstone Elizabethan villa. He spread his message to a national audience in 1841, with his *Treatise on the Theory and Practice of Landscape Gardening, Adapted to North America.*[8]

Downing sought to advise those who had neither Hyde Park's ideal location nor Hosack's capacity to spend. He rejected the established practice of planting trees along straight drives, or as a screen in front of the house, as "ancient" and "mechanical." Instead, he promoted a "modern" naturalistic style. But naturalistic gardening was more problematic in the Hudson Valley than in the English countryside. Persuading estate owners to pay for informal gardens that seemed almost identical to the surrounding countryside was a challenge; and with many kinds of trees available locally for transplanting, such gardens would generate little business for nurserymen like Downing.[9]

Downing thus condemned gardens that replicated valley woodland: they were unimaginative and deceptive "waxworks." Rather, he proposed to plant nonlocal (both North American and Eurasian) species. "One of the chief elements of artistical imitation in Landscape Gardening, being a difference in the materials employed in the imitation of nature from those in nature herself," he explained in his characteristic convoluted style, "nothing can be more apparent, than the necessity of introducing largely, exotic ornamental trees, shrubs, and plants, instead of those of indigenous growth." He recommended that estates in areas with oak and ash "native forest" should be planted with Old World species such as lindens, horse chestnuts, gingkos, and ailanthus (the last being "well adapted to produce a good effect on the lawn, either singly or grouped"), and also with large-flowered southern Appalachian types like tulip trees, magnolias, and dogwoods. Landowners who used "the floral and arboricultural riches of all climates" to establish "a richness and a variety never to be found in any one portion of nature" would achieve the "*beau ideal* in Landscape Gardening."[10]

Within a few years, however, Downing's perspective on species geography changed. In 1841 William Backhouse Astor, John Jacob Astor's oldest son and prospectively the richest man in the country, hired Hans Jacob Ehlers, who had studied forestry in Germany and gardened for the Danish royal family, to lay out Rokeby, his estate about ten miles north of Hyde Park. Downing was deeply concerned about competition from someone with both a formal horticultural education and experience with exotic ornamentals. In the 1844 edition of the *Treatise,* he lashed out at the "professional quackery" of this (unnamed) "foreign *soi-disant* landscape gardener." He argued that while Ehlers's mode of work was "common and appropriate" in level inland areas like much of Germany, it was not adapted to the "bold and lake-like features" of the Hudson Valley. By then known more as a writer and designer than as a plant dealer, Downing eliminated most of his arguments for nonlocal vegetation. Instead, he advocated a picturesque approach that economically employed "the raw materials of wood, water and surface."[11]

Downing shifted his views again when offered an opportunity to work on a more nationally significant stage than the Hudson Valley estates. Asked in 1850 to redesign the National Mall in Washington, D.C., he adopted an eclectic approach. He proposed that the area adjacent to the White House have a formal arrangement that would symbolize the government's power. The grounds of the Smithsonian would become a carefully labeled "public museum of living trees and shrubs," encompassing all the world's species that could survive in the climate of Washington; tropical plants would be maintained in a hothouse a few blocks farther east. Around the Washington Monument (as mentioned in Chapter 4), large trees from all parts of the continent would form a grove of national unity and magnificence. Here he dismissed the ruggedly picturesque, however, in emphasizing that American trees that looked leggy in the wild would have "a graceful and majestic development of head" in this garden setting. The Washington grove would be as superior to a typical American forest as "the best bred and cultivated man of the day" was to a "buffalo hunter of the Rocky Mountains." Visitors to the capital would thus learn about the world, the nation, and the value of culture; and they would get hints on plants they could grow back home.[12]

Two years later, disillusioned by lack of congressional support for his Washington project and disturbed by the ongoing Irish invasion of New York in the wake of the Famine, Downing shifted position again—to a strident horticultural nativism. In early 1852, he editorialized that some races ("notwithstanding the socialistic and democratic theories of our politicians") were superior to others ("or our American Indians, or our Chickasaw Plums, would never have given place to either the Caucasian man, or the luscious 'Jefferson' "). He then emphasized that further improvement required that "a race should be adapted to the soil by being produced upon it": hence, fruit varieties raised from American-matured seeds were as superior to foreign scions as native Anglo-Americans were to immigrants.[13] Two months later he was bemoaning the moral and intellectual capacities of Irish gardeners. Noting that many claimed to have worked for English aristocrats, he doubted whether "Paddy" provided "any other exotics than potatoes and cabbages." More seriously, he complained that alien gardeners were unable to adapt their horticulture, grounded in combat against continued *"damp, wet, want of sunshine,"* to the *"very opposite"* American climate, where the "bugbears are *drouth, hot sunshine,*" and sudden changes in weather.[14] A third essay, mentioned in Chapter 4, was a broadside attack on foreign trees. With the rallying cry, "Down with the Ailanthus!" Downing excoriated this species, which he had previously recommended, because it *"smells horribly"* producing "Ailanthus malaria," and because it *"suckers abominably"*—"if the Tartar is not laid violent hands upon, and kept under close watch . . . we shall have him upon us tenfold in the shape of suckers innumerable—little Tartars that will beget a new dynasty, and overrun our grounds and gardens again, without mercy." His deeper objection to this "petted Chinaman," as well as other foreigners like the silver poplar, however, was a "patriotic" one. He attacked the "national extasy for foreign suckers," and urged his northeastern audience to plant "clean natives" like maples, tulip trees, and (southern) overcup oaks.[15]

The thirty-six-year-old Downing drowned in the Hudson River in 1852 when the boiler of the racing steamboat on which he was traveling exploded. His followers carried his messages in different directions. The *Treatise* was taken over by Henry Winthrop Sargent, who had dropped out of banking to develop a small estate near Fishkill, across the river from Newburgh. Sargent added a long supplement on "the progress of landscape gardening since Mr. Downing's death" to the editions of the *Treatise* he prepared in 1859 and 1872. These texts featured the French parterres and Italian topiary of his friend Horatio Hollis Hunnewell's estate in Wellesley, Massachusetts (see Figure 10.1), as well as his own enthusiasm for "horticultural oddities and freaks," notably specimen evergreens from eastern Asia, Australia, and New Zealand. Readers of these editions of the *Treatise* would see Downing's interests in naturalism and natives as merely stylistic preferences, which had been largely superseded.[16]

Downing's picturesque ideal was publicized, but subverted, by New York celebrity editor Nathaniel P. Willis. In late 1852 Willis hired Downing's assistant, Calvert Vaux, to design Idlewild, a "simple picturesque country house" a few miles south of Newburgh (see Figure 7.2). A few years later, Willis described his rural existence. On the one hand, he promoted an antigarden: he retained the hemlocks on the building site to make the house look like "an old familiar settler almost before the roof was on," and he dramatized the setting through language rather than design—the prominence on the river near his home changed from the prosaic "Butter Hill" to the romantic "Storm King" through Willis's efforts in his book, *Out-doors at Idlewild*.[17] On the other hand, Willis displayed a sophisticated understanding of the changes that

Figure 7.2 Idlewild, Nathaniel P. Willis's country home designed by Calvert Vaux in 1853. This view from water level emphasizes the placement of the house in the midst of established woodland. Illustration from Calvert Vaux, *Villas and Cottages* (New York: Harper and Brothers, 1867), 266. Reproduction courtesy of the Rutgers University Library.

summer people were making in the Hudson Valley. He reported that working farmers were selling their agriculturally inferior riverside hills to city folk and using the profits to buy better land further west. He emphasized the incongruities of a landscape where literary men encountered free-ranging pigs, where Irish immigrants were better neighbors than careless natives, and where wealthy villa owners in old clothing clashed with trespassing fashionably dressed *arriviste* hotel guests. Willis was particularly struck by the realization that he could sit on his veranda, seemingly alone in the woods, yet witness the movements of thousands of people daily along the river. Even more disconcertingly, he could set his clocks to Manhattan time by watching the trains now running regularly along the east bank. The picturesque Hudson Valley, Willis understood, was a product of modern culture.[18]

The Vegetation of Central Park

The plant choice issues that had been raised in the Hudson Valley became more important in the late 1850s when a small group of advocates began to make that region's estate landscape experience part of the lives of millions of Americans. In New York's Central Park, and the many other urban landscape parks built between the Civil War and World War I, thousands could see (in condensed and intensified form) both the kinds of scenic vistas that Hudson River painters had idealized and publicized, and the naturalistic vegetation that valley estate owners and landscape gardeners had cultivated. In these parks, public gardeners continually made choices about plants. Issues involved the selection and siting of species, the arrangement of different morphological categories, and most significantly, the biogeographical appropriateness of plants from other localities, regions, or continents. Both planners and gardeners participated in decisions that were by turns technical, aesthetic, and ideological.

Calvert Vaux made the Hudson Valley landscape central to the American park experience. In 1850 Vaux was a twenty-five-year-old architect practicing in his native London. Downing lured him to Newburgh, where he assisted on both the National Mall and local projects. He adopted Downing's synthesis of Hudson Valley scenery, romanticism, and Americanism; his landscape designs in the years after Downing's death emphasized mature trees and dramatic views. Vaux did not, however, share Downing's professional insecurities, his identification with the valley gentry, or his disdain for foreign laborers. Vaux's biographer, Francis Kosky, emphasizes that this naturalized American was both an admirer of democratic values and a realist who used his status, abilities, and connections to push his ideas forward in public arenas.[19]

In 1857 Vaux moved from Newburgh to Manhattan and immediately inserted himself and his Anglo-Hudson aesthetics into the struggles for control over Central Park, the city's new public landscape project. In April of that year, the Republican-dominated state government seized control of park development from Democratic municipal officials. Vaux successfully campaigned to shelve the plan that engineer Edgar Viele had prepared for the city commissioners, and suggested instead an open competition. He then persuaded Frederick Law Olmsted, the park's general superintendent and a Republican insider, to join him in developing a proposal. Vaux and Olmsted overwhelmed the competition with the Greensward Plan, a highly professional presentation that included a detailed proposal, a ten-foot-long hand-drawn map, and tableaus of strategic viewpoints pairing Brady Studio photographs of the present bleak conditions with Vaux's watercolors of their proposed improvements. Some parts of the park would evoke English traditions, but others—especially where bedrock was exposed—would express the Hudson Valley picturesque.[20]

Most of the Greensward Plan involved grading, drainage, roads, and structures, but the final and most visible phase of park improvement was planting. Over the preceding two centuries, the land for the park had been cleared and recleared of trees, left to hogs, and cleared again during the transition from private to government control. It contained large patches of briers and was filled with "poisonous ivy" so toxic that a surveyor was incapacitated for two weeks.[21] All agreed that the vegetation needed to be improved, but they differed significantly over what improvement meant. The city commissioners, advised by Viele, had proposed that the park be made "an exclusive American Arboretum, not admitting within its inclosure a single vegetable foreigner, naturalized or otherwise."[22] The state commission, which included the prominent importing nurseryman James Hogg, changed direction; it began to plant standard landscaping species, including elm, magnolia, lilac, and Norway spruce, without regard to origins.[23] The Greensward Plan sketched a bifurcated approach. Vaux and Olmsted prepared a detailed design for a forty-acre "arboretum of American trees" along Fifth Avenue from Ninety-seventh Street to the park's northern border (see Figure 7.3). Visitors could walk on curving paths through a carefully labeled, taxonomically arranged collection of the eastern North American trees and shrubs hardy in New York. They would thus appreciate the "great advantage that America possesses" in vegetational diversity, and would see that American trees, when properly cultured, could be beautiful. Vaux and Olmsted emphasized, however, that this arboreal museum would be situated within a larger cosmopolitan naturalistic landscape; the rest of the park would be filled with "all the trees, native or foreign, that are likely to thrive."[24]

Over the next decade, official planting emphases fluctuated repeatedly. The annual report produced in early 1859 repeated the language of the Greensward Plan, but added that "American trees of the stateliest character, standing somewhat openly, are designed to predominate wherever the nature of the surface will permit."[25] The following year the proposal for an American arboretum was replaced by plans for a cosmopolitan one, which would include "specimens of every tree and shrub which can be grown upon the site in the open air." The artificial character of the setting would be emphasized by changing the

Figure 7.3 The Arboretum section of the Greensward Plan for Central Park, 1858, combined informal paths and a numbered, taxonomically arranged composition of North American trees. Bob Lorenzson, reproduction courtesy of Robert T. Augustyn and Paul E. Cohen, New York City Parks Photo Archive.

Greensward Plan's curving paths into rectilinear *alleés*, and adding a 1,000-foot-long reflecting pool (see Figure 7.4).[26] In 1861 the park commissioners contracted for a 200-foot-long, two-level glass conservatory on Fifth Avenue between Seventy-fourth and Seventy-fifth Streets. At the same time, Hamilton Fish, Cyrus Field, and other wealthy New Yorkers were establishing the American Zoological and Botanical Society, which would build a combined zoo and garden modeled on Paris's Jardin des Plantes.[27]

Four years' slog of war, and, more locally, the anti-elite arson and stone throwing of the 1863 Draft Riots, made plans for a Parisian-style arboretum and a giant glass house less appealing.[28] Instead, in early 1865, the commissioners reached back to the previous century's debates regarding the biotic differences between the Old World and the New. Anticipating installation of a menagerie stocked with the standard lions, camels, and elephants, they acknowledged that other parts of the world might "excel in the magnitude of the products of the

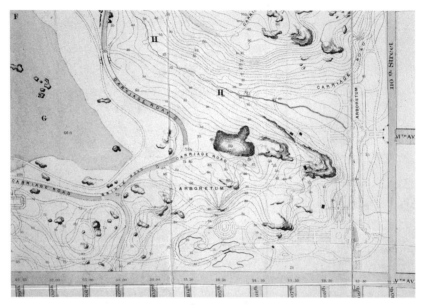

Figure 7.4 The Arboretum section of Central Park, redesigned in a more formal manner. This map appeared in the 1861 *Annual Report* of the Board of Commissioners of Central Park. Collection of the New-York Historical Society, negative #80203d.

animal kingdom." But, they declared with uncharacteristically purple prose, "naturalists accord to our own continent marked superiority of vegetable life." North America's trees "are peculiarly numerous and majestic, its fields luxuriant and prolific, its flowers brilliant and varied." Central Park's vegetation would "mark and emphasize this characteristic of the country, from every river, plain, and mountain side, of whose broad latitudes are derived the bounteous supports of its present growth, and the well-founded promises of its future greatness." The park would become a living monument to the reunited continental nation.[29]

But while commissioners formulated and reformulated master plans, horticulturists were establishing facts on the ground. The most important early influence on Central Park's vegetation was Austrian-American Ignaz Pilat. Pilat had come to New York in the mid-1850s, claiming gardening experience at the Austrian imperial palace of

Schoenbrunn and at the University of Vienna Botanic Garden. He began to work in the park in 1857 and quickly demonstrated his versatility by directing a landscaping crew and, at the same time, coauthoring a report on the plant species growing on the land. Technical competence and bureaucratic skill enabled him to advance gradually in a highly conflicted political setting. While working for Olmsted and the Republican commissioners he rose from foreman in charge of "grubbing and nursery work" to the park's landscape gardener. In 1870, when Democratic Boss Tweed's crony, Peter Sweeny, became city parks commissioner, Pilat advanced smoothly to the position of chief landscape gardener of New York City.[30]

Pilat understood the importance of modeling Central Park's vegetation on that of the Hudson Valley. At the same time he implemented Olmsted's scheme to expand the vegetational meaning of "American" beyond stereotypical northeastern plant assemblages by featuring striking species such as magnolias, yuccas, and prickly pear cacti, and by using masses of large-leafed eastern North American types (pawpaw, catalpa, and skunk cabbage) to produce effects that appeared "tropical."[31] His truly notable work, however, was in naturalizing "foreign vegetables." This can be seen in his most important landscape development, the Ramble (see Figure 7.5).

Vaux and Olmsted had given the hill north of the Lake no special attention in the Greensward Plan. In 1858, however, Pilat and his crews dug and plumbed a stream, rearranged rocks, laid out walkways, and carted in soil and plants. Within a decade, the naturalism of the Ramble was so convincing that even sophisticated visitors such as Horace Greeley believed it was an old, undisturbed area. Art critic and guidebook writer Clarence Cook more knowingly emphasized that this idealized sample of Hudson Valley wildness was, "in almost every square foot . . . a purely artificial piece of landscape gardening."[32] Yet even he did not see the full extent of its artfulness. He explained naively that "an effort has been made to bring into these bounds as many of the wood flowers and flowering shrubs, the native growths of our forests, as would thrive there—foreign flowers and imported shrubs being put in places more seeming artificial." But he then commented approvingly on Pilat's plantings of honeysuckle and of "the blessed dandelions, in such beautiful profusion as we have never seen elsewhere, making the lawns, in places, like green lakes reflecting a heaven sown with stars." He also emphasized the patches of money-wort *(Lysimachia numularia)*: this "pretty, little, vulgar plant, long since exiled from all aristocratic

178 · Fruits and Plains

Figure 7.5 Stereopticon of a constructed naturalistic element in the Central Park Ramble, late 1860s. Reproduction courtesy of the Herbert Mitchell Collection, with thanks to Sara Cedar Miller.

gardens . . . seems to delight in showing how, in this stately garden of the people, it can hold its own by the side of many plants with far finer names and a much prouder lineage." All these species had come from the Old World in the previous three centuries, but Cook either did not know or did not care. Pilat, who had risen from central European obscurity to control the public spaces of New York City, must have been pleased.[33]

The increasing horticultural richness of Central Park can be tracked through plant lists. Rawolle and Pilat's 1857 catalog referred to 285 species, including ferns and mosses. In 1863 the number of species and varieties had grown to almost eight hundred. By the late 1860s, the nonbotanist Cook could assert that in number of species, the park had become the world arboretum imagined a decade earlier. Labels and a guidebook would complete the work. In 1873 the commissioners published a catalog of the park's plants, with a map, listing 1,447 hardy species and varieties.[34]

Natural in the Artistic Sense of the Word

In New York in the 1860s initiatives involving natives and the naturalized were muddled. Two decades later in Boston, the issues were clearly stated if not consistently applied. Disagreements over the importance of planting natives and over what was native pitted against each other two leaders in landscape gardening—Frederick Law Olmsted and Charles S. Sargent. Their views were so strong because projects such as the Arnold Arboretum, the Back Bay Fens, and the Muddy River Improvement were almost completely blank slates that could be shaped toward any desired naturalistic result, and because they believed that these environments would influence public perceptions for decades to come. Drawing largely on the work of Cynthia Zaitzevsky, I discuss the disagreements between Olmsted and Sargent in order to convey the depth of expert interest in the issue of native and naturalized plants in naturalistic settings, and to illuminate their specific concerns. Ecological authenticity was not the issue. The focus was on naturalistic artistry; tensions were ultimately between central European effervescence and Puritan restraint and attachment to the familiar.

The task at the Arnold Arboretum was to design a naturalistic embodiment of a botanical catalog. Within this explicitly artificial framework, Olmsted and Sargent worked harmoniously. The arboretum had been established in 1869 as a joint project of Harvard and Boston to be both a "museum of trees" and a public park. Thirty-year-old Charles S. Sargent became director; his credentials included experience managing the family estate in Brookline, familiarity with both Horatio Hollis Hunnewell and his father's cousin, Henry Winthrop Sargent (Downing's editor), a brief period of study with Asa Gray, and ability to channel donations to Harvard.[35] The weightiest design issue he faced in the 1870s was to arrange the initial tree plantings. With a time horizon of centuries, the arboretum's display would be more permanent than any other kind of museum collection. Sargent planned to arrange specimens taxonomically using the "natural system" of English botanists George Bentham and Joseph Hooker's *Genera Plantarum* as a guide (see Figure 7.6).[36] American natives (by which he meant northeastern species) would predominate, but related types from the Pacific coast and beyond would be placed nearby for comparison and completeness. He wanted trees to have enough space to grow without crowding so that property owners and gardeners could assess what a

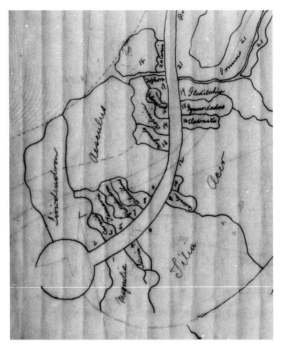

Figure 7.6 This detail from Olmsted's tenth study for the layout of roads and species at the Arnold Arboretum (dated May 4, 1880) shows the first section of the taxonomically ordered path that a visitor would take from the park entrance, with eastern American and eastern Asian genera mixed. Sequence shown is (1) Magnolia, (2) Liriodendron, (3) Cercidiphyllum, (4) Idesia, (5) Tamarix, (6) Tilia, (7) Phellodendron, (8) Ailanthus, (9) Cedrela, (10) [Kohlreutia], (11) Aesculus, (12) Acer, (13) Negundo, (14) Robinia, (15) Halimodendron, (16) Cladastris, (17) Sophora, (18) Gymnocladus, (19) Gleditschia, (20) Cercis, and (21) Prunus. Reproduction courtesy of the Arnold Arboretum, Boston.

---

specimen planting might ultimately look like. Finally, the ensemble of trees should be naturalistically harmonious and parklike, so that visitors would draw pleasure from the arboretum's vistas indefinitely into the future. Sargent hired Olmsted, who had dissolved his partnership with Calvert Vaux and had moved to Boston, to solve these numerous challenges. From 1877 to 1885, the two men collaborated to lay out carriage drives and to situate plantings so that visitors would experience natural taxonomic order, the forms of individual trees, and a harmonious setting. The mutually satisfying result has been maintained to the present.[37]

During these same years Olmsted began a project that was artificial in a more mundane way. In the 1850s, Boston leaders decided to increase their city's land area, raise revenue, and maintain social hierarchy by gradually filling in the Charles River marshland known as the Back Bay and building a residential neighborhood for the prosperous. Engineers determined that a catch basin would need to be dug at the western end of the Back Bay to control both sewage-laden spring runoff from higher ground and seawater that storm tides periodically pushed up the Charles River estuary (see Figure 7.7). In 1880 Olmsted proposed that an artificial valley walled with soil and rocks, and then planted, would be both cheaper and visually more appealing than the alternative of a giant masonry tank. The main challenges were to manage water flow to avoid stench and to select plants that were beautiful but would also survive occasional inundation by polluted or brackish water.[38]

Olmsted's solution was the Back Bay Fens. While the name was an Anglophilic allusion to the marshes adjacent to the other Cambridge,

Figure 7.7 In the 1880s, the Boston Park Department dredged and filled a section of the tidal flats on the south side of the Charles River to create the Back Bay Fens. Area in the midst of construction in 1882. Photograph from Boston Park Department, *Annual Report,* 1883, courtesy of the Boston Public Library.

the design would be modeled, Olmsted explained, on the "original conditions of the locality"—a New England salt marsh. He suggested that a tidal creek could meander through the long, narrow basin. But to avoid twice-daily fluctuations between "slimy mud-banks" and a deep pool, he proposed a tide gate that would reduce water level changes to one foot. The banks would then be planted to be natural "in the artistic sense of the word."[39] He was unsure, however, how to implement this vegetational vision. A local nurseryman experienced with beach planting brought in large numbers of New England seacoast species, including beach plum, bayberry, blackberry, and goldenrod (as well as Oregon holly grape and Japanese honeysuckle), but in the idiosyncratic conditions of the Back Bay Fens, nearly everything died.[40]

Olmsted responded by importing a horticultural hand he could trust. William (Wilhelm) Fischer, like Pilat, was a central European immigrant gardener carrying an impressive but vague résumé (University of Heidelberg, Sans Souci, the Royal Horticultural Society, and Chatsworth). He had worked on Central Park from its beginning and was its "superintendent gardener" in the early 1870s. In 1884 Olmsted persuaded Fischer to come to Boston as his landscape gardening assistant. He encouraged Fischer to select species for the Fens that would display "tufted, ruffled, and mysteriously intricate" foliage but would survive the unusual conditions. Fischer's plantings included local birch, pine, and oak, Midwestern black locust and Kentucky yellowwood, and Old World white willow and ailanthus (see Figure 7.8). Pleased with the result, Olmsted put Fischer in charge of planting the large new Franklin Park. He advised staffers not to "check his spontaneity," and Fischer responded by planting large numbers of flowering shrubs and herbaceous perennials with the goal of showing nature "in holy day dress."[41]

For Charles S. Sargent and his associates, American parks were not appropriate places for central European Catholic "holy day" interpretations of nature. In 1888 William A. Stiles, editor of Sargent's new weekly, *Garden and Forest*, critiqued the plantings in Brooklyn's Prospect Park, which had been designed by Olmsted and Vaux and installed by Pilat two decades earlier. Stiles claimed that flowering shrubs such as forsythia, lilac, and deutzia looked out of place along the borders of the park's "natural woods." Attempts "to force nature, so to speak, by bringing in alien elements from remote continents and climates, must inevitably produce inharmonious results." He asserted that parks existed in order to

Figure 7.8 Taken thirteen years after the photograph in Figure 7.7, this photograph shows how William Fischer's plantings transformed the appearance of the containment basin. Photograph courtesy of the National Park Service, Frederick Law Olmsted National Historic Site.

foster the "idea of repose," and this experience depended on encountering only the nonalien combinations of plants that "we have become accustomed to see."[42]

Olmsted understood that this criticism applied as much to his current work in Boston as to his past project in Brooklyn; his responses were both textual and horticultural. A careful rebuttal in *Garden and Forest* focused, not on the specifics of park plantings, but on Stiles's connections among nativity, familiarity, and repose. He asked readers to consider an agricultural old field filled with a mix of human- and bird-planted species. A naïve observer, he asserted, could not distinguish plants of "foreign ancestry" (with some exceptions like gingkos) from "old native stock." On the one hand, American trees such as catalpas or honey locusts were odder-looking than European maples or elms; on the other, many foreign-origin plants (including buttercup, mint, and Cherokee rose) were neither "disreposeful" nor "inharmonious." He then argued for the positive value of Old World plants: they were often more effective in redeeming ground that was "dull, dreary, forlorn, and tamely rude" than "any plants natural to it," and foreign species such as white willow and paulownia supplied "vivacity, emphasis, [and] accent."[43]

Olmsted's tactics on the ground in Boston were much less confident. He ordered Fischer to subordinate foreign plants to natives, to pull up flowers, and generally to be more "modest" in his plantings.[44] These actions did not mollify Sargent. He pushed his view that (with a few exceptions) only familiar and plain New England species should be planted in Boston-area parks. The issue came to a head around a project to beautify the Muddy River, which formed part of the boundary between Boston and Brookline. Olmsted's firm had the contract for plantings but needed approval from both municipalities' park officials. In April 1892, as a member of the Brookline commission, Sargent demanded to see Olmsted's plant lists. He crossed out foreign species (except the restful weeping willow [*Salix babylonica*] and the familiar oriental plane tree [*Plantanus orientalis*]), eliminated some plants whose pre-Pilgrim ranges had not extended into New England (Kentucky coffee tree [*Gymnocladus dioicus*], Carolina silverbell [*Halesia carolina*], and Carolina allspice [*Calycanthus floridus*]), and even rejected some local species that flowered too intensely (Eastern redbud [*Cercis canadensis*] and virgin's bower [*Clematis virginiana*]).[45] When Olmsted's staffers refused to accept these changes, Sargent walked away from the contract and organized the plantings on the Brookline side of the stream on his own. A few years later, when the senior Olmsted and William Fischer retired, Sargent became the Boston park system's authority on both landscape design and plant selection.[46]

The Problem of Styles

This intramural dispute over park vegetation was overtaken in the 1890s by much broader discussions. Advocates of Italian and French formalism directly challenged the perspective shared by both Sargent and Olmsted. Interest in Japanese gardens generated deeper, if more ambiguous, issues. By the 1910s, American naturalism had become merely one among a number of choices from a style menu. Within that context, fine distinctions about nativity and familiarity declined in significance.

Charles Platt's *Italian Gardens* (1894) stimulated a major rethinking about estate design in the United States. Platt argued that gardens were parts of architectural plans, and that they should be designed primarily to harmonize with buildings and to extend living space out of doors. He thus subordinated vegetation to structures and proportions, and he focused on the shapes rather than the species of plants. Platt gained visibility when Charles Sprague (Charles Sprague Sargent's relative and neighbor) hired him as landscape designer after having dismissed Olmsted's partner, Charles Eliot. The predominance and character of the new style was announced by Sargent's son-in-law, French-trained architect Guy Lowell, in *American Gardens* (1902). In this lavish guidebook, "American" referred to the location, creation, and ownership of properties, not to the appearance of the gardens or to the species planted.[47] Formal landscape design gained permanent public visibility through the reconstruction of the National Mall in Washington. In the first decade of the new century, architect Charles McKim removed both the naturalistically arranged trees and carriage drives that Smithsonian leaders had established around their building in the 1850s as a tribute to Downing, and the elaborate display of American and exotic species that William Saunders had planted on the Agriculture Department grounds two decades later. He instead created spaces that were monumental, symmetrical, linear, and clean. Elms were planted as blocks of greenery and as shade providers, not as specimens or accents.[48]

Emulation of east Asian models was less widespread than interest in Italian or French formalism, but it was more important from a vegetational perspective. Chapter 5 noted the extent of American interest in Japanese plants and plantings during the second half of the nineteenth century. Importers struggled to satisfy enthusiasms for Japanese conifers, plums, peonies, lilies, and, above all, chrysanthemums. Chrysanthemums dominated American flower shows in the 1890s; interest was so great that Cornell's agricultural experiment station briefly planned annual reports on new models.[49] Charles S. Sargent, who was crossing off foreign plants from the Muddy River Improvement in 1892, changed perspective gradually over the next two decades, as Arnold Arboretum collector Ernest Wilson sent back a multitude of new Asian ornamental species. Many of these (including species of *Juniperus, Malus, Prunus,* and *Viburnum*) spread across garden landscapes in greater Boston and elsewhere with the name *sargentii*.[50]

East Asian plants and gardens—Japanese in particular—were subversive of the categories used during the previous decades. Americans knew intellectually that Japanese plants were exotic. Similarly, they could learn that Japanese chrysanthemums, bonsai, and gardens were constructions that required planning, skill, and labor. Yet Japanese gardens with Japanese plants offered experiences that were intuitively congruent with those of the native American wild garden landscape. They were aesthetically the opposite of Tuscan or French formality: they featured irregularity, informality, harmony, and above all, the *Garden and Forest* ideal of "repose." Moreover, as Asa Gray had established, east Asian plants were more closely allied to those of New England and the Carolinas than were those of western Europe or California, and the diversity of species in the two regions was also similar; they were an odd mix of the exotic and the familiar.

The rise of formal and Japanese styles together undercut the premises of American naturalism. Such gardens demonstrated that naturalistic landscapes were merely expressions of one rather old-fashioned taste, and were not an unproblematic idealized American nature. This multiculturalism was particularly evident in official Washington, where the three kinds of gardens existed in separate-but-equal juxtaposition. The Capitol grounds, designed by Olmsted in 1874, were a rapidly maturing naturalistic landscape park populated largely by North American species. The National Mall consisted of a two-mile-long formal lawn with straight drives and symmetrical rows of elms. Finally, as noted in Chapter 6, the new Tidal Basin was fringed in an orientalist manner with Japanese flowering cherries. The Brooklyn Botanic Garden, planned in the early 1910s, juxtaposed the three styles more intensively: the fifty-acre site included a formal alleé, a naturalistic brook with a "local flora section," and "a real Japanese garden, planned and constructed by Japanese."[51]

The war in Europe briefly made this multiplicity of styles problematic. With the different Old World countries occupying an inconclusive slough of destruction during 1915 and 1916, American leaders sought to articulate what was distinctive and better about their own nation. University of Massachusetts professor Frank Waugh sought to advance this cultural project within the area of landscape gardening, but his claims were deeply confused. On the one hand, he forthrightly advanced the premise that styles in landscape gardening were national, or "more strictly speaking, racial." On this basis, he proposed that the American garden spirit should be found in the "native landscape" that had been experienced by the pioneers. On the other hand, he dismissed what he called the "native plant cult," both because it was unworkable for a continental nation ("one gardener would accept any species native to America; another insists on plants from his own state; the garden maker of real convictions accepts nothing but what grows naturally on his own farm"), and because "it wouldn't be good democratic Americanism, either, for the great bulk of our citizens are derived from foreign stocks." He also suggested that only "an intolerable Puritan" would exclude cultivated species such as the lilac and the apple because they were nonnative. Waugh's resolution of these tensions entailed embracing the idea that American landscape gardeners could employ an indefinite number of leitmotifs. These could be topographic (the "river motive"), geographic (Connecticut), botanical (sunflowers), historical (Lookout Mountain), or literary (Shakespeare). It was a more contrived list than any botanical garden could encompass. These issues disappeared in 1917, however, when the United States joined the allied gardening nations of England, France, Italy, and Japan in their fight against the Central Powers.[52]

Prairie Spirit

Northeasterners' struggles to garden landscape were recapitulated, in a shorter time span and with greater seriousness, in Illinois and Wisconsin. Interest in replicating familiar Anglo-Hudson scenery competed with desires to evoke the regionally distinctive prairie. Landscape historians have focused on the pre–World War I innovations of the Danish German immigrant Chicago park designer Jens Jensen and the American horticulturist Wilhelm Miller. I suggest, however, that Jensen's and Miller's "prairie style of landscape gardening" drew so much from German and Olmstedian naturalism, and placed so much emphasis on shrubs and trees, that it contained little that was distinctive. The truly important development occurred, not on Chicago parklands or North Shore estates in the 1910s, but in southern Wisconsin in the 1930s, where Aldo Leopold planted a vast wildflower garden.

Chicago Improvement
In the second half of the nineteenth century, Chicago boosters promoted prairie as one of the elements underlying the greatness of their city. But while they honored early settlers' accounts of the beauty of the unbounded landscape of grasses and flowers, they themselves valued it primarily as a grid that generated wheat, corn, livestock, and, in the city itself, house lots. In searching for scenery, wealthy Chicagoans

188 · Fruits and Plains

congregated near Lake Michigan, along Lake Shore Drive and on the ridges north of the city.

Chicago land was a major challenge for public gardeners. In 1869 the city hired Olmsted and Vaux to plan parks on the far South Side. Their report, submitted two years later, focused almost completely on the area near Lake Michigan. It suggested that a landscape of islands and lagoons in what is now Jackson Park would possess an intricacy that would complement the simple grandeur of the water view. They had little to say about the area more than a mile inland that would become Washington Park. If it were "rolling prairie," it might have possibilities; but this land was completely flat and did not have soil that could support the "great spreading trees" that were the "distinctive glory of all park scenery." Their sole recommendation was for "a large meadowy ground, of an open, free, tranquil character."[53] These plans were shelved when the center of Chicago burned later that year. In succeeding decades, the Midwestern landscape gardeners H. W. S. Cleveland and O. C. Simonds dealt practically with land that was flat, windswept, and subject to climatic extremes. They suggested that vegetation echo the horizontal lines of the land and the lake, and they reinforced the common theme that planting should begin with demonstrably hardy local tree and shrub varieties.[54]

Chicago landscape gardening changed significantly after 1900. Build-out increased public interest in the large park areas on the west side of the city. Leaders of major enterprises such as Sears Roebuck, International Harvester, and the Board of Trade were laying out estates. And prosperous Chicagoans who went looking for scenery on Sundays in their new automobiles saw that the Prairie State consisted of mile upon mile of featureless farmland. The artists, designers, and business and civic leaders who met periodically in the center of the city for dinner and conversation at the Cliff Dwellers Club—whose name linked the vistas from its office tower location with those from southwestern pueblos—reimagined the problem of the prairie as one of regenerating the regionally distinctive topographical and vegetational beauty limned by early promoters.[55]

In later years, Jens Jensen became the icon of this movement, and was seen as its prime mover. Prior to the 1910s, however, his interests were heterogeneous. Raised in the Danish German province of Schleswig, he migrated to the United States in 1884 following service in the German Imperial Guard. He worked his way up in the Chicago Parks Department to become superintendent of the west-side Humboldt Park in

1896, where he was particularly enthusiastic about the landscaping value of Russian olive. Jensen criticized uneducated immigrants' preference for elephant rides over "natural scenery and sylvan beauty," and he attacked politicians for allowing ads to be displayed in park concerts during the "sacred melodies of Wagner." After losing his municipal position in 1900, he argued that the United States should emulate Germany by putting parks in the hands of professionals.[56] As a founder of the Cliff Dwellers, Jensen comfortably situated a wide variety of landscape features within the context of the Midwestern heartland. He claimed that his garden designs were inspired by both prairies and the Zuni. He wrote lyrically about prairie inspirations for his creation of a long view, using lawn and shrubbery, on the estate of Harry Rubens in Glencoe. More boldly, he characterized the prairie river, formal rose garden, and conservatory palms that he installed in Chicago's west parks as all interpretations of Illinois landscape. On the other hand, when asked in 1909 to plant the grounds of Chicagoan Chauncey Dewey's cattle ranch in the Flint Hills of Kansas, Jensen submitted a design consisting of trees, shrubs, and perennials, including forsythias, peonies, and chrysanthemums.[57]

Wilhelm Miller was the individual who explicitly claimed that there was a "prairie style of landscape gardening." He was unsuccessful, however, in articulating its contours or promoting its significance. German on his father's side and a New Englander on his mother's, Miller received the first American Ph.D. in horticulture, from Cornell, in 1900. He worked for Liberty Hyde Bailey on the *Cyclopedia of American Horticulture* and on *Country Life in America,* and then edited the Manhattan-based *Garden Magazine,* where he highlighted irises, peonies, and other suburban favorites. He became an advocate in a 1911 book that advised owners of suburban properties to end "literal imitation of English gardening ideas," which were unsuited to this continent's climatic extremes, and to instead use tough native species. A year later, having moved to Chicago to create a "landscape extension" program in the University of Illinois Horticulture Department, he began to promote "the 'Illinois Way' of beautifying the farm" and the "prairie spirit in landscape gardening."[58]

Some of Miller's advice was ordinary naturalistic uplift: farmers should move pigsties away from roads and should replace "gaudy" specimen plantings with lawns framed by shrubs and trees. His broader vision was that landscape gardening could revive the Illinois pioneers' prairie spirit. This entailed that "at least ninety percent" of planted species should be native to the Prairie State; it did not, however, mean prairie plants. Featuring the park and estate designs of Jensen and O.C. Simonds, Miller suggested that prairie could be "idealized" by placing locusts or junipers on roadsides to frame views of wheat fields. A group of horizontally branched trees, a terraced rose garden, or a "long view" down a strip of lawn framed by shrubbery would "conventionalize" prairie. Prairie could be "symbolized" in urban settings with foundation plantings of prairie rose *(Rosa setigera)*. While the spirit was prairie, the materials were the landscape gardener's mainstay trees and shrubs.[59] Given the breadth of Miller's conception and the unexceptionality of his examples, eastern reviewers questioned whether the prairie spirit had any real content. Miller's more immediate problem was that he had used his university agricultural experiment station circular to urge Illinoisans to fight municipal graft, abandon party-line voting, and oppose "military, economic, and social" war. The university terminated his contract in 1916; unsuccessful in an attempt to establish himself as a consulting landscape architect, he faded from the scene.[60]

Original Wisconsin

Aldo Leopold, Norman Fassett, and Theodore Sperry were the developers of a real prairie style of landscape gardening. Between 1935 and 1940, they transformed about twenty-seven acres of old pasture in Dane County, Wisconsin, a few miles southwest of Madison, into a naturalistic garden of grasses and wildflowers that they called a prairie. This act of historical naming enabled them to resolve the problem faced by landscape gardeners from Downing to Miller. They planted a landscape that was distinguishable from, and an improvement upon, the common vegetation around it, but which was plausibly naturalistic.

The University of Wisconsin Arboretum began as a provincial Olmstedian park project. In 1911 the private Madison Park and Pleasure Drive Association hired the young Massachusetts landscape architect John Nolen to prepare a comprehensive plan for the improvement of their city. Among Nolen's recommendations was the idea that the city and the university should emulate Boston and Harvard's partnership of the 1870s by establishing an arboretum-park on the shore of Lake Mendota, west of the city and the university campus.[61] That suggestion went nowhere. The arboretum idea was revived in the late 1920s, however, by local boosters seeking to transform a failed suburban development on the small and marshy Lake Wingra, a few miles southwest of the city. They argued that the state and the university should fund a park, arboretum, and wildlife refuge as part of the ongoing initiative to establish a conservation professorship for Madison-based forester and game manager Aldo Leopold. The university approved this plan in 1932, appointed landscape architect William Longenecker to the position of executive director, and asked Leopold to take on the arboretum's research directorship as one of his professorial duties.[62]

Disagreements arose immediately over issues of plant choice. Longenecker envisioned a landscape park containing systematically and ecologically ordered displays of all the perennials, shrubs, and forest trees that might prove hardy in Wisconsin. Visitors to the arboretum would be inspired to beautify their own properties, and would learn what different ornamentals and woodland trees looked like and which were worthwhile. Leopold wanted to send the visiting public a different message. He was uninterested in what he considered merely "a 'collection' of imported trees." Instead he wanted to show how much the state's vegetational quality had declined since the 1830s, and to provide a vision for improvement in the future. Advised by botany professor Norman Fassett, he proposed that the arboretum should be "a reconstruction of original Wisconsin." It would be "a bench mark, a datum point, in the long and laborious job of building a permanent and mutually beneficial relationship between civilized men and a civilized landscape." This disagreement was resolved by dividing the arboretum into areas controlled by either Longenecker or Leopold.[63]

For Leopold and Fassett, original Wisconsin was an essentially steady state, consisting of forest, wetland, and prairie, that had existed prior to Anglo-American settlement. (They passed over the major presence of Indians in Dane County during the Woodland Period, evident in the number of mounds—over one thousand, more than anywhere else in the United States.) Creating replicas of these plant communities on a few hundred acres would require a number of different kinds of effort. Sections with trees could redevelop on their own if there were fire suppression and culling of undesirable species. The right mix of wetland vegetation depended largely on steam dredges that could change the monotonous marsh into a more varied landscape of islands and lagoons. Shoreline areas with different slopes and soil compositions could then be planted with cattails and pondweeds that would attract wildfowl.[64]

The real gardening challenge, however, was to create a "Wisconsin prairie" (the present-day Curtis Prairie). The basic prerequisite was labor. In 1934 the arboretum received a windfall when the state established a work relief camp for transients on its grounds. Then, when complaints arose about the behavior of these migrants and hoboes, the university persuaded the National Park Service to take over the camp and use it for the Civilian Conservation Corps (CCC) (see Figure 7.9). The CCC recruited a more tractable pool of young local men, and its involvement enabled the university to hire the young National Park Service plant ecologist Theodore Sperry as foreman. "Camp Madison" averaged about two hundred residents during the second half of the 1930s, at a cost to the federal government of more than two million dollars.[65]

The first step in the creation of a Wisconsin prairie park was to clear existing old-field growth. Tree control was a straightforward matter of destroying saplings, but was complicated by Fassett and Sperry's interest in leaving one large tree standing to evoke early settler accounts of "oak openings"; each year laborers had to pull up a crop of squirrel- and

Figure 7.9 This photograph from the late 1930s shows University of Wisconsin Arboretum horticultural director William Longenecker with tanned and happy Civilian Conservation Corps (CCC) boys planting sod they had trucked to Madison. Reproduction courtesy of the University of Wisconsin-Madison Archives, Steenbock Library, Madison.

bird-distributed oak seedlings. The major problem was quack grass. Sperry and his workers sought to eliminate this Old World pasture mainstay and agricultural weed by plowing deeply, harrowing to dry out the rhizomes, and then replanting with clover to smother remaining growth. Irritating plants such as nettles and thistles were also a concern, without regard to their geographic origin. Finally, Leopold sought to suppress high-density populations ("thickets") of plants that were too common, such as goldenrods and asters.[66]

Once the ground was cleared, the major issues involved plant choice. In principle, Fassett and Sperry's palette could include any of the species associated with prairies in or near Wisconsin during the previous century. A present-day list of such plants totals between 340 and 550. But prairie gardeners in the 1930s were neither capable of nor interested in cultivating such a diverse flora. Sperry's planting list from 1935 to 1939 consisted of about fifty species. In both his exclusions and featured species, his goal was to plant an assemblage that would not be confused with common or despised pasture.[67]

The largest category of excluded species consisted of the dozens of plants that were small, had inconspicuous flowers, or were visually generic. There was minimal interest in devoting labor and space to vegetation that added little to the field's visual composition. More straightforwardly, Sperry did not replant the nettles and thistles that had been removed when the land was cleared, nor did he introduce additional species with similar properties. While some of the more memorable native species that people encountered on Wisconsin prairies were greenbrier *(Smilax lasioneura)*, prickly pear *(Opuntia macrorhiza)*, and poison ivy *(Toxicodendron radicans)*, they were not part of the arboretum plantings. The most interesting group of exclusions was of species poisonous to livestock. Prairie larkspur (*Delphinium carolinianum* subsp. *virescens*), sundial lupine *(Lupinus perennis)*, and death camas *(Zigadenus elegans)* were all visually impressive Wisconsin natives. But the prosperous rural citizens whose sensibilities Leopold wanted to touch would not have appreciated a field filled with seed-bearing specimens of the weeds they had worked for a century to eradicate.[68]

Sperry wisely emphasized familiar species that would, under proper cultivation, provide a spectacular mass display. His most frequently planted species was turkey-foot grass (*Andropogon gerardii*, now commonly called big bluestem). The most-planted forbs were stiff sunflower *(Helianthus rigidus)* and three species of *Silphium* (including compass plant and rosinweed). Others included blazing star *(Liatris)*, prairie goldenrod *(Solidago rigida)*, prairie rose *(Rosa carolina)*, prairie bush clover *(Lespedeza capitata)*, prairie coneflower *(Lepachys pinnata)*, and prairie painted cup *(Castilleja sessiliflora)*. They were either large (big bluestem, compass plant, and stiff sunflower could all grow ten feet high in a good summer), had conspicuous flowers (blazing star, rose, coneflower), or unusual characteristics (indicated in names such as compass plant and painted cup). While Wisconsinites might know these plants, they would have seen them only in small populations or in fields browsed by livestock. At the arboretum, by contrast, they were able to display their capabilities and to reinforce each other visually as elements of a multiacre garden. People who visited this landscape, especially in the peak summer vacation months of July and August, would experience a wonderful wildflower garden in the style of a prairie. It was both easy and pleasant to imagine that this was original Wisconsin.[69]

CHAPTER EIGHT

The Horticultural Construction of Florida

The preceding chapters have examined temperate and humid regions of North America that were, in climate and vegetation, at least superficially similar to northwestern Europe. Efforts to replicate European culture in those regions had begun in the 1600s. Horticulturists did not appear on the scene until two centuries later; when they did, their goals were to improve or to fix existing vegetation. In one part of the United States, however, horticulturists took the lead in settlement and created something quite new from scratch. The Florida Peninsula, and especially the territory south of the 27th parallel—from Jupiter Inlet on the Atlantic Ocean to Charlotte Harbor on the Gulf of Mexico—was a region with an unusual geographic and climatic history, and with an idiosyncratic ecology, where small changes could have much larger effects than in most other parts of North America (see Figure 8.1). Essentially unpopulated by Anglo-Americans prior to 1880, it was the final settlement frontier in the contiguous forty-eight states. Horticulturists imagined a new kind of garden culture based on tropical species, a few of which they found growing there spontaneously. This vision made them leaders in the rapid development of south Florida during the next half century.[1]

Exploring the horticultural construction of south Florida is important for understanding the history of a major American region that has received little attention by comparison with its western analog, southern California. In the context of this book, the value of examining Florida horticulture comes from the extent to which it

196 · Fruits and Plains

Figure 8.1 East Florida, with locations discussed in the text emphasized. The base map, from the *Encyclopaedia Britannica*, 9th edition (1878–1879), displayed the state's few railroads as dark lines. Overlay by the Rutgers University Cartography Laboratory.

recapitulated, with greater intensity and concentration, the tensions among desires for the exotic, enthusiasms for the native, and fears of the alien that have appeared in previous chapters. I begin with a sketch of the region's biogeographic peculiarity, and then review the horticultural efforts of Euro-Floridians prior to the United States' takeover in 1821. I then describe American efforts to understand and to alter the peninsula's biotic makeup. In the 1830s Henry Perrine sought to make south Florida a center for tropical plant production. During Reconstruction Harriet Beecher Stowe planned to restore Florida to the Union by persuading large numbers of Yankee gardeners to buy farms, supply the North with winter produce, and vote Republican. Agriculture Department botanists worked to fill south Florida with citrus groves and garden ornamentals. In the twentieth century privately patronized scientists delineated a native Florida flora, and they extended landscape gardening ideals to the Everglades. The last part of the chapter deals with invading pests: I contrast Floridians' acceptance of annoying biotic competitors with Northern federal officials' efforts to control exploding populations of new species such as the water hyacinth and citrus canker. In 1929 Florida was the place where Charles Marlatt demonstrated that, with near-totalitarian power, government scientists could exterminate a major alien pest. The question was whether this model could then be extended to the rest of the continental nation.

A Climatic Island

Topographically, the southern half of the Florida Peninsula is an integral part of North America. Biogeographically, however, it is a tropical island less than 5,000 years old. During the Pleistocene glaciations, the lowering of the oceans caused the peninsula to double in width. With a climate similar to present-day Oklahoma, the land was covered by temperate grasses and scrub plants driven down from the north. Then, from 18,000 to 3,000 years ago, rising oceans flooded much of the peninsula, and warmed and moistened what remained. The modern terrain and flora of south Florida—the Everglades, in particular—are about 5,000 years old, a millennium younger than the traditional age of the Garden of Eden.[2]

The consequence of this prehistory was that south Florida was quite un-Edenic in its vegetation. While Caribbean islands were rich in endemics, the Florida mainland south of latitude 27°N contained only those tropical species that had been carried there during the Holocene by birds, currents, or hurricanes, or by the humans who had been living there since 12,000 BP. These plants presumably arrived, died out, and arrived again numerous times. Many of the previously common temperate-zone plants disappeared due to warming, efficient tropical species, and fire-using humans, but some refugees from New Jersey and Ohio held on.

Many areas in south Florida were dominated by single species, notably sawgrass *(Cladium jamaicense)*, palmetto *(Sabal palmetto)*, and slash pine *(Pinus elliottii)*.[3]

The human presence on the peninsula declined dramatically during the first three hundred years of European influence. The pre-Columbian population in the eastern part of the state may have been more than 100,000; by 1760, epidemic disease and warfare reduced it to less than 1,000.[4] The Spanish abandoned their initial ambitious missionary program for a minimal effort to extract products that required little labor (i.e., cattle and horses) and to maintain enough military presence to keep the British away from the Florida Straits. The one enduring Spanish change in vegetation was the introduction of orange trees; by the early 1800s, these had spread so widely that many believed they were indigenous to the peninsula.[5]

British imperial officials began to advertise Florida as a site for tropical agriculture soon after they took over the territory in 1763, offering large grants of land along the St. Johns River to promoters. The Spanish maintained this policy when they regained control in 1783.[6] But the isolation of these settlements from the rest of the world, the prevalence of insects and diseases, and the vulnerability of plantation houses to raids from Seminoles (Creek Indians who had migrated south from Georgia and Alabama during the preceding century) or from white Georgians prevented the creation of a viable agricultural economy adapted specifically to Florida's climate. The Georgia and Alabama planters who came in quantity after 1821 were interested primarily in the cotton lands of the Panhandle region. Small groups of Seminoles, escaped slaves, and the military dotted the four-hundred-mile-long coast south of Jacksonville.

An Invasive Species from New Jersey

Northeastern horticulturists were the first Americans to focus on the unique climatic and geopolitical position of Florida. Their idea was that the region could become what otherwise seemed an oxymoron—a United States tropical garden. They hoped to grow cold-intolerant plants whose products could be marketed profitably in the North, and to create bucolic winter destinations for invalids, the elderly, and people of leisure. Between 1828 and 1840, physician and diplomat Henry Perrine struggled heroically to carry horticultural improvement from New York to south Florida. He has generally been viewed as an uncontroversial if quixotic improver. His importance grows if the depth of thought and the particular passions involved in his project are foregrounded. For Perrine and his patrons, horticulture stood at the center of American conquest of south Florida. It provided a rationale for invasion by white "native Americans"; it would make the area worth having; and it required clearing the ground of existing inhabitants. Those inhabitants responded violently.

Henry Perrine was born near Princeton, New Jersey, in 1797. He studied medicine with both Samuel L. Mitchill and David Hosack (discussed in Chapter 3) at the College of Physicians and Surgeons in New York.[7] After graduating in 1819, Perrine migrated to west-central Illinois where he established a medical practice, married, and fathered two daughters. His rise to ordinary provincial prosperity was interrupted, however, when he accidentally drank a near-fatal dose of arsenic. The long-term neuropathy caused by the poison made him acutely sensitive to cold; sending his wife and children to his in-laws' home in upstate New York, he moved his practice to Mississippi. The climate there was still too "variable" for his nerves, which were sensitized further by his reliance on medicinal mercury and ergot. After wintering in Havana in 1826, he established himself permanently in the tropics by obtaining the position of U.S. consul in Campeche on Mexico's Yucatan Peninsula.[8]

When in Cuba, Perrine discussed the possibility that local tropical crops could survive in south Florida and that Americans' industriousness and "free institutions" would provide a competitive advantage.[9] He thus took interest in Treasury Secretary Richard Rush's 1827 circular (mentioned in Chapter 5) that asked diplomats to send back exotics. Perrine was enthusiastic about all kinds of tropicals, but saw particular promise in sisal (*Agave sisalana*), the fiber-producing relative of the yucca that grew readily in Yucatan. He believed that sisal could be grown in south Florida, only 300 miles to the northeast, and that it could be as versatile and profitable as cotton.

By early 1829 Perrine was sending both information and potted sisal plants to S. L. Mitchill. He overcame Mexicans whom he considered ignorant collectors, and American naval officers who would not provide adequate care during transport, but he was stymied by the lack of a government-affiliated botanic garden to shepherd the plants through Customs. His specimens froze on the docks in New York.[10] In late 1831, temporarily back in New York, Perrine suggested to the House Agriculture Committee that the government should compensate him for his expenses in gathering the plants requested by the secretary of the treasury by providing him land for a tropical experimental garden in Florida. He featured his relations with experts S.L. Mitchill, John Torrey (his fellow medical student), and nurseryman William Prince, and he displayed his Democratic Party connections by datelining his proposal "Tammany Hall." The Florida Territorial Legislature supported Perrine's proposal and Congress published it, but the plan went nowhere.[11]

In 1834 Perrine tried again, sending from Campeche a series of proposals and explanations to Congress, the secretary of state, and various agricultural publications. He explained that his earlier difficulties in sending viable material were not due to the tenderness of the plants themselves; rather, he reported with anger and triumph, the seemingly friendly but incompetent Mexicans had been boiling his seeds and drying out his seedlings because they believed he was "robbing the country of all its valuable plants." For Perrine this dishonesty was part of a larger pattern of Caribbean peoples' antagonism toward the United States.[12] The best response would be home production of tropical products. He explained to Congress that south Florida was *"the only tropical territory on the globe in which tropical vegeculture can be pursued by the best species of the human genus under the best Government in the world."* The "colored species" was "indolent, ignorant, immoral, imbecile, and, consequently, poor"; on islands such as Jamaica, "the black cloud of fanatical emancipation" was leading to a dramatic decline in economic productivity. "Native American" success in extracting profits from high-value tropical species would dampen secessionist impulses in the South by restoring prosperity to the South's "ruined fields." Those states would then "emancipate and transport [to other countries] all their colored laborers," and so the region would ultimately be populated solely by whites.[13]

Again, Congress took no action. The Jackson administration's interest in Florida in the 1830s centered not on tropical plants, but on the Seminoles. In 1829 Jackson had decided to move all southeastern Indian tribes west of the Mississippi River. Forcing this policy on the Seminoles was difficult because the federal presence south of latitude 27°N was so minimal. It was important to north Florida plantation owners, however, because slaves could easily escape to Indian refuges, and then, even worse, mount "piratical" raids on their former residences. From the administration's perspective, a colony of horticulturists and their slaves at the southern end of the peninsula would increase both the complexity of the escapee situation and the hostility of the Indians.[14]

In March 1837, however, the new Van Buren administration appointed the South Carolian, Joel Poinsett, as secretary of war. Perrine decided that this was the time to act. Poinsett had been ambassador to Mexico when Perrine first went to Campeche, and he shared Perrine's enthusiasm for tropical plants (he introduced the showy red *Euphorbia pulcherrima*, which Americans later named the Poinsettia, from Mexico into the United States). Perrine resigned his consular position, moved to Washington, and in the autumn laid out his scientific arguments with lists and drawings of plants, analyses of soils, weather data, and an annotated copy of the plant geography chapter of John Lindley's *Introduction to Botany*. Submitting testimonials and copies of earlier documents, he detailed all the problems about which he was not complaining, argued for the value of tropical products, suggested that his garden's plants could eventually be grown farther north, and denied accusations that he was a land speculator.[15]

Perrine's project became deeply entwined in the Seminole War debates. In January 1838, congressional opponents of the war complained bitterly about the army's "perfidy and treachery" in jailing Seminole leader Osceola when he met them under a flag of truce. In February, after a series of bloody and inconclusive battles near present-day Palm Beach, the field commander, General Thomas Jesup, advised Secretary Poinsett that the Seminoles' lands had little agricultural value and that the way to end this "most disastrous war" was to leave the Everglades to the Indians.[16]

Perrine's memorials presented the opposite position. The heart of his project was the belief that "Tropical Florida" could play a unique role in the nation's agricultural economy. Its development depended on insuring the safety of the white species and their (unmentioned) slaves, who would clear the land, make the climate salubrious, and learn to grow tropical crops. The result would be "the creation of a dense population of small cultivators and family manufacturers" and a Riviera-like winter resort for invalids. The alternative to this idyllic future was continued warfare with "savage Seminoles" and "murderous fugitives" (escaped slaves), and a landscape of "pestilential swamps and impenetrable morasses, [which] will become the fortresses of the worst portions of the black and piratical inhabitants of the adjoining West India islands."[17]

In March Poinsett informed General Jesup that the government was committed to the expulsion of the Seminoles from Florida, and he appointed the more aggressive Zachary Taylor as his new commander. That same month, the House printed a ninety-nine page compilation of Perrine's documents and the Senate prepared a 142-page report, supplemented by twenty-four engravings of tropical plants. The law that passed in early July, granting Perrine and his associates a township (36 square miles) of land of his choosing south of 26°N, was a federal affirmation of support for the settlement of the entire Florida Peninsula by native Euro-Americans.[18]

Poinsett promised Perrine that the land he picked out (about fifteen miles south of modern downtown Miami) would be protected by an army base, but he was vague about timing. Perrine decided to move forward with a preparatory nursery on an outlying island. Ten-acre Indian Key, twenty-five miles south of the Florida mainland, was protected by a small navy facility a mile away on Teatable Key, and its owner, the adventurer-entrepreneur Jacob Houseman, was glad to cooperate with the first person authorized to acquire a large tract of mainland property. On Christmas Day 1838 Perrine moved to Indian Key with his family. His imagination expanded as Houseman's slaves began hacking temporary nursery beds from the scrub-covered limestone for the plants that would soon arrive. He wrote a detailed report on the logistics and economics of silk culture in Florida, and he urged the secretary of the navy to acquire a "precious cargo of living plants" from Mexico immediately: he explained that recent French threats against that country meant that an American expedition would not be faced with the usual "Creole animosity to foreign enterprise."[19]

Perrine had placed himself, however, in an untenable position. While his long-term plans were linked to the War Department, he was now dependent on the kindness of the navy. While that organization would not cooperate in acquiring Yucatan plants, its local sailors showed a worrisome amount of interest in Perrine's teenage daughters. The most vexing problem was that the regional naval command in Key West, whose mission was to control smuggling, required that all foreign shipments be routed through that official port of entry. As a result of that detour nearly all the plants sent to Indian Key arrived dead.

In early 1840, approaching a second spring with no significant plant deliveries, Perrine roundly denounced the combination of "murderous red men" on the Florida mainland, his "organized enemies" in Key West, and "the inimical government of the United States at Washington," who together were obstructing his enterprise. His partner, Houseman, belittled the expenses and casualties incurred by the military by proposing to the Florida legislature and to Congress that he would be willing to "catch or kill all the Indians of South Florida, for two hundred dollars each." In August navy commanders sent the active force on Teatable Key to the southwest coast of the mainland in hopes of finding and eliminating the southernmost group of Indians. Their intelligence, however, was faulty. A day after the sailors left, more than one hundred Indians attacked Indian Key by night. Houseman and his wife swam to safety, and Perrine hid his family in a crawl space, but the horticulturist himself was killed, the island's supplies and buildings were looted or burned, and the slaves left with the Indians. Indian Key was never rebuilt, and even after the Seminole hostilities ended in 1842, Americans made no coordinated efforts to establish themselves in the Miami area. Perrine's sisal plants, on the other hand, spread rampantly in their new home.[20]

Nutmeggers and Citrus

Northerners—soldiers, bureaucrats, reformers, and adventurers—experienced Florida in large numbers for the first time during the Civil War. Some grasped the peninsula's climatic uniqueness, sociopolitical blankness, and horticultural potential. They imagined a state that was not, like its neighbors, grounded in cotton, oligarchy, and rebellion. Rather it would be a tropical outpost of New England, combining a garden landscape, production for national markets, and conditions for living easily during northern winters. The centerpiece of this vision was the orange. Yet the meaning of citrus culture ranged from the harmonious and integrated villages of antebellum New England imagined by reformers to the replication of that region's proprietary enterprises, company towns, and commercial leisure. In Florida in the 1870s, one end of the spectrum was occupied by the abolitionist, author, and orange grower Harriet Beecher Stowe; the other by investor, diplomat, and orange grower Henry Sanford. Examining the activities of these two individuals, raised less than twenty miles from each other in Connecticut (the Nutmeg State), displays the potential that Yankees saw in Florida during Reconstruction, and the limitations of these visions.

The recent *Beechers, Stowes, and Yankee Strangers* begins with its authors' perplexity that in 1865 the integrationist New Jersey minister John Swaim was sending reports from Jacksonville to his former congregants in Newark that were more about the beauty and settlement potential of Florida than about race and spirituality. Swaim and other reformers were boosters because, from their tenuously held wartime base in the state's northeast corner, they imagined a transformation that was simultaneously demographic, political, and experiential. At the end of the war Florida's population was only about 165,000, of whom fewer than 32,000 were voting-age males. Nearly 48 percent of those men were "colored," and some fraction of the whites consisted of rebels ineligible to vote. Northern liberals could thus reasonably imagine that a few new people could combine with the former slaves to make Florida Republican and to develop an economically progressive and racial harmonious society.[21]

Republicans immediately began to encourage Northern "colonists" when they gained control of the state government in early 1868. Vermont educator John S. Adams, appointed to the new position of immigration commissioner, prepared a pamphlet that dismissed King Cotton as a "social tyrant" who relegated his "votaries" to "the vast solitudes of remote plantations," and produced oligarchy, ignorance, and stagnation. He urged "strong immigration of new men with new views and new desires" who would initiate "an absolute and entire change in the methods of agriculture, and an increase in the diversity of occupations." The result would be the creation of villages that would contain the proven Northern means of social progress—"the Church, the School, the Press, and Post Office."[22]

The exemplary "new man" was Harriet Beecher Stowe. Her son Frederick, an army veteran and an alcoholic, had bought a cotton plantation south of Jacksonville at the end of the war. Harriet came south in the winter of 1867 to manage his household and insure that he stayed sober. The plantation failed and Frederick Stowe left for the West and disappeared, but Harriet and her husband, Calvin, bought thirty acres of waterfront property with an orange grove on the St. Johns River near the village of Mandarin (named for the orange), latitude 30.16°N. For the next fifteen years, Stowe and her family wintered in Mandarin and a nephew managed their grove. Stowe established contact with other Northerners, including Harrison Reed, the state's Republican governor from 1868 to 1873. In 1869 her brother Charles, a minister who was under fire in Massachusetts for heresy, moved to Florida and was soon appointed state superintendent of schools.[23]

Just as Henry Ward Beecher had evangelized for both God and gardening in Indianapolis in the late 1830s, his sister Harriet envisioned a combination of community, reform, and fruit in Florida. *Oldtown Folks,* a novel she completed in Mandarin in 1869, was an evocation of the virtues of the preindustrial New England culture that Stowe had experienced in Litchfield, Connecticut, in the 1810s. *Palmetto-Leaves* (1873), a collection of articles she had written for Henry's new magazine, the *Christian Union,* sketched how that kind of life could be transplanted to the South and rejuvenated. Stowe preached from a text of cucumbers: gardeners near Mandarin who transcended "ordinary culture" through generous applications of intelligence and manure grossed up to $850 per acre by growing vegetables, strawberries, and peaches for spring shipment to New York.[24] New England village institutions, family and neighborliness, and simple rural pleasures such as picnicking, could grow upon this base of truck farming. Experiencing a softer climate, Yankees would lose their compulsion to rush through the seasons; with lusher greenery, bugs, and sand all around, the tidiness that Stowe and her sister, Catherine, had just outlined in their guide to housekeeping, *The American Woman's Home,* would inevitably be relaxed. Stowe was a realist in acknowledging that Florida had a malaria problem, but she emphasized that the disease was absent from November through May. New Englanders could be part-time Floridians, tending gardens, voting, and educating both colored and white natives in liberal Protestantism, yeoman prosperity, and mutual tolerance. The result would be a new and balmier version of preindustrial Connecticut.[25]

A different Nutmegger strategy for Florida came, not from ministerial Litchfield, but from the industrialized Housatonic River Valley. Henry Sanford, born in 1823, received a large and steady income from his inherited interest in an uncle's tack factory. Living primarily in Europe after 1849, he circulated in cosmopolitan society, participated in international business ventures, and pursued a diplomatic career. His most important activity (while officially ambassador to Belgium) was coordinating countermeasures against Confederate agents in Europe. By donating cannons to the Minnesota militia he obtained the honorific title of "General."[26]

After the war General Sanford decided to participate in "*reconstruction after my own manner.*" He saw the South primarily as a field for profits, but also hoped that he could aid his friends and "help along our repentant southern brethren." In 1867 he purchased an orange grove near St. Augustine from President Lincoln's private secretary, John Hay. Three years later, traveling up the St. Johns River with his wife to see his property for the first time, he became an enthusiast for Florida development. He purchased more than 12,000 acres on Lake Monroe, the upriver terminus of navigation. His idea was to create, not Stowe's old-style New England village, but a modern proprietary town. In Sanford (latitude 28.8°N), as it inevitably was named, Sanford built a wharf, lumber mill, slaughterhouse, and resort hotel; he established a real estate company that advertised the desirability of purchasing town lots and country properties on what was called, with pseudo-Hispanic nostalgia, the "Sanford Grant."[27]

The linchpin of this development project was citrus. Sanford believed that oranges could, like tacks, be mass-produced at a high level of quality and sold profitably in ever-increasing numbers. He hoped to overcome the main antebellum constraints on Florida citrus production—intermittent freezes, inadequate transportation, sour or bland fruit, and scale insects—by locating his groves at the southernmost place that riverboats could reach and by testing a large number of introduced varieties for hardiness, flavor, and pest resistance. His ten-acre experimental garden was, by itself, larger than nearly any orange grove in Florida in the early 1870s; his main grove, Belair, included ninety-five acres of oranges and fifty of lemons. Sanford hired a large number of African Americans from the cotton region around Tallahassee to construct and operate this frontier property. He planned a permanent community for these workers and their families, offering low-interest mortgages and donating land for a church and school.[28]

A significant number of Northern colonizers embraced some combination of Stowe's and Sanford's plans for Florida. In 1875 former governor Reed founded *The Semi-Tropical,* an upscale monthly that promoted Northern immigration. Reed provided news and editorials that emphasized Florida's development and political harmony. The newly created Florida Fruit Growers Association sent in reports on its activities. Charles Beecher submitted to Reed an imaginative account of Florida in 1976, with airplanes, a circumferential railroad, and a gardenized Everglades. These Yankees saw a reasonable prospect of transforming the environmental and political landscape of Florida over the next generation.[29]

Their vision, however, was not realized. White Southern resistance, poor management, and the limitations of geography stood in the way. Sanford began to experience difficulties as early as 1870 when whites shot at the black workers who had been "imported" to the general's "Yankee nest." With more irony than wisdom, he responded by bringing in Swedes; their church was threatened with arson. He held his outpost with the aid of armed guards and federal marshals.[30] His longer-term problem was his inability to manage such a large and novel venture. He plunged into projects such as mills and hotels without considering cash flow, he was unwilling to delegate authority over the specialized enterprises, and he was not sufficiently interested in the project to spend enough time on site. As his debt mounted in the 1880s, he lost control of his land, and development of Sanford slowed dramatically.[31]

The largest obstacle, however, was implicit in the title of Reed's magazine: both the visions of Stowe and Sanford, and the climate of northeast Florida, were only "semi-tropical." St. Johns River towns consisted of vernacular frame structures with touches of Victorian gothic similar to those in both rural New York and Colorado. The mix of tourists, invalids, and entrepreneurs resembled travelers on the Mississippi. From a New England perspective, January and February along the St. Johns River were nonwinter months, but the vegetation—a mix of conifers, oaks, maples, and gums, many without leaves—did not differ radically from that of the North. Stowe's emblematic palmetto, and the "golden jessamine" *(Gelsemium sempervirens)* she considered characteristic of the area, could both be found as far up the coast as North Carolina, and orange trees—albeit in tubs, with indoor winter storage—were common Northern estate fixtures. This was not the lush floral world of the tropical conservatories built in the previous quarter century, with their palms, bananas, and bromeliads.

The region's real problem, as it turned out, was that the weather was not even reliably semitropical. In 1894 and 1895 northwest winds brought in temperatures low enough to destroy essentially all the citrus trees in the area. The Sanford Swedes abandoned oranges and reinvented their town as the prosperous but distinctly untropical "celery capital" of the United States.[32]

Flamboyant and Royal

In the four decades around 1900, nearly four centuries after Ponce de Leon reconnoitered the coastline from Cape Canaveral to the Keys, Euro-Americans finally began to settle the southeastern edge of the Florida Peninsula. The population of the area south and east of the northern end of Lake Okeechobee (present-day Martin, Palm Beach, Broward, and Miami-Dade counties), which was less than 1,000 in 1890, grew to 17,500 in 1910 and more than 210,000 two decades later. The settlement of south Florida was rapid in part because it had been retarded during the previous three decades by political turmoil. But the area changed so dramatically because its major developer, Henry Flagler, combined resources and know-how more effectively than the long line of Florida real estate promoters, from Ponce to Sanford, who had preceded him.

Horticulture was integral to the Florida that Flagler and his followers shaped. The region would not be a utilitarian farmers' frontier, but rather a new kind of place whose sensuous beauty would draw the richest Americans. Flagler projected a landscape of luxury hotel gardens, estates, and orange groves; its appearance would be truly tropical. It would be a place where the exotic fit naturally within a cosmopolitan American framework. This idea was embodied in the arboreal names central to Flagler's major south Florida development: the Royal Poinciana Hotel in Palm Beach paired the Madagascarian flamboyant tree with the pantropical coconut. The broader development of south Florida was shaped by a loose network of horticulturists, including former forestry professor John C. Gifford, retired Smithsonian naturalist Charles T. Simpson, farm machinery magnate Charles Deering, and New York Botanical Garden botanist John K. Small. The presiding figure was David Fairchild, who guided both the Agriculture Department and his amateur associates in introducing Americans to the avocado, bamboo, cajeput, dasheen, and the rest of the tropical plant lexicon.

Henry Flagler was the cocreator of Standard Oil.[33] For fifteen years beginning in 1867 he worked side by side with John D. Rockefeller to build the integrated system of extraction, transport, and marketing that made their company so profitable and powerful. In the early 1880s he decided to withdraw from daily management of the oil business; after spending parts of two winters in St. Augustine (latitude 29.9°N), one of the northeastern coastal towns that attracted seasonal vacationers, he became interested in local development. He appreciated the old Spanish city's exotic associations and moderate weather, but he was disappointed with its poor hotels and lack of diversions. He decided to transform St. Augustine into a winter Newport—an exclusive resort for the new national business elite.

Producing a beautiful setting in St. Augustine during January and February was difficult. The local vegetation was dormant, and a major priority was to eliminate all suggestions of wetness to reassure visitors that there was no malaria. The young architects, John Carrere and

Thomas Hastings, solved these problems brilliantly through design. Flagler's Ponce de Leon and Alcazar Hotels, which opened in 1888, were "Spanish-Moresque palaces" that identified, with Gilded Age expansiveness, the early-eighteenth-century military structures of St. Augustine with Iberian styles of the 1300s. The grounds were then planted with exotic-looking Mediterranean and Mexican plants—olives, citrus, the tougher palms, and agaves—that were satisfyingly green but were associated with arid rather than swampy environments (see Figure 8.2).[34] Florida's other seasons and riskier places could be experienced through art. The landscape and flower painter Martin Johnson Heade had moved to St. Augustine in 1883 and began to produce both evocative renderings of Florida swamplands and portraits of the region's signature flowers—the magnolia, Cherokee rose, and orange blossom. Flagler adorned his hotels with Heade's landscapes, and installed the artist as a tourist attraction in a studio attached to the Ponce de Leon.[35]

Figure 8.2 Courtyard plantings at St. Augustine's new Hotel Ponce De Leon in the late 1880s featured yucca, sisal, palmetto, Canary palm, and other arid-themed species. "Photographs taken in St. Augustine, March 1888," Diary and Album Collection, P. K. Yonge Library of Florida History, Department of Special and Area Studies Collections, George A. Smathers Library, University of Florida, Gainesville.

The profitability of the St. Augustine hotel venture on its own was unclear. Flagler, however, was able to bring his deep reservoirs of Standard Oil money, along with his experience in system building and in manipulating legislatures, to the problem of Florida development. He invested in the railroad that ran from St. Augustine to Jacksonville to ensure that it would provide access to his hotels, but within a few years gained monopoly control of regional rail construction. By the early 1890s his Florida East Coast Railway had obtained grants for more than two million acres of public land from the compliant state government.[36]

Flagler personally scouted ahead of his railroad engineers for flagship properties. In 1893 he visited Lake Worth (latitude 26.7°N) and realized that the barrier island separating this open lagoon from the ocean was preadapted as a luxury winter resort. Nearly 250 miles south of St. Augustine, it abutted the Gulf Stream; the moving tropical waters moderated temperatures, enabled winter sea bathing, and kept down insects. Island guests could be insulated from the resort's support infrastructure and its black workforce, both located on the mainland. Moreover, the area was scenically distinctive because, when a Spanish ship foundered offshore fifteen years earlier, local salvagers had appropriated and planted its cargo of thousands of coconuts. These paradigmatic tropical trees enabled promoters to give the island the distinctive identity of Palm Beach.[37]

The Palm Beach aesthetic blended New England familiarity and tropical cosmopolitanism. Flagler built a five-hundred-room Colonial Revival hotel modeled on Catskills or White Mountain resorts. He eschewed Yankee white clapboards, however, for bold "Flagler yellow," and he named the hotel the Royal Poinciana for the red-flowered tree introduced into the Caribbean a century earlier. The Royal Poinciana grounds and golf course were planted with tropical vegetation, including a coconut grove, banana trees, agaves, and poincianas (see Figure 8.3). These plants, many of them with botanical labels, were central to the Palm Beach experience. They were both signals of the island's reliably warm climate and invitations to guests to indulge their tropical fantasies.[38]

Planting Miami

In February 1895, a few days after the "great freeze" destroyed citrus groves throughout the Florida Peninsula, Miami booster Julia Tuttle

Figure 8.3 Grounds of the Royal Poinciana Hotel, Palm Beach, around 1900, featuring rows of coconut palms. Reproduction courtesy of the Historical Society of Palm Beach County, Palm Beach.

sent Flagler a spray of orange blossoms to show that her small town (latitude 25.8°N) had escaped. Flagler supposedly responded by agreeing immediately to extend the Florida East Coast Railway beyond its Palm Beach terminus, south to Biscayne Bay. Thus, at least in promotional folklore, the development of Miami was a response to flowers.[39] Horticulture was not easy on the Miami ridge, however. Planting on the bare coral limestone required large amounts of dynamite and swamp muck. The number of local species was small. The more difficult problems were to determine what plants—whether imported from the Everglades, the Caribbean, or elsewhere—would survive, which kinds had value, and how they could be made to flourish. Amateurs, nurserymen, and government scientists worked incrementally on these issues from the 1880s into the 1920s and beyond.

South Florida horticulture began, not on the Atlantic coast, but in the more accessible and less rocky territory around Tampa Bay. In 1881 Pliny Reasoner, the teenage son of an Illinois nurseryman, set up a center for fruit and ornamentals in Manatee (latitude 27.5°N). His main business was mass production of grafted citrus varieties, but he was also interested in ornamentals and novelties. He acquired unfamiliar types from metropolitan dealers, the Agriculture Department, and collectors. Locally, he worked with Illinois-born former navy carpenter and amateur conchologist Charles T. Simpson. In 1885 the two men collected together along the southwest Florida coast; their most important find was a group of royal palms *(Roystonea regia)* about fifteen miles southeast of present-day Naples. Simpson later commented that the sight of these hundred-foot-tall trees, familiar in Cuba but only rumored in Florida, had made him weep for joy; Reasoner dug a supply of smaller specimens, brought them back to Manatee for propagation, and renamed his business the Royal Palm Nurseries.[40]

Advancing horticulture in south Florida involved difficulties unique to the region. In late 1884 a "vigilance committee" began to harass African Americans and Northerners around Manatee. When they killed the town's Illinois-born postmaster, a Yankee posse that included both Reasoner and Simpson tracked them down and brought them to trial; some were convicted. Threats of reprisal induced Simpson to sell and move to Nebraska. Reasoner, however, who was younger, bigger, and an employer, not only held on but prospered. He developed a catalog-based mail-order business, prepared a one-hundred-page report on Florida fruits for the USDA, and coordinated the horticultural display at the Florida Subtropical Exposition in Jacksonville. In September 1888, however, he died suddenly of yellow fever. His younger brother Egbert continued the Royal Palm Nurseries, but without Pliny's degree of commitment to public service and innovation.[41]

In the decade after Reasoner's death, USDA plant scientists took over guidance of south Florida horticulture. The initial stimulus came from New York senator and Republican boss Thomas Platt. In 1891 he pushed the USDA to investigate citrus diseases prevalent in the area north of Orlando (where his wife owned an orange grove). Vegetable Pathology Division head Beverly Galloway sent staffers Walter T. Swingle and Herbert J. Webber to Eustis, a few miles east of Mrs. Platt's grove, to create the Subtropical Research Station. They rapidly expanded from the study of blights to the search for blight-resistant varieties and then for varieties that were generally of higher quality. Swingle established ties with the state horticultural society and by 1894 was proposing an experiment station in south Florida to Flagler's representatives. In 1897 the USDA leased a small parcel of land three miles south of Miami; Swingle imagined development of a major tropical botanical garden that would combine work on plant pathology, breeding, and introductions.[42]

This scheme stalled, however, in 1898. With the American takeovers of Puerto Rico, Hawaii, and the Philippines, the importance of advancing tropical horticulture in Florida was less self-evident. Swingle and Webber moved to other projects elsewhere, and the south Miami garden limped along as a test plot. Private individuals with federal connections took the initiative. One was John C. Gifford, who had been an assistant professor at Cornell under former USDA forestry chief Bernhard Fernow. He left the academic world in 1903 at age thirty-three, when the Cornell department was suddenly downsized, and moved to the artists' colony of Coconut Grove, seven miles south of Miami, where he combined real estate development with gardening experiments. A second innovator was Pliny Reasoner's former collector, Charles T. Simpson. He had been rescued from Midwestern rural isolation in 1889 by an offer to become an assistant in the Smithsonian Institution's Department of Mollusks. He retired in 1903 and purchased a nine-acre bay-front property three miles north of Miami that he began to develop as a garden.[43]

Gardenizing south Florida involved major changes in the land and its vegetation. Simpson emphasized how geologically new the area was and how limited the repertoire of plants; much of the land, he explained, was "a waste of weeds and sawgrass." Gifford initially planted arid-climate species such as agaves.[44] Gradually, however, the men learned to manage soil and water, and they experimented with the "plant immigrant" seeds available from the nearby USDA garden. Gifford had an enduring impact when he introduced the northern Australian cajeput tree *(Melaleuca quinquenervia)* in the hope that it would both supply the region with lumber and would dry up swampy areas, thereby lowering the risks of malaria and yellow fever.[45]

Simpson became Miami's major garden guide (see Figure 8.4). His *Ornamental Gardening in Florida* (1916) sketched the importance of building up soil with manure, compost, and legumes, the need to poison the herbivorous local land crabs, and the range of plants that could be grown. Simpson emphasized species diversity rather than design coherence. He maximized this quality by creating "condensed" representations of different kinds of landscapes. An excavated pool mimicking a sinkhole would support a large number of ferns; a dead tree in a low, shaded spot could be used as a base for dozens of orchids and other epiphytes.[46] These informal garden rooms were connected by junglelike paths. Simpson's garden consisted primarily of "native plants," but this category included any species that Simpson found growing in south Florida on its own: local sea grapes and mangroves, palms and cacti from the Keys, orchids from the Everglades, Jerusalem thorn from west Texas, West Indian sea hibiscus, Central American yellow elder, pantropical coconut palms, and Australian casuerinas. He supplemented these species with confessedly exotic eucalypts. South Florida homes,

Figure 8.4 Charles T. Simpson in his garden in Lemon City, on the north side of Miami, in the 1920s. Negative X-287-6. Reproduction courtesy of the Historical Museum of Southern Florida, Miami.

he predicted, would soon be "shaded and embowered with the glorious vegetation of the equatorial regions."[47]

Around 1920 David Fairchild took the lead in these initiatives. In 1912 he visited Miami and described it as "one big botanic garden" where "plant amateurs are sure to develop." Four years later his wife, Marion, drew on her prospective Bell family inheritance to buy a ten-acre waterfront parcel in Coconut Grove as a winter home. Fairchild planted much of the property with fruit trees, but he surrounded the house, which he gradually transformed from a typical frame structure into a rambling series of Asian-Deco pavilions, with palms, cycads, philodendrons, and other tropical ornamentals. He gave the property an exotic Indonesian name, "The Kampong."[48] Fairchild's more public project was to establish permanent plant introduction facilities near Miami. In 1921 he organized a campaign to turn part of a surplus Army base five miles south of Coconut Grove into a USDA tropical experiment station. Six years later he began to advocate for the creation of an arboretum open to the public. The goals of what came to be called Fairchild Tropical Garden were to display and to breed ornamentals that would flourish in south Florida.[49]

One element within this horticultural movement was the preservation of natural areas. The paradigmatic local site was the hammock thirty miles southwest of Miami that longtime Coconut Grove resident Kirk Munroe, a prominent writer of boys' adventure books, had named Paradise Key. This area of raised ground just inside the Everglades contained a diverse flora including the region's only wild royal palms. In 1905 Munroe's wife, Mary, teamed with John Gifford's wife, Edith, to induce the Florida Federation of Women's Clubs to call for state preservation of the hammock. Nine years later club leaders, together with Henry Flagler's widow as a donor, persuaded the legislature to create Royal Palm State Park.[50]

Preservation of vegetation closer to the city was more problematic. Most Miami pioneers made their livings directly or indirectly from real estate and therefore considered the loss of established plant communities the price of progress. Only a few had both the sensibility and the wherewithal to choose privacy and scenery over profit. Charles Deering, chairman of the board of International Harvester (the company that had enabled the plowing up of the prairie), was deeply interested in history, beauty, and nature. His two-hundred-acre waterfront estate, Buena Vista, had a panoramic view of Biscayne Bay looking out toward north Miami's barrier islands. The property included a Mediterranean mansion and formal garden with palm alleés, but also mangrove-lined canals, alligator pools, a monkey colony, and an aviary stocked with exotic tropical species that Deering hoped to naturalize. He consulted Simpson, who lived nearby, on plants and planting, and in return subsidized publication of Simpson's book; he was enthusiastic about collecting and distributing plants, especially cacti, from Florida, other parts of North America, and beyond.[51]

Deering despaired in the years after 1913, however, as he saw his beautiful view being destroyed by former auto dealer and highway promoter Carl Fisher, who was transforming Miami's barrier strip into the vulgar resort of Miami Beach. He decided to retreat southward, to the land that Henry Perrine had claimed seventy-five years earlier. Cutler Hammock, a few miles south of the property that would become Fairchild Tropical Garden, contained old trees and a mangrove swamp, was close enough to deep water for a yacht basin, and was protected from encroachment because it looked out only on tiny Chicken Key (which Deering purchased), and was separated from the railroad and highway by a slough. The major problem was that the site had been subdivided and platted with roads, and included a public dock. Deering bargained with multiple sellers over ten years, persuaded the county to demap the streets, and built a new dock and access road on the south edge of his property. He preserved his privacy by building a seven-foot-high concrete wall between his estate and the road.[52]

Deering moved palm trees south from Buena Vista to frame his new view, but he left his monkeys, Muscovy ducks, and exotic shrubs behind. He decided instead to create a naturalistic representation of the plants native to south Florida. For guidance on how broad a palette of species would be authentic, and which associations were naturally harmonious, he turned to John Kunkel Small. Small, a botanist at the New York Botanical Garden, had been collecting in Florida since 1901 with particular attention to the "isolated and virginal" plant associations found in places like Paradise Key. He sought to discover the few remaining unknown species in Florida and, on a more workaday level, to determine the ranges that Caribbean species occupied. He could thus assure Deering that the royal palm, mahogany, and many more plants that were common in Cuba or the Bahamas but were rare in Florida, were in fact Florida natives and not introductions.[53] In 1916 Deering outfitted Small, William T. Simpson, and two local plantsmen for the first of a number of cruises through the Everglades and the Keys to collect trees, cacti, orchids, bromeliads, and other garden material; Small also brought back photographs of notable wild plant communities. For the next decade, he provided private guidance to Deering and his gardeners on issues that included methods of fire protection, replacement of the old public road with what was called "the new hammock," and creation of a naturalistic group of royal palms on Chicken Key (Figure 8.5 shows an example of Small's work). By the early 1920s Deering's estate was by far the most significant native plant garden in south Florida.[54]

It was a private and transitory pleasure, however. Deering donated his El Grecos to the Art Institute of Chicago, but he did not take the testamentary steps that would give his garden institutional permanence. The estate passed to his heirs on his death in early 1927. (They sold it to Miami-Dade County in 1985, and it is now a park.) The estate's enduring legacy passed through John K. Small. In 1904 Small had commented blandly that the settlement of south Florida would soon eradicate many of the native species. Deering provided Small the opportunity to tour the region repeatedly and to witness the effects of fire, drainage projects, and roads on vegetationally notable locales. Deering also demonstrated that there were alternatives to the acceptance of development and eradication.

Figure 8.5 In 1921 John K. Small photographed this naturalistically pruned and planted vista on Charles Deering's estate at Cutler. Reproduction from the John K. Small Collection (SM 1546), Florida State Archives, Tallahassee.

In 1929 Small published *From Eden to Sahara: Florida's Tragedy*, which overlaid tropical travel writing with a jeremiad warning that "the future of North America's most prolific paradise seems to spell DESERT." Not only were the animals and plants being wiped out, but the soil was "being drained and burned and reburned until nothing but inert mineral matter is left." Before-and-after photographs showed lands "vandalized" and "ravished" by fire, tree cutting, excavation, and road building. Small's conclusion was that "steps for protection of selected areas should be taken at once by the state and federal governments." In a separate article, he joined the call, spearheaded by Coconut Grove landscape architect Ernest Coe, for creation of a "Tropical Everglades National Park."[55]

The idea of an Everglades park became a realistic possibility after the hurricanes of 1926 and 1928, the collapse of the Miami real estate bubble, and the Depression combined to halt south Florida development. As in the West, park promoters fought hunting, agricultural, and mining (in this case, petroleum) interests who did not want to be permanently excluded from the proposed area. In one respect, however, making the Everglades a park involved a problem not faced in the West. National parks had been created primarily to preserve scenery—mountains, geysers, canyons, and glaciers. The Everglades, by contrast, was flat and featureless, and its water views did not compare with those from Miami Beach or the railroad to Key West. The congressional committee that reviewed the park proposal in late 1930 wanted to know, "from the yardstick . . . of a showman," what "the spectacular" was in south Florida. Witnesses including Small, Coe, National Park Service director Horace Albright, and nurseryman Harlan Kelsey all responded by emphasizing the "strangeness" of the mangrove forests and the hammocks with their palms, ferns, and orchids. For Park Service consultant H. C. Bumpus, the clearest argument for the value of the Everglades lay in the amount of money being spent to imitate it. As a former director of the American Museum of Natural History, his examples were indoor ones. Small, however, could apply the new yardstick fashioned by Charles Deering in explaining that the Everglades should be appreciated as "a naturally made botanic garden."[56]

Totalitarian Power over Pests

Florida horticulturists focused on the new, the tasty, and the beautiful, whether found overseas or in the Florida wild. These encounters took place primarily during the citrus-ripening months from November into April, when nature was reasonably well behaved. Maintaining these experiences, however, depended on a year-round support system. During the hot and humid half of the year, from May through October, engineers and laborers worked to keep the wrong aspects of nature from seizing the upper hand. Hurricanes were not a significant problem during the development years from 1885 to 1925. The main challenges involved pest problems more varied and severe than in the rest of the United States. In Florida predators and parasites were not challenged by freezing temperatures. Remote wetlands provided secure refuges. Ecological niches that had opened during the dramatic transformations of Florida during the Holocene—from sea level rise to resort construction—offered new arrivals greater chances for spreading explosively. Florida natives from both the Cotton Belt and the Caribbean experienced weeds, insects, crabs, and reptiles as ordinary irritants. Northerners, by contrast, sought to control invading threats to development. Florida became the showcase for the national government's ability to defend America.

Water Hyacinths

The Anglo-American struggle to penetrate Florida in the nineteenth century was, in large part, a battle against pests. Early army campaigns were resisted as much by saw grass, saw palmettos, and mosquitoes as by Seminoles. While the St. Johns River provided the best pathway into eastern Florida, its upper reaches were blocked by weeds—both snags and trees growing horizontally from the banks. By the 1870s the Army Corps of Engineers made the river navigable as far as Lake Monroe; with steady water levels and no competing railroads, it remained the main artery for tourists, fishermen, farmers, ranchers, and lumbermen.

In the early 1890s, some sections of the St. Johns began to fill with floating plants soon identified as the water hyacinth *(Eichhornia crassipes)*. In early discussions the spread of this decorative species, with its shiny leaves, curious air sacs, and pastel blue flowers, was attributed to particular ornamental gardeners who had obtained it either from the New York hothouse trade or from a display at the 1884 Cotton States Exposition in New Orleans.[57] Such a view reflected a limited perspective, however. The Royal Palm Nurseries offered the water hyacinth as a regular catalog item as early as 1888. More significantly, the species had spread widely from northern South America during the previous century; hurricanes, birds, or travelers would have brought it to Florida within a short time after its discovery in the St. Johns. The knowledgeable John K. Small in fact believed that it was an indigenous plant that had long been growing unnoticed in the "wilderness of the interior of peninsular Florida." The water hyacinth was thus a biogeographically different phenomenon from, for example, the gypsy moth or the melaleuca (see Chapter 9), where a species moved across a major ecological barrier due to the actions of a single person.[58]

The water hyacinth became a public problem within a very particular context. Palatka, located on the west bank of the St. Johns about sixty miles south of Jacksonville, had prospered after the Civil War as a shipping center for citrus, cotton, and cattle, and as the place where travelers transferred from ocean-going ships to the small steamboats that could manage the shallows of the upper river. In the early 1890s, however, the town suffered major setbacks. Flagler's new Florida East Coast Railway (FECR) moved citrus and tourists more rapidly and farther than did river steamers. The problem more specific to Palatka was that it had been bypassed: while the FECR took control of the town's existing rail connections by bridging the St. Johns with a hastily built trestle and drawbridge, its trains stopped at an unpopulated junction four miles to the east. The most immediate issue was that the new trestle's bracing trapped objects floating downriver. In 1893 water hyacinths began to accumulate behind the bridge. Stimulated by fertilizer and manure washed from upriver farms and ranches, as well as by Palatkan sewage, the plants approached their remarkable capacity of doubling in mass every week. They soon grew into "acres and hundreds of acres" of tightly interlaced plants. The following year water hyacinths enveloped the town's wharves, blocking sailboats and tangling propellers. Even paddle wheelers had difficulty churning through the rubbery vegetation (see Figure 8.6).

This local controversy became a federal issue because Palatkans claimed that the bridge trestle and the water hyacinths together constituted an "obstruction to navigation." This categorization placed the problem within the domain of the Army Corps of Engineers. The corps had developed dramatically during the two preceding decades as the implementer of the Rivers and Harbors Acts that annually dispersed federal dollars around the country to dredge channels, clear snags, and

Figure 8.6 Water hyacinths in the St. Johns River, Palatka, Florida, around 1895. Reproduced from Herbert J. Webber, *The Water Hyacinth and Its Relation to Navigation in Florida*, U.S. Department of Agriculture, Division of Botany, Bulletin 18, 1897. Reproduction courtesy of the Rutgers University Library.

build jetties. It also held authority over structures erected in navigable waterways.[59] For both Palatkans and the engineers, the problem initially was the FECR's cheaply built trestle, not the water hyacinths. In April 1895 the local representative of the corps reported that plants periodically blocked the town wharf, but that rebuilding the trestle would substantially resolve the problem. A year later town leaders petitioned the secretary of war to require the FECR to construct "a proper bridge with wide, open spans" so that water hyacinths could pass through, and army engineers supported them by noting that the FECR's omission of water-level braces from plans they had filed with the War Department may have violated the law.[60]

Chief of Engineers W. P. Craighill was uninterested in confronting the most powerful corporation in Florida, however. He requested a third round of reports, and the focus suddenly shifted from the bridge to the weeds. Staffers referred prominently to an article in the *New York Sun* in which steamboat owner J. E. Lucas described how water hyacinth mats were so thick that men could walk on them, and that they had to use axes and saws to free imprisoned boats.[61] This interpretation of the problem led the army engineers to enter the field of weed control. In 1899, when entomologists and nurserymen were battling over the propriety of federal involvement with pests (see Chapter 6), the army obtained a $35,000 appropriation to fight the water hyacinths.[62]

After an initial unsuccessful attempt to crush the plants, the corps turned to chemistry. Most of the arsenical insecticides used in the late nineteenth century had the unfortunate side effect that they could kill plants as well as bugs. The Louisiana-based Harvesta Chemical Compounding Company decided to exploit this liability by selling a mixture of salt, saltpeter, and arsenic as an herbicide. Army engineers reported that this proprietary product killed water hyacinths, and so they purchased a used party boat, installed a steam-powered pump, and in 1902 sprayed 2.9 million square yards of water hyacinths with more than 240,000 gallons of herbicide. They projected that the St. Johns could be kept clear if funding was increased tenfold.[63]

This program was controversial from the start. Cattlemen complained that their animals died after browsing on sprayed water hyacinths. The corps responded that the probable cause was saltpeter, and therefore shifted to a mixture of arsenic acid and sodium bicarbonate. In 1905, however, Florida Representative Robert Wyche Davis inserted an amendment to the Rivers and Harbors Bill prohibiting the corps from using chemicals toxic to animals against the water hyacinth in

Florida. In a tense exchange on the House floor, bill manager Theodore Burton forced Davis to take responsibility both for future stream blockage and for a permanently lower level of Rivers and Harbors spending in Florida.[64] A year later army engineers collaborated with a veterinarian to determine whether arsenic-covered water hyacinths were, in fact, dangerous. They reported that cattle fed with such plants died within four or five days, and that there was no way to get the animals to avoid the poison by adding what were thought to be unpalatable substances.[65] A few years later they developed a solution that fit much more comfortably into their traditions. The "hyacinth elevator" used a scoop mounted on the front of a boat to dredge plants from the water surface; a conveyor belt lifted and then dumped the plants onto the shore, where they dried out and died. Under certain circumstances, however, the corps continued to spray with arsenic until new herbicides became available in the 1940s.[66]

Citrus Canker

Oranges were more important to Florida than steamboats. In the last third of the nineteenth century citrus was one of the state's major industries. By comparison with their competitors in California, however, Florida growers were slow to organize either a system for common protection against pests, or the capacity to identify and investigate new problems. They finally took their first steps in these directions in the mid-1910s, in response to the appearance of citrus canker, but they depended heavily on Northern federal officials and scientists for leadership.

Throughout the nineteenth century, east Florida promoters assumed that their balmy and well-watered region was the natural citrus garden for the Northeast. The problem of shipping perishable fruit to the North was chronic but would ultimately be resolved. Floridians were thus shocked in the years after 1900 when, just as their transportation infrastructure was maturing, their oranges were losing out to those of Southern California—a desert area three times as far as Florida from New York markets. In 1890 Florida produced about three times as many oranges as California; twenty years later the proportions were reversed.[67]

The reasons for this change were that culture mattered more than nature, and that Florida's natural advantages were cultural impediments. Citrus growers succeeded in Southern California because the need to irrigate entailed large capital investments and intense cultivation of relatively small properties. Grove conditions were disciplined by growers' organizations, banks, and the state horticultural board. The Southern California Fruit Exchange organized picking, grading, and cross-country transport to maximize penetration of national markets. After 1906 the University of California's Citrus Experiment Station in Riverside provided a permanent cohort of scientific guides for growers.[68] In humid Florida, by contrast, oranges grew on their own or with minimal care. Trees could get through most winters in most counties, and fruit was harvested and sold by everyone from large-scale entrepreneurs to Yankee retirees and uneducated Florida "crackers." The consequences were that varieties were not standardized, quality was not regulated, and harvesting schedules were up to individual growers. Shippers did not begin to cooperate to avoid production gluts and famines, or to pressure railroads on rates and delivery times, until the 1910s. The state had neither a permanent citrus research establishment nor a grove inspection system. Florida citrus was thus, by comparison with California's, a bulk product of variable quality and fluctuating availability. Profits were lower, new ventures were chancier, and pests more likely to establish themselves unnoticed.[69]

These issues became significant when a new kind of scab appeared in Florida orange groves in 1912. What came to be called citrus canker had arrived from Japan on imported nursery stock. It spoiled the appearance of fruits and sometimes damaged enough leaves to make trees unproductive. Canker flourished in Florida because the state's climate was, in contrast to California's, similar to that of the Southeast Asian home of both the orange and the disease.[70] The state's horticultural leaders, just organized as Florida Growers' and Shippers' League, sought unsuccessfully to gain control over the problem. The local scientists they funded to study the canker misidentified the cause as a fungus. Those scientists reasonably recommended that infected trees should be burned and groves quarantined, but no one could compel compliance: two prominent growers went to court in highly publicized cases and prevented interference with their trees.[71]

Both scientific leadership and money came from Washington. In early 1915 the young Bureau of Plant Industry pathologist Clara Hasse determined that citrus canker was in fact a bacterium (now *Xanthomonas axonopodis* pv. *citri*); it was carried from tree to tree by winds and insects, but moved between groves primarily on infected fruit, stock, or workers' clothes. That year Congress appropriated $25,000 to aid states in controlling the disease. The availability of federal money induced Florida to establish a state plant board with regulatory powers. State plant commissioner Wilmon Newell, an Iowa-born entomologist, developed a long-term program funded primarily by the federal government to inspect groves and burn suspect trees. By 1919 citrus canker was under control.[72]

The Mediterranean Fruit Fly

In September 1929 Charles Marlatt, chief of the Plant Quarantine and Control Administration (PQCA), remarked to his longtime friend and sometime adversary, David Fairchild, that unfortunately he was not a "dictator" who could employ millions of men to "destroy all the fruit in Florida." Given the American system of government, "all we can do is to come as near to [those measures] as we can." These musings about totalitarian power were stimulated by his bureau's current campaign in Florida against the Mediterranean fruit fly, or medfly. During the previous decade Marlatt's staffers had struggled inconclusively in the North against such insects as the gypsy moth and the Japanese beetle, and they confronted citizens who were sometimes resistant to plant quarantine. In Florida, by contrast, peninsular geography enabled entomologists to hold defensible lines, and the human population was too small, shallow-rooted, and dependent on the federal government to get in the way.[73]

The medfly was first noticed in Florida on April 6, 1929, when State Plant Board nursery inspector J. C. Goodwin found unfamiliar maggots squirming in a grapefruit he cut open for dinner at his home in Gainesville. Since the fruit had come from a tree at the Federal Bureau of Entomology facility in Orlando, state entomologists immediately contacted their federal colleagues, and the two groups began to find other infested groves in that area. They were unable to identify the larvae, however, and so they sent samples to Washington. When Smithsonian insect taxonomist J. M. Aldrich declared them to be Mediterranean fruit flies, PQCA chief Marlatt immediately boarded a train for Florida. He met with fellow entomologist and plant commissioner Wilmon Newell to plan strategy, and the two men together explained the situation to a select group of growers, bankers, and newsmen. On April 15, the Florida Plant Board announced simultaneously the insect's arrival, an emergency state appropriation of $50,000 to fight it, and initial quarantine rules.[74]

Both money and strategy were under the control of Marlatt. He asked President Hoover and congressional leaders to make an emergency transfer of $4.25 million in unused pink bollworm funds to medfly control. This measure was rapidly approved, with east Florida representative Ruth Bryan Owen (daughter of William Jennings Bryan) explaining that the nation had a duty to fight this "landing party" of an "invasion of insects" like it would any alien enemy.[75] The Agriculture Department appointed Newell a federal agent, imported grove inspectors from Texas, and named a monitor to oversee state spending. Uniformed state militia paid by the USDA patrolled the boundaries of a twelve-county restricted zone. They searched baggage and hand luggage of train passengers, and they blocked roads, searched automobiles, and sprayed their interiors with arsenic. At least one person who tried to hide fruit under his car was jailed for sixty days (see Figure 8.7).[76] Within the quarantine zone, planting and shipping of fruits and vegetables were prohibited. In a smaller "infested area" state officials hired crews with federal money to enter properties, strip trees, spray with lead arsenate and an arsenic-molasses "bait," and uproot black-eyed peas, melon vines, tomato plants, and other summer annuals. The PQCA began research to determine what plants, especially wild species, were susceptible to medfly infestation, and which extermination techniques were most effective. The program was inspected and

Figure 8.7 In April 1929, members of the Florida National Guard, paid as laborers by the federal government, manned roadblocks, searched automobiles for fruit, and sprayed them to kill medflies. Reproduction courtesy of the Florida Department of Plant Industry Library, Gainesville.

endorsed by two blue ribbon commissions of entomologists and agricultural science administrators.[77]

The Florida Plant Board alerted the public about the dangers of the fly and urged them to cooperate. Posters displayed the insect and its damage, and urged prompt reporting of sightings. University of Florida horticulture professor H. Harold Hume roused growers by declaring that the fight against the medfly was the latest stage in "man's warfare against the living things of earth." While admitting that the insect had spread around the world, he emphasized that it had never before "met a bunch of red blooded Americans." Newell and his staff countered skeptics who asserted that the medfly had long been present in Florida, or who focused on the singularity of the circumstances of its discovery to suggest that it was an escapee from the Bureau of Entomology laboratory, or, worse, that state entomologists had released it intentionally to keep their agency from being abolished.[78]

The entomologists and their allies believed that with sufficient resources victory was possible and would have enduring consequences. In fall 1929 Marlatt suggested a 500 percent increase in federal funding for quarantine and eradication. Prominent science journalist Paul De Kruif dramatized the efforts of "the famous Doctor Marlatt" and "big Newell," and he emphasized that defeat of the medfly depended on getting more money from an indifferent Congress. At the annual meeting of the Society of Economic Entomologists in December 1929, Newell explained that the medfly campaign was their profession's chance to do something "really big," like the building of the Panama Canal or the development of radio. If they could persuade Congress to provide more money and then succeed with eradication, the next generation would wonder why their ancestors had been "such fools" as to let foreign pests invade the United States.[79]

As Florida Goes, So Goes the Nation?

The medfly disappeared from Florida in 1930. The effect of the medfly campaign on the significance of economic entomologists in national affairs was, however, precisely the opposite of what Newell predicted. The national plant quarantine system that Marlatt had laboriously constructed over the previous quarter century crumbled in the first half of the 1930s. It was undermined by events on the ground, by scientific criticism, and by national policy changes.

A high-stakes confrontation between the federal government's scientific and political leadership seemed to be taking shape in the winter of 1930. Marlatt had managed Congress successfully for two decades, and had easily induced Will R. Wood, the Indiana Republican who chaired the House Appropriations Committee, to endorse the initial emergency action against the medfly. But Wood resisted Marlatt's call for a large increase in funding only a few months after the first appropriation, and he resented the entomologists' argument that Congress would be to blame if their campaign failed. In December 1929 Wood terminated federal funds for medfly eradication, field inspections, and militia checkpoints. More dramatically, he reasserted congressional authority by providing a public platform for critics of the PQCA. Congressional hearings in Orlando in February 1930 featured witnesses ranging from H. L. Frost, a sophisticated Massachusetts entomologist turned Florida citrus grower, to self-described "Florida cracker" T. M. Arnold, who described how he threatened the State Plant Board foreman and his "niggers" with a ".38 Smith & Wesson special" to keep them from digging up his plants and spraying near his livestock. Wood and his colleagues grilled Newell and other government "bugologists" on issues of waste, fraud, and abuse.[80]

If the medfly population had exploded in spring 1930 in the wake of the federal campaign's abrupt termination, the entomologists would have been vindicated. Anticlimactically, however, no infestations occurred in 1930, and in November the quarantine was lifted. While the entomologists could claim that they had stopped the insects, the congressmen had the last words about judgment, money, and power.[81]

Two prominent academic entomologists—Glenn Herrick of Cornell and Gordon Ferris of Stanford—raised more principled objections to plant quarantine. They argued in science policy forums such as *Science, Scientific Monthly,* and *New Republic* that the Plant Quarantine Act had neither slowed the entry of foreign pest species nor kept those that had arrived from spreading; that economic entomologists had produced "hysteria" by exaggerating the dangers posed by the medfly and other pests; and that the PQCA was a bloated bureaucracy that stifled dissent, disunited the states, and isolated the United States from the rest of the world.[82]

Finally, the Depression and the New Deal shifted national capabilities and concerns away from those emphasized by quarantine advocates. In 1932 the financial crisis forced the Bureau of Plant Quarantine (successor to the PQCA) to abandon its decade-long fight against the corn borer, and to acknowledge that the lines maintained against the gypsy moth, Japanese beetle, and pink bollworm had all been overrun. At the same time, bureau pathologists realized that the quarantine on nursery stock had done nothing to keep out the great plant pest of the new decade, Dutch elm disease: that fungus had arrived, unimpeded, in timber imported from France to both Ohio and New Jersey for furniture veneer. Henry A. Wallace, who became agriculture secretary in 1933, was more interested in varietal improvement, revegetation of eroded land, social uplift, and international trade and friendship, than in crop protection, and he reorganized bureaus so that they would contribute to the department's central aims.[83]

Within this context, the Bureau of Plant Quarantine's chief, Lee Strong, announced a fundamental reexamination of the principles underlying Marlatt's Quarantine 37. He acknowledged that Marlatt's strategy to reduce the risk of pest introductions by reducing plant imports would ultimately "result in the stopping of all commerce," and he noted that even comprehensive plant inspection would not have prevented the entry of chestnut blight and other major recent pests. The new policy emphasized international cooperation and "horse sense." In 1935 narcissus bulbs, banned in 1926, were again allowed into the United States; the Massachusetts Horticultural Society celebrated the event as comparable to the end of Prohibition.[84]

Federal entomologists and their allies reasonably viewed Florida as a horticultural construction. Citrus groves formed the basis for the economic prosperity and social relations of the region stretching from Miami to St. Augustine. Defending the groves from invading insects was important enough to justify dictatorial measures. Quarantine was possible because the peninsula was effectively isolated by its seacoasts and its climatic uniqueness. The vegetable farmers and laborers who manned the area during the summer months could be controlled by a relatively small force of federal and state inspectors. The work of these officials, and the tenuous hold that the medfly had in the area, led to the disappearance of a major insect pest.

This achievement could not, however, be generalized to other parts of the United States east of the Rockies. Those states produced a multiplicity of agricultural products, and had few boundaries that were more than arbitrary political lines. War against new pests in the Northeast and Midwest would have been much more intensive and extended than the fight against the medfly in Florida. People were an additional problem: millions of farmers and gardeners planted what they wanted, and motorists resisted roadblocks. Finally, national leaders in the 1930s were more concerned with reviving commerce and with strengthening nations to the west of Germany than with minimizing risks for particular groups of farmers. The renewed availability of Dutch bulbs symbolized that, Charles Marlatt to the contrary, Americans were participants in a cosmopolitan horticulture.

CHAPTER NINE

Culturing Nature in the Twentieth Century

This chapter carries my themes to the present. I examine three developments and their interactions during the last century. The first is the trajectory of horticulture: it diffused widely throughout the country, but in the process became itself diffuse—losing coherence and ultimately influence. The second is the increasing importance of experts working to manage noxious insects, plant diseases, and weeds, and their coalescence during the last two decades around the problem of invasive species. Finally, I discuss efforts to go beyond the interdiction and eradication of undesirable organisms to the restoration of prior biotic communities. These three developments were connected in that, as horticulture proper declined, the ecologists involved with invasive species and restoration took over the horticulturists' characteristics. This occurred most obviously on the practical level, where ecologists embraced horticultural methods of plant management. The more subtle but profound changes were in levels of awareness of the historical depths of biotic disturbance in North America, of the extent of ecological alteration, and of the inevitability of management. My goal is to move beyond general arguments about nature in order to engage with the particular perspectives and initiatives of plant culturists.

From Horticulture to Gardening: Onward and Upward?

In the decades between 1900 and 1930, horticultural leaders promoted a broad vision of their enterprise. Liberty Hyde Bailey and Wilhelm Miller's four-volume *Cyclopedia of American Horticulture* (1900–1902), as well as Bailey's presidential address at the founding of the American Society for Horticultural Science in 1903, reaffirmed the long-standing vision that horticulture encompassed fruits, flowers, and trees, and also the improvement of both private gardens and public landscapes.[1] A wave of new illustrated magazines—*Country Life in America* (1901), *House and Garden* (1901), *Horticulture* (1904), and *Garden Magazine* (1905)— enticed prosperous Americans to follow horticultural paths. Catalog merchants such as W. Atlee Burpee, Stark Brothers, and Conrad-Pyle supplied seeds and plants to retail customers nationwide. During World War I the National War Garden Commission made vegetable growing a patriotic duty. The publication of *Standardized Plant Names,* coordinated by printer-reformer-rosarian J. Horace McFarland, facilitated clear communication among scientists, nurserymen, landscape gardeners, and amateurs.[2]

These possibilities seemingly expanded further with the great change in Americans' experience of landscape in the first half of the twentieth century—the gasoline-powered colonization of new suburbs. With more automobile commuters and larger plots of ground, the cultured landscape became the American norm. Suburbanites interacted regularly with lawns, vegetables and fruit trees, flowers and shrubs, and woodland or scrub.[3] On the model of McFarland's American Rose Society, they advanced "special plant" societies devoted to, among others, the dahlia, sweet pea, gladiolus, iris, and peony.[4]

In the mid-1920s an alliance of botanists, foresters, nurserymen, garden club women, park advocates, and the Washington elite publicly affirmed the national importance of horticulture by persuading Congress to create the National Arboretum. This new and autonomous enterprise within the U.S. Department of Agriculture would support scientists and plantsmen working on the full spectrum of cultured woody species from pulpwoods to rhododendrons, and would coordinate introductions, breeding work, and varietal standards. Located on high ground two miles east of the Capitol, abutting the grand parkway then being planned to connect Washington with Baltimore and the Northeast, the arboretum would be a national attraction. Visitors could stop there for an initial panoramic view of the Seat of Government and of the capital's monuments. They would experientially grasp the importance of horticulture for the improvement of their homes, cities, and the nation as a whole.[5]

These displays of unity and progress masked significant difficulties, however. Horticulture, as envisioned by men such as Bailey and McFarland, depended on cooperation among botanical garden scientists, land grant college professors, landscape professionals, commercial seed and plant dealers, and amateurs. For a variety of reasons, however, many of these alliances broke down between the 1920s and 1950s. The botanical gardens suffered from problems involving mission, location, and leadership. The land grant horticulturists, to a significant degree, "ran away with" horticulture; landscape gardeners, by contrast, ran from it by repositioning themselves as architects. Efforts in the late 1940s to reunite horticulture were futile; the response of amateurs and plantsmen was to combine under the less-ambitious identity of "gardening."

The major difficulties that botanic gardens faced in the first half of the twentieth century involved urban geography. As museums of plants, they placed great emphasis on permanency, but they were situated in places where pollution levels, population densities, and potential patrons were rapidly changing. The Arnold Arboretum was protected by its proximity to wealthy Brookline and the parks that buffered it from poorer and more industrial areas. The New York Botanical Garden, by contrast, was bordered increasingly by non-luxury apartment buildings in the 1910s. In addition, its wealthy supporters were dispersing into Westchester County and beyond. New York patrician, naturalist, and racist Madison Grant worked to counter this last trend by spearheading the construction of the Bronx River Parkway, which would enable Westchester County estate owners and residents of wealthy villages such as Scarsdale to reach the garden without experiencing traffic jams or dispiriting views of factories. This initiative, however, was not sufficient. Patronage by estate owners declined with the Depression, and the garden, as a New York City institution, was not able to draw tax revenue from suburbanites. The dilemmas associated with location were most acute at the Missouri Botanic Garden. Surrounded by St. Louis, it was heavily affected by both coal smoke and an unsympathetic city government. In 1925 the garden's trustees sold 40 percent of the 125-acre garden to private developers in order to create a 1,600-acre arboretum and tropical plant facility thirty-five miles to the west. For the next four decades, the garden's leaders vacillated between their desire to leave the city behind and their fear that such action would interfere with public support and government funding.[6]

Major gardens also suffered from lack of creative leadership in the critical decade of the 1920s. The Arnold Arboretum experienced an interregnum when Charles S. Sargent, who led the institution for a half century, died in 1927 at age eighty-six; he had groomed no successor who could participate comfortably, as he did, in both Boston society and Harvard science. The New York Botanical Garden missed a more significant opportunity. In 1919 the aging Nathaniel Britton hired the young ecologist Henry Gleason as assistant director. Gleason was scientifically prominent for advancing the argument that what many considered "natural plant communities" (notably, tallgrass prairie) were in fact historically contingent associations of species. He thus had the potential to develop a garden research program linking the evolutionary, climatic, and human history of North America with contemporary plant geography and horticultural potential. Britton, however, pressured Gleason to focus on the garden's established project of cataloging all New World plants, and by 1922, the younger man felt the need to write his superior for permission to use vacation time for ecological fieldwork. By the time Britton retired in 1929 Gleason was deeply involved in taxonomy. In succeeding decades the Missouri Botanical Garden's director, Edgar Anderson, belatedly made the integration of ecological and human history in North America central to research and outreach.[7]

The missed opportunity to develop historical biogeography at the New York Botanical Garden was particularly unfortunate because, as I explained in Chapter 6, the botanical gardens' existing public *raison d'etre* for research—finding and introducing novelties—was losing momentum and status in the 1920s. Quarantine 37 made private expeditions in search of live plants unfeasible; the more serious problem was the Federal Horticultural Board's stigmatization of garden leaders as frivolous promoters of ornamentals who were unconcerned about the risks of introducing pests. After 1930 the gardens were not the major institutions—uniting explorers, scientists, and gardeners—that were imagined a generation earlier.

The fate of the National Arboretum indicates how marginalized botanic gardens became. As a legislatively mandated institution it could not die, but development of the site was deferred, first by the Depression and then by World War II. When it finally opened in 1954, the successors of the USDA researchers who had earlier supported it were moving from offices on the National Mall to the 12,000-acre facility of the newly consolidated Agricultural Research Service in Beltsville, Maryland, about seventeen miles to the northeast. They could reach the city quickly on the new Baltimore-Washington Parkway. This scenic road transitioned to a commercial strip just before reaching the arboretum. Instead of the monumental entrance planned in the 1920s, a large berm was built to shield the arboretum's interior from traffic sights and noise. The facility's invisibility to tourists became more complete as the population of the area separating it from the Capitol transitioned from white to black. With a modest entrance on a cul-de-sac it became an enclave for scattered projects such as azalea development, and for interested Washingtonians with cars.[8]

The trajectory of land grant college horticulture was much simpler than that of the botanic gardens, but was also problematic for intergroup cooperation. The colleges, with support from state horticultural societies, had created professorships in horticulture in the last three decades of the nineteenth century. Horticulture professors affirmed common identity and established a national network through creation of the American Society for Horticultural Science in 1903. But most of these scientists emulated their colleagues in other agricultural disciplines by focusing almost completely on problems of commodity production. Horticultural science was about railcar loads of orange juice concentrate and Red Delicious apples, not endless varieties of strawberries, tomatoes, roses, or chrysanthemums. In 1926 Liberty Hyde Bailey warned the horticultural scientists that they were losing touch with much of their former subject matter and with many of their potential supporters. The hardheaded younger generation honored rather than heeded their old visionary leader, however; they preferred dealing with a few growers' associations and commodity brokers to interacting with amateurs seeking novelty and ornamentality.[9]

Landscape professionals underwent the most dramatic shift, abandoning their identification with horticulture. Downing, the field's pioneer, had been the creator of *The Horticulturist*. Both he and Olmsted called themselves landscape gardeners, and, as I mentioned in Chapter 7, they were deeply interested in plant choice issues. With the decline of American naturalism and the rise of the architecture profession, however, focus shifted to design. The professional training program that Frederick Law Olmsted Jr. established at Harvard in the early 1900s was in landscape architecture, not landscape gardening. The faculty that Liberty Hyde Bailey recruited to develop an "outdoor art" program within Cornell's agriculture college pushed for two decades to free their students from plant science requirements and to become the Department of Landscape Architecture in the School of Architecture. Their major tool for affirming their competence to change identity was their record in preparing students to win the competitions of the architect-dominated American Academy in Rome. Within such a formalist framework, horticulture was an ancillary activity at best.[10]

With the return to peace and prosperity in 1945, some leaders of commercial, scientific, and amateur horticulture recognized that fragmentation was a problem, and they tried to rectify it. They were unsuccessful. Rose importer Robert Pyle believed that disparate interests could be reunited through a "distinctly American" umbrella organization. The American Horticultural Council or, more informally, United Horticulture, was created in 1946 to coordinate existing groups and to organize national surveys, annual congresses, and a central office that would distribute newsletters, guide local groups, advocate public improvements, and coordinate with institutions in other nations (see Figure 9.1). Dozens of societies and clubs affiliated with the council, but few were sufficiently interested in its organizational charts and large plans to provide financial support. Within two years, Pyle realized that what he had thought was a "virile seed" was making little growth. Neither plugging along nor a management consultant's report solved these problems, and in 1960 the council merged with the American Horticultural Society, which was primarily a publisher of magazines and guidebooks.[11] The interests involved had been too disparate for too long.

Amid these changes, amateurs gradually shifted their identification from horticulture to gardening. The rhetoric of horticulture had emphasized activities that extended from home grounds to food markets, society meetings, public exhibitions, and parks. Gardening, by contrast, was a private activity, pursued for its own sake within circles of family and friends. It was, in the language of the early twentieth century, a "hobby." It could involve fruits and vegetables, but its primary objects were shrubs, perennial borders, and rock gardens, arranged as accents within sweeps of lawn.[12]

The transition from horticulture to gardening involved substantial social changes. The northeastern horticultural societies were male-dominated urban organizations, centered around regular, textually documented meetings held at permanent locations. By contrast, the garden clubs that boomed from the 1910s onward were female suburban networks grounded in automobility and hospitality; they left few records of their activity. They federated at the state and national levels, and their leaders could mobilize around such major causes as the National Arboretum; but like other women's clubs that relied on oral communication, they were chronically underorganized. Few of their initiatives involved long-term commitments.[13]

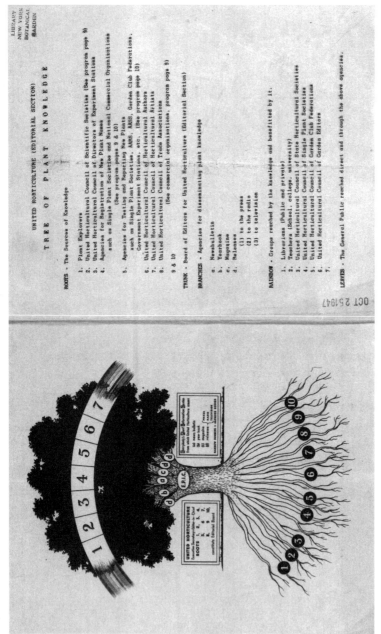

Figure 9.1 The American Horticultural Council's Tree of Plant Knowledge, 1947. The trunk, "everybody's plant information center" (EPIC), also the board of editors of United Horticulture, would draw knowledge from roots 1–10, and distribute it through branches a–d. The results would emanate, as the rainbow, to 1–6, or 7, and, alternatively, to the leaves or general public. Reproduction from the American Horticultural Council Collection, vertical file, courtesy of the LuEsther T. Mertz Library, New York Botanical Garden.

The experiential elements of gardening were more important than the social aspects, however. In the absence of united horticulture, how did gardeners acquire and use information and plants? Property owners turned to personal contacts, garden clubs, and landscapers for local knowledge and materials. Newspaper columns, magazines, and guidebooks were significant. But the most widespread sources of information and material were large commercial interests—marketers of fertilizers, pesticides, and, especially, seeds and plants. Established dealers such as Stark Brothers and Burpee's supplied new and improved fruit trees, tomatoes, and marigolds; upscale catalogs—notably, Wayside Gardens and White Flower Farm—guided buyers toward rare new roses and azaleas.

The essays that Katherine Sargent White published in the *New Yorker* between 1958 and 1970 exemplified the new lower level of culture associated with gardening. White acknowledged and validated prosperous gardeners' dirty secret: they read "the catalogs" not only for seeds and plants, but also for information, for attitudes, and for private pleasure.[14] White focused almost exclusively on flowers and a few vegetables. She ignored civic, park, and landscape gardening and, for the first eight years of her series, did not discuss the botanical gardens and societies that had long guided serious amateurs. She then incongruously introduced the New York Botanical Garden's *Wild Flowers of the United States* into her homely world of mail-order businesses as "the million-dollar book" (underwritten by Laurence Rockefeller).[15] Her collective title, "Onward and Upward in the Garden," expressed unsentimental irony about the progressivist pretensions of her Unitarian forebears and of horticultural enthusiasts more generally. Suffering from chronic and worsening health problems at the isolated Maine home she shared with her husband and gardening audience, E.B. White, she instead emphasized the annual cycle of growth, beauty, frost, and hope of renewal.

How Pests Became Invasive Species

While horticulturists were fragmenting and losing influence, pest controllers became increasingly important, if sometimes controversial, figures. In the late 1940s, entomologists, plant pathologists, and weed scientists adopted common technological approaches to control. A generation later, weeds joined insects, molds, and blights as organisms subject to border interdiction. Finally, in the 1990s, pest controllers recruited ecologists as allies under the new banner of the fight against invasive species.

From the 1890s to the 1930s, entomologists and, to a lesser extent, plant pathologists, had followed strategies that combined border interdiction with inorganic poisons such as arsenic, lead, sulfur, and copper. They had advanced professionally through their official positions as horticultural inspectors and their expertise in devising spray regimes. Weed controllers were left behind in this transformation. Plants were the focus of the Plant Quarantine Act, but only as carriers of insects and diseases—not because they themselves might be undesirable. Weeds were subject only to fitfully enforced state regulations. Tractors increased the amount of mechanical power that farmers could direct against weeds, but inorganic herbicides were too undiscriminating for general use—as demonstrated in Florida in the early 1900s by the controversy over the army engineers' poison campaign against the water hyacinth. With little of the entomologists' and phytopathologists' modern chemical science, weed controllers developed neither a journal nor a professional society; between the 1920s and 1940s, they organized themselves only on the regional level into "conferences" that emphasized bureaucratic coordination of efforts to manage the major weeds of major crops.[16]

Chemists' and plant physiologists' development of organochemical poisons in the first half of the 1940s created new possibilities for pest control and new identities for controllers. The consequences of DDT for entomologists are well known. In the 1930s, these scientists were under fire for fostering the routine use of the cumulative poisons arsenic and lead on fruit, and for promoting intrusive and ineffective quarantines. During the next decade, however, new kinds of insecticides such as DDT spread from the military to the civilian sector. These organochemicals were more effective and less toxic than arsenicals. They increased the supply of cheap, bug-free fruit, and pushed back mosquitoes and such suburban nuisances as the Japanese beetle and the gypsy moth. But the criticisms that had been raised about arsenicals—acute poisoning of workers, harm to nontarget animals, persistence in the environment, creation of resistance, and uncertainties about chronic low-level exposure—were directed from the beginning at DDT. When insecticide safety became a major public issue in the wake of Rachel Carson's *Silent Spring*, entomologists counterattacked, but they also adapted to the restrictions imposed on them, as they had in the 1930s.[17]

Organochemical herbicides appeared with greater suddenness than insecticides, and became controversial in a more politically dramatic way. Hormonal weed killers (notably 2,4-D and 2,4,5-T) were developed rapidly and secretly during World War II. USDA researcher L. W. Kephart provided the first public hints about the new chemicals, to the first North Central Weed Conference, in November 1944. A year later, the words of this "prophet" were fulfilled with a coordinated set of reports on the effectiveness of hormonal herbicides on major crop weeds. The new agents acted selectively, decomposed rapidly, and, as analogs of plant hormones, had minimal acute effects on animals. Agricultural uses ranged from ordinary weed control to the defoliation of cotton plants to facilitate mechanical picking. 2,4-D killed ragweed on roadsides, and, with a change in wetting agent, was also effective against water hyacinths. By 1950 weed controllers were celebrating the extent to which herbicides had elevated the status of their field. Supported by industry and government, they finally established their own journal and national society, and began to plan for graduate programs and departments in what came to be called "weed science." They were moving slowly toward parity with the other spray disciplines.[18]

In the late 1960s, weed scientists suddenly leaped ahead of entomologists—but in the unwanted category of controversy over dangers their chemicals posed to human health. Prior to 1966, the major complaints about hormonal herbicides involved wilted roadside wildflowers and lost animal habitat. The chemicals became a public issue when prominent scientists opposed to the Vietnam War (notably, Harvard biochemists John Edsall and Matthew Meselson, and Yale botanist Arthur Galston) argued that the military use of Agent Orange (a mixture of 2,4-D and 2,4,5-T) and other herbicides *prima facie* comprised illegal chemical warfare, and in addition that herbicidal destruction of mangroves on the Mekong Delta would damage locally important fisheries. These arguments, while well publicized, persuaded few outside the antiwar movement. In 1969, however, a systematic federal pesticide testing project indicated that 2,4,5-T caused both tumors and birth defects in laboratory animals.[19] Following an exposé in the *New Yorker* and a Senate hearing at which Galston testified, the government cut back the spraying program. In addition, the new Environmental Protection Agency put weed killers on its regulatory agenda. Toxicologists soon found that 2,4,5-T's problems came from the presence of dioxins as synthesis by-products. Vietnam veterans kept Agent Orange in the public eye for a generation, and herbicides surpassed insecticides as exemplars of the dangers that could lurk in agricultural chemicals.[20]

Another, and ultimately more important, form of parity between weeds and bugs, and their associated scientists, came out of the Agriculture Department. In the 1950s and 1960s, responsibilities for quarantines were dispersed among different divisions of the scientist-dominated Agricultural Research Service. In 1971, however, the Nixon administration's policy of separating the regulatory functions of government from those of research and development led to the consolidation of agricultural quarantine operations in a new and autonomous Animal and Plant Health Inspection Service (with the entomologically redolent acronym APHIS). Francis J. Mulhern, the service's first administrator, pushed to make the agency managerially coherent and bureaucratically secure. He realized that while his border control portfolio covered invertebrates and both animal and plant diseases, it did not include plants per se. In 1973 he thus pushed forward long-stalled proposals to establish federal control over the movement of weeds. A year later the Noxious Weed Act authorized the Agriculture Department to exclude and control plants "of foreign origin" that were "new or not widely prevalent in the United States." More than sixty years after the passage of the Plant Quarantine Act the federal government finally acquired taxonomically inclusive power over the movements of species across the nation's borders.[21]

How vigorously and comprehensively this power would be implemented was problematic. The Noxious Weed Act gave APHIS control only over species that the service specifically blacklisted. This approach gave inspectors and importers clear criteria and thereby minimized controversy; its weakness was that it was reactive and hence often too late. The opposite, so-called "whitelist" approach, combining general exclusion with named exceptions entering under permit, had been pioneered with Quarantine 37. In 1977 the Carter administration issued a broad executive order (11987) restricting "the introduction of exotic species into natural ecosystems" and mandating the secretary of the interior to coordinate implementation. But with no funding, tensions between the Interior Department and the USDA, and different perspectives within the USDA, this initiative went nowhere.[22]

Academic ecologists with environmental interests were uninterested in these mundane plant and wildlife management issues. *Conservation Biology: An Evolutionary-Ecological Perspective*, a collective manifesto conceived in the late 1970s by environmental scientist Michael Soulé, placed no emphasis on problems resulting from the introduction of species into North America. The most notable discussion of the subject around 1980 was by Florida State University ecologist Daniel Simberloff. He presented a mournful literature review that emphasized the merely anecdotal character of nearly all research on the survival and effects of introduced species, and he questioned whether meaningful scientific generalizations or predictions were possible in this area.[23] In the early 1980s, the activist international Scientific Committee on Problems of the Environment (SCOPE) encouraged ecologists in Australia, Britain, South Africa, and the United States to organize conferences that would highlight "biological invasions" as a problem. While scientists on the Commonwealth's temperate biogeographic islands sketched a major issue that required immediate action, the Americans who conferred in 1984 generated no unified perspective. Washington State University botanist Richard Mack provided a compelling history of the introduction, spread, and deleterious consequences of brome grasses in the West, and Cornell entomologist David Pimentel, a veteran of the insecticide controversies, surveyed problems of introduced pests in agriculture and forestry. But Simberloff's lead paper emphasized that the field lacked scientific credibility. *Ecology*'s review of the proceedings contrasted Pimentel's "glibness" and Simberloff's caution.[24]

While federal departments maneuvered and academic scientists looked elsewhere, local managers—notably, in south Florida—responded actively to vegetational changes.[25] Aquatic weeds were a continually growing problem. During the first half of the twentieth century, hydrological managers in south Florida extended a network of freshwater canals throughout the region. Suburban properties fronting on these new bodies of water were planted with gardens and lawns. With conditions similar to those in the St. Johns River and its tributaries at the end of the nineteenth century, many canals soon filled with water hyacinths. The plants blocked flow gates and generated stagnant conditions inimical to people and to the introduced fish that comprised a major element in the state's leisure industry. In 1946 scientists working for the Army Corps of Engineers showed that 2,4-D killed water hyacinths. Eleven years later, the corps obtained funding for a comprehensive aquatic weed control program in Florida.[26]

Suppression of the water hyacinth opened opportunities for subsurface weeds unaffected by herbicides. One of these was alligatorweed *(Alternanthera philoxeroides)*, which had reached Florida from South America around the same time as the water hyacinth, but had not heretofore caused major problems. The other was a plant that Florida botanists initially considered a southern relative of Canadian waterweed *(Elodea canadensis)*, and thus named *Elodea densa*. They belatedly learned that it was an Old World pantropical, *Hydrilla verticillata*, which had been introduced from Sri Lanka through the tropical fish trade. In 1969 Florida required aquatic plant dealers to obtain permits for new species from the state's Department of Pollution Control; a few years later, Orlando-area congressman Lou Frey Jr. was a major backer of the Noxious Weed Act. Both hydrilla and the long-pervasive water hyacinth were placed on the initial APHIS blacklist as "new or not widely prevalent" noxious plants of foreign origin.[27]

To the west and south of Miami the issues were not aquatic plants but trees and shrubs. In the first half of the twentieth century the northern Everglades had been drained, cleared, fertilized, and planted. Soil deterioration and changed water policies then led to significant farm abandonment during the next three decades. Opportunistic species occupied these tropical old fields and began to spread south toward Everglades National Park. By the late 1960s, park managers were particularly concerned about three species: Brazilian pepper *(Schinus terebinthifolius,* previously Florida holly), long grown throughout the Caribbean as a medicinal plant; Australian pine *(Casuarina equisetifolia)*, which had been carried around the world in the mid-1800s to stabilize exposed coastal areas; and John Gifford's *Melaleuca quinquenervia*. While no foreign species were desired in the national park, these stood out because they formed monostands on cleared land.[28]

Melaleuca was particularly problematic due to its fire properties. As Gifford had hoped, the trees dried up the soil around them; in addition, their exfoliating bark formed an ideal fuel. Then, when a fire came, they not only survived but also dispersed protected seeds onto surrounding uncovered, enriched soil. In the 1970s park officials, state weed managers, and scientists organized a series of workshops on melaleuca. They broadened their reach in 1984 by creating the Florida Exotic Pest Plant Council, whose purpose was to include "weeds of natural areas" in state and federal weed management systems in Florida.[29] One tactic was to have melaleuca declared a federal noxious weed. The USDA was reticent about such action because the council's open-ended definition of weeds would dilute USDA programs targeted at fields, roadways, waterways, and rangeland, and because a USDA claim over "natural areas" would encroach on the authority of the Interior Department.

The department acquiesced, however, when south Florida congressman E. Clay Shaw Jr., advised by state weed controller and Pest Plant Council leader Don C. Schmitz, threatened to block appropriations and introduced legislation to make the declaration by fiat.[30]

Foreign pests became a real federal issue in 1990. Concern arose, however, not in one of the executive departments or in an elite advisory body such as the National Academy of Sciences, but rather in the less-prominent Merchant Marine and Fisheries Committees of Congress, and then in the Office of Technology Assessment (OTA). For two decades, the OTA had struggled to balance goals of providing usable advice to Congress and facilitating informed science activism. Those two concerns were evident in the development of *Harmful Non-Indigenous Species in the United States* (1993), the report prepared for the House committee.

The OTA report was the endpoint of a sequence of events that began with the discovery of zebra mussels *(Dreissena polymorpha)* near Detroit in June 1988. These common European freshwater mollusks had reached the Great Lakes because the completion of the St. Lawrence Seaway in 1959 had made the lake system accessible for the first time to ships that carried fresh water, as ballast, from Europe to America, and because the pollution controls mandated by the 1970 Clean Water Act had enabled Lake Erie to support shellfish. The zebra mussel spread rapidly through the lakes during the next year, generating headlines when a colony clogged the unshielded water supply inlet pipe in the western Lake Erie town of Monroe, Michigan.[31]

The zebra mussel became a federal concern because it involved interstate, international, and marine transport issues. But the taxonomic scope and bureaucratic locale for research, regulation, and control were unclear. Bills introduced in 1990 through the Merchant Marine and Fisheries Committees proposed a task force representing the Departments of Commerce and the Interior, the Environmental Protection Agency, the Coast Guard, and the Corps of Engineers, to monitor and control "exotic species" or, alternatively, "non-indigenous aquatic nuisances" on the Great Lakes (specifically, the zebra mussel and the tubenose goby [*Proterorhinus marmoratus*]). Interior Department officials resisted that approach, however. They argued that the problem was broader than aquatic nuisances, and they wanted an initiative that would be more substantive (and presumably better funded) than a call for interagency cooperation.[32] Committee leaders were not in a position to legislate more broadly, but did not want to lose the issue of nonindigenous nuisances to other committees and interests. They therefore referred it to the OTA. Phyllis Windle, an experienced staffer trained in ecology, was named project director, and in 1993 the OTA published its substantial report on "harmful non-indigenous species."[33]

The explicit goal of the OTA report was to help legislators. It summarized the state of knowledge about the "pathways and consequences" of introductions of mollusks, plants, fishes, plant pathogens, insects, and vertebrates; it assessed the risks, costs, and benefits associated with different control policies; and it sketched jurisdictional issues and organizational problems in the Interior Department, in APHIS, and in the rest of the USDA. Finally, the report presented legislators with a series of legislative and regulatory fixes that they could consider. The importance of these "policy options" within the congressional setting was evident in the inconclusive hearing that reviewed the report. Administrators of the different departments, engaged in the endless labor of bureaucratic turf protection, focused entirely on these action items. They accepted small recommendations, but defended their agencies' core missions and touted their major accomplishments.[34]

The real impact of this project, as OTA staffers understood, could lie more in its preparation than in its recommendations. Windle identified about two hundred people in universities, federal and state agencies, environmental organizations, and trade groups who were involved with harmful nonindigenous species, and put them into working contact with each other. They created a national advisory panel and organized regional stakeholder meetings that raised consciousness and facilitated networking. They provided research seed money in the form of contracts for specialized draft reports. Interested individuals saw a new front for scientific activism and came to the fore. University of Wisconsin ornithologist Stanley Temple, for example, had complained broadly about inaction regarding exotics a few years earlier with little effect; he coauthored the contract report on vertebrates.[35]

The most remarkable trajectory was that of Florida weed controller Don Schmitz. After organizing a helicopter visit by the OTA staff (and ABC News) to melaleuca-rich sites in the Everglades in 1991, he was made a member of the project advisory panel, and he participated in the presentation of the final report to President Clinton's science advisor. Directed by his superiors to prepare an analogous report on peninsular problems, he recruited Daniel Simberloff, the state's most eminent ecologist, with a passionate argument that personal accomplishments such as "awards, book chapters, and plaques will likely end up in a land fill one day," but that preserving and maintaining the environment would be a lasting legacy. Over the next four years the self-described "pragmatic hack state biologist" with a master's degree from the University of Central Florida collaborated with the Harvard-trained Distinguished Professor of Biological Sciences on a major collection of papers and a forceful polemic in the National Academy of Science's *Issues in Science and Technology*. Schmitz also worked with Windle and ecologist James Carleton on petitions to Vice President Gore calling for the creation of a permanent federal commission that would coordinate "the war against invasive exotic species."[36]

The collaboration between Schmitz and Simberloff was emblematic of a broader reconciliation between government pest controllers and environmentalist academic ecologists, a generation after *Silent Spring* and Agent Orange. Pest controllers identified the targets of opportunity—feral cats, zebra mussels, purple loosestrife, kudzu, salt cedar, or melaleuca—and did the work of stripping soil and of poisoning, chopping, and burning pests. Ecologists framed the fight against exotics within the capacious context of biodiversity maintenance, and they defended active methods against animal protectors, pesticide skeptics, and tree huggers. The pest controllers' long-standing simple public message—that undesirable organisms should be understood through the model of plague—came to the fore.[37]

At the end of the 1990s, newspapers, magazines, and books delivered the message that the United States was threatened by alien invasion and that the nation needed to respond vigorously—through increased public awareness, coordinated control measures, eradication campaigns, and improved border protection.[38] As the OTA panel emphasized, the choice was between "life out of bounds" and "life in balance." During this decade, some ecologists questioned reliance on the concept of balance, and emphasized the normality of chaotic and sometimes dramatic fluctuations. Philosophers and skeptical ecologists began to note conceptual weaknesses in invasion biology.[39] These developments, however, were insignificant. The problem of harmful nonindigenous species was not one of scientific theory but rather of horticultural practice. As the OTA staff noted in its conclusion, the world had become a garden to be managed by people willing to undertake responsible action.[40]

The resolution of the terminology issue confirmed this perspective. During the previous decades, the categories of concern—"pest," "nuisance," "weed," "noxious," "invasive," "exotic," and "alien"—varied by taxonomic jurisdiction and expressed different managerial values. In the 1990s, various general terms were introduced to the public. "Alien species" was easily comprehended, and its association with space monsters, while frivolous, was attention-getting and conceptually relevant. Careful activists, however, were dissatisfied with the term's connotations: on the one hand, it caused difficulties among crop promoters because it made foreignness *per se* the issue; on the other hand, it reinforced linguistic links between human immigrants and noxious pests.[41] "Exotic" was problematic because it was associated either with positive aspects of the rare, strange, and beautiful or, more vulgarly, with burlesque. "Introduced," favored by Simberloff, was too broad and its connotations were too positive. Windle's "harmful non-indigenous" was too bureaucratic. In early 1999 the Clinton administration passed over the "invasive exotic species" used by its ecological advisors in favor of the simpler "invasive species." Executive Order 13112 diffused this language through the creation of the Invasive Species Council and a series of other initiatives to coordinate federal responses to these organisms. The term combined the idea of "biological invasions"—established in biogeography largely in the contexts of Australia, New Zealand, and isolated islands—with Americans' long history of battles against aggressive garden plants such as mint and crabgrass. It reached back through the boll weevil and the Rocky Mountain locust to the Hessian fly.[42]

The Problem of the Prairie, Revised

Beyond efforts to exclude harmful nonindigenous species and to suppress those that were establishing themselves, one more step was possible—vegetational restoration. Conservationists in the twentieth century wrestled with the problems associated with lands that people and their domesticated animals and plants had used and, to varying degrees, used up. Forests were cut and abandoned. Ranchers put livestock on private and public grasslands, leaving when vegetation was consumed, inedible plants came to predominate, or grazing became unprofitable. The most notable tracts were farms, where preexisting plant populations were replaced for decades or centuries by nonlocal crop species and their companion weeds, until changes in soil quality, climate, or markets made the cycle of plowing, planting, weeding, and harvesting unprofitable. Euro-Americans never abandoned land completely. But when effective control shifted from economically self-interested individuals to philanthropists or public bodies, new questions about plant management arose.[43]

In the eastern woodland, vegetational restoration was more of a phenomenon than a problem. Forest promoters in the late nineteenth century had advocated replanting, using European larch, Norway spruce, Scotch pine, or Eastern white pine. By the 1920s, however, scientists recognized that forests grew back quickly, albeit with changed species composition, through complex successional sequences. Both Yale silviculturist James Toumey and Harvard Forest director Richard T. Fisher celebrated the "resiliency of the New England forest" and embraced "ecological forestry."[44] With labor expensive and profits slim, forest management involved minimal changes in species composition. In the 1940s, and again beginning in the 1990s, Harvard Forest scientists emphasized that vegetational composition was a contingent consequence of historical events, but that what resulted was true forest.[45]

In large parts of the South, by contrast, replanting was advanced as an emergency response to a widespread crisis. The central concern of the USDA's Soil Conservation Service in the 1930s was to stop erosion; its leaders were utilitarian about plant choice and were enthusiastic about novelties. Operating from the Carolina Piedmont to the Oklahoma Dustbowl, the service distributed "native" Russian olives, grasses imported from Turkestan, American locusts, and, in the Southeast, kudzu.[46] This East Asian legume could be established on sound soil above the gullies running through poorly maintained upland, and would then creep down, root in the sterile subsoil, and trap runoff silt. Nitrifying the soil and decaying into its own compost, it would gradually restore the land to fertility and eventually to its original level (see Figure 9.2). Kudzu could be controlled in fields by livestock, and on roads and other rights of way by convict or low-wage labor. In the third quarter of the twentieth century, however, both free-ranging hogs and chain gangs became less prominent on the Southern landscape. As a consequence, kudzu spread, and by the 1970s had become the widely hated "vine that ate the South." But the issues associated with kudzu were not new ones. Like Johnson grass, the USDA's earlier hope for renovating the South, kudzu was a way to use the self-propagating capabilities of vegetation to brake a long-term environmental decline resulting from neglect. Frederick Law Olmsted, Liberty Hyde Bailey, and other Yankee horticulturists would have considered Southerners' labors to keep kudzu under control salutary indicators of the slowly rising level of the region's culture.[47]

The truly difficult issues associated with vegetational restoration can best be seen by turning once again to the tallgrass prairie. This landscape was as fundamental to the national economy and to American identity as the eastern woodland. But prairie was harder to define than

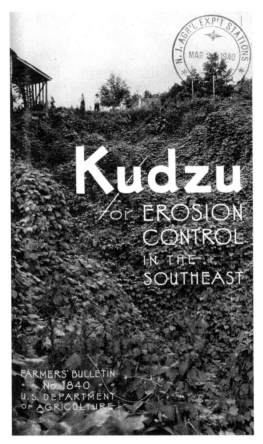

Figure 9.2 The title page of this 1939 U.S. Department of Agriculture farmers' bulletin offered a dramatic angle on the ability of kudzu to fill gullies. Reproduction courtesy of the Rutgers University Library.

forest. Forests consisted of trees, and when the trees were cut and the land plowed and planted, the forest was gone. Cutting plants on a prairie, or even plowing it and growing wheat, was not the same kind of change. For farmers, prairie was flat, treeless land that was sufficiently watered and deep-soiled for normal cultivation; the change from wild grasses and forbs to domesticated grass monocultures—whether tropical American corn or western Asian wheat—was not a major one. Ranchers had a different conception: they minimized the importance of flatness and soil depth, and instead emphasized land that was covered with palatable perennial grasses and forbs (thus including bluegrass and clover, but excluding noxious or poisonous weeds, regardless of origin). A third definition was adopted by the Wisconsin scientists discussed in Chapter 7; for them, prairie was an association of numerous nonwoody plant species that contained no foreign-origin species and relatively few noxious weeds.

Farmers, range managers, and ecologists learned that while most forests regenerated themselves, maintaining prairies involved choice, thought, and labor. In spite of their previous wide extent, North American tallgrass prairies were not a stable state in the twentieth century. Without continued intervention, they lost ground to woody plants. This new problem of the prairie was the reverse of that posed in the nineteenth century. Rather than asking why prairie had no trees, and whether they could be grown, prairie managers sought to keep trees out, and resented their seemingly relentless advance. The new problem of the prairie resulted in substantial rethinking of the relations between culture and nature.

From the Wisconsin Arboretum to Chicago Wilderness

As I noted in Chapter 7, prairie restoration arose within botanical and conservation circles at the University of Wisconsin in the 1930s. For Aldo Leopold and his associates, the central task was to re-create the impressions emphasized by early Wisconsin settlers through the use of species selected from old plant lists. By the end of the decade, Theodore Sperry and the Civilian Conservation Corps (CCC) had produced a visually distinctive landscape consisting almost completely of native nonwoody plants. It was a momentary flowering, however. In 1939 funding for the CCC camp ended. Horticultural director William Longenecker hoped to continue the development of the ornamental plantings, but the scientists did not want to devote the arboretum's now-minuscule funding to what they considered public park projects. They proposed to cut maintenance, limit public access, and utilize the area for ecological investigations, primarily of animals. Within a few years, the arboretum's laboriously constructed prairie plant community was invisible to any but the most discerning. Other species predominated—not the agricultural weeds that Leopold had emphasized in the mid-1930s, but, on the one hand, Kentucky bluegrass that had persisted without much notice during the transition to prairie, and, on the other, shrubs and trees such as elderberry, dogwood, and oak, which birds and squirrels were introducing from neighboring woods.[48]

In the early 1940s the orchid-culturist-turned-ecologist William B. Curtis began to think about ways to restore the arboretum's prairie. He wanted first to make the land passable, and then to suppress bluegrass and woody plants without destroying Sperry's laboriously planted perennials. Unlike Sperry, he could not rely on gangs of young men to chop, weed, and haul, nor could he depend on gasoline to run mowers because of the war. A decade earlier, Kansas State College agronomist A. E. Aldous had demonstrated the value of using fire to clear prairie rangeland, but had warned that burning in early spring eliminated bluegrass. Curtis understood that such a regime of plant management could alleviate many of the arboretum prairie's problems. Incinerating dried canes would make walking possible. Fire would kill off much of the bluegrass, and it could also break up accumulated thatch sufficiently for summer-growing prairie forbs to emerge. It would keep down elderberry bushes and dogwoods, and slow the spread of oaks without killing the scattered specimen trees. Burning was not risk free: the heavy smoker Leopold died of a heart attack in 1948 while firing the property he celebrated in *Sand County Almanac*. But burning was so successful that after the war it became both a regular part of arboretum prairie management and itself a subject of scientific analysis (see Figure 9.3).[49]

A few other Midwestern institutions (most visibly the Morton Arboretum, established thirty miles west of Chicago by salt magnate Joy

Figure 9.3 In the late 1940s, staffers and students equipped with torches, beaters, and sprayers began regular controlled burns to maintain the University of Wisconsin Arboretum prairie. Reproduction courtesy of the University of Wisconsin-Madison Arboretum Archives, Madison.

Morton, son of the Arbor Day founder) planted prairies over the next two decades. In 1968 Knox College in Galesburg, Illinois, hosted a symposium on prairie and prairie restoration, framed both as a response to the impending environmental apocalypse of overpopulation, overconsumption, and pollution, and as a "Happening" with practical advice on planting.[50] Prairies became emblematic regional environmentalist projects: in the 1970s, the Fermi National Accelerator Laboratory, a few miles west of the Morton Arboretum, replanted parts of the bluegrass pasture (and bison preserve) above its buried atom smasher with a prairie mix.[51]

The most consequential initiative was Nature Conservancy activist Stephen Packard's campaign, beginning in the late 1970s, to establish "oak savanna" (grasses and forbs with scattered burr oaks) in Cook County's less-developed parks. These "Chicago Forest Preserves" had been managed for the previous three generations to encourage tree growth. Packard, however, organized grassroots volunteer groups to remove trees and shrubs (including Midwestern white ash and black locust, and introduced common buckthorn [*Rhamnus cathartica*]), to burn fields, and to plant mixes of grasses and forbs based on Wisconsin precedents and on mid-nineteenth-century plant lists. Then, in 1994, the conservancy mobilized "a virtual who's who of conservation and planning groups" to advance the "Chicago biodiversity initiative," or "Chicago Wilderness," which would gradually transform large sections of Cook County parks from woodland to savanna.[52] Some ecologists were skeptical that the Chicago suburbs had been oak savanna prior to white settlement, and they did not think such a plant assemblage was stable over time in any case. The real opposition came, however, from people living near parkland. They preferred trees and shrubs to grass, did not want nesting birds and other suburban wildlife incinerated, were concerned about the dangers posed by grass fires, smoke, and herbicides, and were generally angry at the Nature Conservancy's top-down approach. After months of debate in 1996 about nature, the native, and restoration, the county halted the project.[53]

The Long View from Konza Prairie
These efforts to establish particular mixes of species on long-cultivated land in what was considered the boundary region between prairie and forest were visionary and exhilarating. But the more important scientific and environmental questions concerned the restoration, not of species,

but of function—of community stability or ecosystem self-regulation—on unplowed "true prairie" far from eastern woodland influences. The center for these initiatives was in the Flint Hills of Kansas. They culminated in the 1970s in the development of the Konza Prairie Research Natural Area.

In the last quarter of the nineteenth century, syndicates organized in Chicago, New York, and Great Britain acquired large expanses in the sixty-mile-wide zone of grass-covered hills running the length of Kansas between the 96th and 97th meridians. They pushed out the scattered small farmers, fenced the unplowable upland with the newly invented barbed wire, and established a system for receiving cattle from Texas and the Southwest in the spring, fattening them on the land's mix of bluestem grasses, Kentucky bluegrass, and palatable forbs, and then shipping them as demanded to stockyards in Kansas City. Flint Hills ranchers worked to maintain the long-term "health" of their grasslands—to keep up yields, clear out old straw, keep down poisonous and noxious weeds, and suppress woody plants. Their tools were, on the one hand, regulating the season and time of grazing and the number of cattle per acre, and, on the other, hiring men to cut, pull, burn, and (after 1945) spray undesirable species.[54]

Scientists at Kansas State College, located in Manhattan in the northwest corner of the Flint Hills, sought to make this work more effective and economical. In the 1930s A. E. Aldous, mentioned above, advised ranchers about the best seasons for mowing nuisance species such as coralberry, sumac, ironweed, and goldenrod, and for grubbing toxic plants such as larkspur, locoweed, and death camas. He also reported on extensive experiments confirming the view of "old-time" stockmen that regular burning, especially in late spring, would remove straw and keep down the proportion of weeds.[55]

In the 1950s the National Park Service sketched a different vision for part of this area. Park service director Conrad Wirth, seeking to expand his system beyond the Mountain West, drew on Interior Department proposals developed in the 1930s for the Great Plains to outline a 57,000-acre fenced Tallgrass Prairie National Park about ten miles northeast of Manhattan. A series of studies and brochures celebrated the "highly refined" and stable "natural type" of vegetation that made up true prairie, and proposed to stock the park with native grazers such as bison, elk, and pronghorn. A circuit road and horse and hiking trails would enable the public to see a remnant of prairie as it had been experienced by Indians and the first white settlers. More remote areas would provide scientists opportunities for research.[56]

Regional leaders (including news outlets controlled by President Eisenhower's Nebraskan interior secretary, Fred Seaton) pushed the park proposal vigorously in the late 1950s and early 1960s, but local ranchers opposed it. Their primary concerns were personal and economic, but their arguments extended to the agroecological—to the overlap among the categories of range, grassland, and prairie. At a Senate hearing held in Manhattan in August 1963, ranchers testified with pride that they had, with support from Kansas State agronomists, kept the bluestem grasslands beautiful and in good condition for nearly a century. They were skeptical about reintroducing wild animals. On the one hand, "mangey" bison and elk were a less-impressive sight than a herd of fine cattle; on the other, controlling numbers and grazing schedules would be extremely difficult. An unmanaged population would increase and forage so intensively that the prairie would be filled with bare spots, range weeds, and nonlocal woody species like Osage orange. Too few grazers would be worse: the result would be a buildup of straw, and then, with bored park visitors flicking cigarettes from their cars, fire. Ranchers recounted stories from the old days about catastrophic prairie fires, and argued that families on the proposed scenic road (not to mention hikers on back trails) would be trapped if thousands of acres suddenly went up in flames. This argument was taken seriously by western senators skeptical about the park service's plans, and the project was shelved.[57]

The tension between range and park formed the background for what became Konza Prairie. Lloyd Hulbert was a Michigan Quaker and Cornell graduate student who, as a conscientious objector during World War II, served in Montana as a Forest Service smoke jumper and range manager. His dissertation research at Washington State College concerned ways to accelerate succession on rangelands from introduced bromes such as cheatgrass to indigenous perennial species. He joined the Biology Department at Kansas State College in 1955, and two years later, led a faculty committee that proposed development of a natural prairie area of about 3,500 acres for field studies. The committee's understanding of "natural" was conflicted: on the one hand, the prairie would be an "ecological benchmark" that would not include "human activities, fire, and grazing"; on the other, since "grasslands are unnatural if all fire and grazing are excluded," the group proposed to introduce bison and pronghorns and to explore how much intentional burning would be necessary to keep the land "in a condition we think of as natural."[58]

Kansas State administrators were unenthusiastic about any large and visible venture that was not clearly part of their agricultural mission, and declined to support Hulbert's proposal. Four years later they made certain he did not use his position as a Kansas State professor to testify on behalf of the Tallgrass Prairie National Park.[59] Undaunted, Hulbert revived the idea of a "prairie research area" in 1968. He inventoried all the land within a twenty-five-mile radius of Manhattan and identified a 5,000-acre tract that the university could acquire. He suggested that permanent support for the project would come from the federal government under the United Nations Educational, Scientific and Cultural Organization's (UNESCO) new International Biological Program. Once again, however, university leaders did not act.[60]

In the face of this local disinterest Hulbert, like proponents of ranching a century earlier, turned north and east for capital. He had been an active member of the Nature Conservancy from the late 1950s, but had received in return only a few contacts and many exhortations to increase local membership. In the late 1960s, however, wealthy New York birder and conservancy leader Richard Pough persuaded the Minnesota-born 3M Company heir and art collector Katherine Ordway to fund an initiative to collect prairies—small fields first, but eventually, Ordway imagined, tracts large enough that "we can go out on and not see *anything* else—no houses, no power lines, *anything*." The conservancy thus became interested in Hulbert's expertise in both prairie ecology and the Flint Hills land market. They recognized that partnership with Kansas State would make removal of land from the tax rolls more acceptable locally, and could help secure a preserve's long-term future. They did not make a commitment on the scale that Hulbert had proposed in 1968, however. In late 1971 the conservancy purchased a nine-hundred-acre strip fronting on Interstate-70, about eight miles south of Manhattan. This land was then donated to the Kansas State Endowment on the condition that it be maintained forever as the Konza Prairie Research Natural Area.[61]

Establishing the "pre-white-man bluestem prairie ecosystem" on this property visible to motorists passing each day on the Interstate required judgments that would reconcile parkland and rangeland values. Both the tract's small size and lack of money meant that bison and pronghorns could not be imported to maintain the vegetation and to form a subject for research. Instead Hulbert focused on fire, as both a phenomenon to be studied and as a maintenance tool that was economical and consistent with the Nature Conservancy's rules. He attributed pre-white-man fire to both Indians and lightning, but focused his attention entirely on the latter. And while he acknowledged that most lightning fires occurred in late summer, he developed plans to burn in the late spring. Hulbert's predecessor A. E. Aldous had demonstrated that late spring burns were optimal for getting Flint Hills prairie to green up well. Hulbert imagined comparisons of annual, biennial, and longer-interval controlled burning to determine what protocols would maintain high-quality grassland. Species composition and vegetational production on these natural prairies could be compared with longitudinal data from grazed lands to answer much-discussed questions about the deterioration of the ranges.[62]

Nature Conservancy leaders were sufficiently impressed by Hulbert's development of Konza that they agreed in 1976 to expand the size of the natural area dramatically. Ordway funded acquisition of the 7,200-acre Dewey Ranch, which abutted the conservancy's strip on the north, plus additional property that brought the total area to more than 8,600 acres. She wanted to insure, however, that the priorities would be preservation and ecology, not park development or range research. The conservancy thus shifted from donation to a lease, outlined specific conditions that could trigger reversion of control from Kansas State to the Nature Conservancy, and created a scientific oversight board with a non-Kansas majority. This expansion and reorganization made Konza a natural candidate for inclusion, beginning in 1981, in the new National Science Foundation Long Term Ecological Research (LTER) program. Headquarters were established on the part of the Dewey Ranch where, seventy years earlier, Jens Jensen had planned to plant peonies.[63]

The major goal of the LTER project, Hulbert explained in 1985, was to learn why the "natural prairie ecosystem" was self-sustaining, productive, and long-lasting. The research design divided the tract into a number of watershed-based areas; each was then subjected to a particular regime involving periodic burning, mowing, or grazing (initially by cattle, but after 1987, when an electrified perimeter fence was in place, by the long-anticipated bison). Hulbert's expectation was that different treatments would result, over years and decades, in different types of prairie. He noted, for example, that while the agronomists' practice of annual late spring burning encouraged large grasses that "looked great," it reduced populations of smaller plants. He also expected to determine what species of grazers, numbers of animals, or schedules of grazing or burning would put too great a stress on the prairie and lead to a decline in its health.[64]

Konza did not, however, ripen into the peaceable kingdom that Hulbert, as an ecologist and a Friend, anticipated. For two decades the reasons for the stability of the natural prairie ecosystem eluded the Konza LTER team. After 2000, they realized that they were, in fact, tracking its disappearance. Whether left alone, burned every three or more years, or browsed by bison, grasses were giving way to woody plants. Some of these spreading species, such as buckthorn and Russian olive, were Old World horticultural introductions that could be interpreted as alien invaders. Others, however, had been brought only a few hundred miles (notably, Osage orange from Oklahoma and black locust from Missouri). The most important types were already there. Juniper, American elm, honey locust, and, especially, the previously unremarkable rough-leaf dogwood *(Cornus drummondii)*, did not retreat except under the intense range management practice of annual late spring burning (see Figure 9.4). Emulations of lightning fires and grazing were ineffectual in preserving sweeps of grassland. In 2005 Konza

Figure 9.4 This view of Konza Prairie in 2003 featured a honey locust tree protected from bison and fire by a thicket of roughleaf dogwood. The wider spread of woody vegetation, described by scientists as "a serious threat to the remaining tracts of tallgrass prairie," can be seen in the background. Alan K. Knapp, reproduced from *BioScience* (2005): 246, by permission.

ecologists looked outside their system for answers to this exemplification of the new problem of the prairie. They attributed the spread of woody plants to the large-scale changes in plant distribution in the Flint Hills and beyond, which had taken place during the preceding century, and to the recent "change in state" associated with global warming.[65]

The issue of the instability of prairie was being addressed from two other perspectives during the decades of the Konza LTER. The first of these was the old argument that prairies had been created and maintained largely by the Indians. For most of the twentieth century, this perspective, while advanced by a few geographers, anthropologists, and foresters, was ignored, attacked, or even suppressed.[66] The first time it was advanced at the North American Prairie Conference was in 1984. Historian Stephen Pyne's specific and evocative narrative of prairie fire experiences contrasted with the laconic hypotheticals advanced immediately after him by Lloyd Hulbert. Pyne produced a series of books that compiled evidence for the extent of human modification of landscapes through fire over millennia. He smoothly transitioned from recognizing Indians' impact on Midwestern vegetation to celebrating them as gardeners; for Pyne, William Cronon, and other environmental historians, the tangled wilderness that Euro-American settlers experienced was the consequence of the Indians' disappearance, not the setting in which they had lived. The final step, promoted by some ecological restorationists, was to promote modern controlled prairie burning as a restoration of these salutary Indian practices. In an inversion of J. Sterling Morton's springtime Arbor Day, restorationist William R. Jordan III proposed a ritualized autumnal burning season throughout the Midwest to suppress woody plants.[67]

The more radical view was that prairies were not Indian gardens, but the consequence of cycles of exploitative foraging, collapse, and makeshift recovery reaching back to the Pleistocene. In the 1960s paleontologist Paul S. Martin began to advance the argument that the extinction of the large North American herbivores (including mammoths, mastodons, gomphotheres, ground sloths, camels, horses, and short-faced bears) was the result of a 500-year "blitzkrieg" of predation by newly arrived humans around 12,000 years ago. In 1982 Martin collaborated with plant ecologist Daniel Janzen on a paper that attributed the restricted ranges of some of the most distinctive large-fruited American trees—including honey locust *(Gleditschia triacanthos)*, Kentucky coffee tree *(Gymnocladus dioicus),* and Osage orange *(Maclura pomifera)*—to the disappearance of the large-mouthed seed dispersers with whom they had coevolved. The broader implication was that nearly all North American ecosystems lacked, on an evolutionary timescale, some of their major components. These lacunae were particularly consequential in the Midwest, where removal of processors of high-bulk dry browse facilitated woody plants, fuel buildup, and then recurrent fires, started either by lightning or by people.[68]

From this evolutionary ecological perspective, the land covered by tallgrass prairie had been damaged for 12,000 years. The eighteenth-century suspicions that the New World was biologically incomplete were not totally mistaken. Indian management of vegetation through fire was less something to celebrate than to acknowledge as a makeshift second-best. The implication of the survival of sylvic "ghosts of evolution" like the Osage orange, however, was that other, less prominent, species and associations might also be holding on, and that the introduction of herbivores similar to those eliminated at the end of the Pleistocene could reestablish the vegetation of the "American Serengeti." In 1999 Martin and David A. Burney proposed in wildlife activist Richard Foreman's *Wild Earth* to "bring back" the elephants, camels, and cheetahs whose close evolutionary relatives had long populated the Midwest. The burros and horses that had spread from Spain through Mexico in the 1500s were the vanguards of this restorational process. Six years later, this call took the form of a sober proposal to *Nature,* signed by eleven scientists, including Martin, Foreman, and Michael Soulé.[69]

The idea that exotic mammals generally seen only in zoos should be roaming Nebraska spotlighted the ambiguities in commonsense distinctions between wild nature and parks, between American and Old World species, and between the Pleistocene and the present. But the proposal was paradoxical only when animals were at the center of attention. If the focus was on vegetation, "re-wilding" became an ordinary, if ambitious, form of naturalistic horticulture. Like ladybugs, elephants were garden helpers, controlling weeds and enriching soil fertility. They represented an improved way to culture valuable associations of North American prairie plants.

CHAPTER TEN

America the Beautiful

> O beautiful for spacious skies,
> For amber waves of grain,
> For purple mountain majesties
> Above the fruited plain!
>
> —KATHARINE LEE BATES, "AMERICA THE BEAUTIFUL," (1895, 1913)

Kitty Bates grew up on the bleak southwest corner of Cape Cod, deforested for decades to maximize sheep grazing. As a student at the new Wellesley College in 1876, by contrast, she looked out onto a suburban campus filled with rhododendrons, azaleas, and lilies, and she could see and visit the famous arboretum and topiary garden of H. H. Hunnewell on the other side of picturesque Lake Waban (previously Bullard's Pond) (see Figure 10.1). In "The College Beautiful," published in 1887, Bates lyrically celebrated the blue skies, grasses, flowers, different kinds of trees, and dimples on the lake of the campus where she then both taught and lived. Her more famous poem projected the same garden aesthetic onto a larger landscape. Bates counterposed the skies she experienced on a ride up Pike's Peak in 1893 against the Kansas wheat fields whose scenic quality had impressed her on her railroad trip west. The Rockies were the backdrop for the vast "fruited plain" that began in Ohio. The beauty of this well-planted land would, Bates hoped, draw God's continued grace.[1]

Through the allusion in my title, I emphasize the centrality of Bates's kind of horticultural perspective to interpretations of America. This book, sketching national history from a vantage point that foregrounds prosperous people, cultured plants, and noxious pests, is a new variant on those interpretations. The plot line is a simple one. In the eighteenth century, Americans interested in both nature and nation-building recognized that plants and pests were major issues in settlement and in the transition from colonial to postcolonial status.

Figure 10.1 An 1876 stereopticon showing Wellesley College students enjoying the native-plant topiary garden that H. H. Hunnewell had established on his estate, which he had named Wellesley. The newly built college is visible directly across Lake Waban. Reproduction courtesy of the Wellesley College Archives, Wellesley, MA.

From the early 1800s into the twentieth century, a cohesive network used both governmental and nongovernmental organizations to advance American horticultural independence. These efforts culminated, between 1910 and 1940, in the divisions between horticultural cosmopolitans such as David Fairchild and J. Horace McFarland, and nativists such as Charles Marlatt and Aldo Leopold. The aftermath, up to the present, has combined the generalization of garden values with confusion about their status.

The Americans involved in horticulture were sophisticated people filled with both good will and prejudice. They struggled with phenomena and with desires that were more complex than they could grasp. They knew, on the one hand, that people had massively transformed cultured plants and landscapes during previous centuries, and were continuing to do so. On the other hand, they believed that both they and the plants in which they were interested either were or easily could become native to America. They self-consciously manipulated varieties and introduced species, yet they understood that organisms transformed themselves and that certain species had arrived from parts unknown and had spread spontaneously. They wanted both to advance themselves competitively and to be seen as contributing primarily to the common good. The richness of these experiences, the conundrums horticulturists faced, and the reasonableness—if not necessarily the rationality—of their actions have emerged, I hope, through the details of their enthusiasms and their struggles. There are various ways to move from these details to a general picture of the past and to interpretation of the present. I make that move by considering the implications of my account for science, America, and culture.

This story confirms that practical issues of biogeography, plant improvement, and plant protection were central to the development of American science. Both Pennsylvania patriot Charles Thomson's plant introduction manifesto in the late 1760s, and Samuel L. Mitchill and David Hosack's schemes in the 1820s for permanent partnership between the New York Horticultural Society and the federal government were major early national scientific initiatives. The Hessian fly, which engaged American investigators, diplomats, and, ultimately, a federal-private commission, was the first postcolonial public scientific issue.

In the nineteenth century horticultural amateurs, nurserymen, and their organizations were significant participants in science. Gardeners and nurserymen were deeply involved in the work of selection, hybridization, and improvement. Collectors and taxonomists compiled materials for improvement projects. From John Lowell to J. Horace McFarland, amateurs pushed innovation and engaged with policy issues. The Massachusetts Horticultural Society was a major force in plant improvement from the 1830s onward; the Arnold Arboretum and the New York Botanical Garden were in large part institutionalizations of these interests.

Ultimately, of course, federal scientific bureaucrats predominated. By the early 1900s, the aptly named Bureau of Plant Industry and the Bureau of Entomology established themselves in Washington and used both land grant college professors and state bureaucrats to extend their reach. And just as breeders transformed themselves into geneticists in the early decades of the twentieth century, pest controllers became guides within ecology at the century's end. Contemporary work on invasive species is a renewal of a long scientific tradition whose visibility was temporarily eclipsed by the predominance of academic physical scientists during the cold war.

By "America" I mean the nexus of environment and nationality in the United States. John Mitchell, Guillame Raynal, and other eighteenth-century geographical commentators made the reasonable argument that the environmental peculiarities of America could keep the United States from developing economically and politically to a level that was comparable to the western European nations. Jefferson was not unusual in either his pre-Revolutionary fantasy that American Piedmont wine would generate private pleasure, income, and a congenial local social structure, or in his post-Independence argument that wheat could be the foundation for both export-driven prosperity and for a neo-British sociopolitical system. The antagonism of the American environment toward wine grapes, and the weakness of American wheat in the face of the ecologically imperialist Hessian fly, threatened these national prospects.

In the early nineteenth century, cotton fulfilled the vision that the United States could grow upon a plant-export economy, but it accelerated the South's movement toward an agricultural-social model that was neo-Asian rather than European. Crop introduction projects such as silk, sisal, strawberries, and citrus were either emulations of that economic success or efforts to turn the country away from the culture that cotton had created. Most remarkably, plant-centered reformers promoted some introductions—notably Johnson grass in the late nineteenth century and kudzu in the 1930s—with the expectation that they would participate in national efforts to fix the South. These plants were not mindless mistakes, but rather a heritage that Americans have received from their forebears trying to intervene strategically in difficult situations.

On the other hand, Americans feared alien agents that could bring national environmental destruction. The Hessian fly, through both the unique circumstances of its arrival and the host it affected, was a compelling exemplar of the dangers that foreign pests posed to America. In the nineteenth century the situation was confused by the existence of both foreign pests such as the Canada thistle and western American plant enemies like the Colorado potato beetle. But the devastation of such emblematic American trees as the chestnut and the white pine in the early 1900s by Old World organisms refocused national attention on the ways that alien pests could simultaneously destroy a natural resource, alter the American environment, and impoverish national identity. The plant quarantine battles in the years around World War I entailed deep divisions about the direction of the nation and the power that the federal government could exert over people in order to protect plants. These issues came to the forefront again in the 1990s with the fight against invasive species. J. Horace McFarland's belief that the basic question in horticulture and plant quarantine was "What is America?" becomes relevant in the context of the spread of Brazilian pepper and melaleuca in Everglades National Park and the use of herbicides to exterminate them.

The Everglades situation exemplifies how concern about alien species has been linked with beliefs about native plants and native landscape. The existence of natives was a commonsense perception in the nineteenth century but the term was used with little consistency. One problem was that Yankees, Virginians, and Californians all considered themselves native Americans. The category of native plants included various apples, new grapes, prairie forbs, and sequoias. Nativity could be assessed locally, regionally, nationally, or continentally. Finally, belief in nativity involved both continued adaptation to changing political boundaries and idealization of an original flora that had largely been effaced by land clearing and the nursery business. The importance of Paul Martin's modest proposal to restore the prairie with cheetahs and elephants comes, on the one hand, from its insistence that thinking about American landscapes should incorporate human and evolutionary history, and, on the other, from its stark challenge to the restoration ideals, discussed by Olmsted, of familiarity and repose. Stumbling forward toward the past involves radical and unpredictable change.

As a conclusion I return, not surprisingly, to culture. I have used this term obsessively and in ways that might, at least initially, have seemed odd. But practical clarification of this concept is important both for understanding the history of American horticulture and, more broadly, for thinking about the interactions between people and other organisms. I emphasize culture in order to critique both cultural theory and the usual academic hierarchy of sciences, and then, more constructively, to point toward a meaningful usage grounded on the experiences of historical actors.

During the last generation, cultural theory has been an amorphous but important academic entity. But the enterprise has been scholastic in its reliance on a dense circle of authorities radiating out from Michel Foucault and Jacques Derrida, and has inflated heuristic into theory at the expense of attention to empirical work on the diversity of experience. I therefore have considered it worthwhile to gesturally disenchant that field's terminological center. Liberty Hyde Bailey and other horticulturists possessed a nuanced conception of culture. They dealt with problems of high culture and low culture, purity and hybridity, the native and the cosmopolitan, and the difficulties of achieving consensus about taste. They were participants in an imagined community that extended indefinitely in space and time. For horticulturists, however, these issues were routine aspects of their interactions with plants, manures, insects, and farmers, not new insights into theory. My expectation has been that sympathetic study of these actors, centrally involved with culture and with organisms, opens a new strategic salient for understanding change.[2]

My more specific critical goal in emphasizing culture has been to develop an alternative to the inchoate but widespread view that the history of breeding, plant introductions, and pest exclusions should be understood as a history of applied science—of technologies derived from basic principles. I emphasized above that horticulture was an important part of American science, that it involved men of science, and that it utilized concepts and techniques from plant physiology, biogeography, selection theory, community ecology, and genetics. But it is important to maintain focus on the extent to which it was an art, or form of *techne*, and not a science on the academic model. Nineteenth- and twentieth-century culturists were building on traditions of skill that could be traced back into the ancient world. Explanations were interesting and sometimes helpful, but were subordinate to the artisanal goals of creating plants and gardens.

Culture differed fundamentally from more canonical arts such as painting or metalworking, however, because it involved not only traditions of skills but also material chains of living things. These populations of organisms, on the one hand, had been shaped by culturists over the course of millennia; on the other, they were entities engaged in Darwinian strategizing, with gardeners functioning as their evolutionary servants. Culturists were thus, first of all, procurers who moved plants around the globe, gave them ideal conditions for growth, trained their forms, and sometimes surgically merged different individuals. They were also facilitators who could select mutant buds and transform them into independent plants, or who could manipulate plants' sexual partners and reproductive behavior. They then waited for new generations to mature, and selected for propagation individuals that met their standards of taste. Sometimes, as in the case of strawberries, the results were those desired, or better. Other plants, such as pears and sequoias, were either uninteresting or nonviable. Frequently plants moved beyond their caretakers' control. Americans' desires changed, but plants such as Russian olive and melaleuca would be, apparently, forever.

These activities have gained renewed significance during the last generation. Grass developer Jack R. Harlan, who had literally grown up within the Washington agrobotanical community that George Vasey had established (his father Harry was the USDA's barley collector and breeder in the interwar years), accelerated agricultural scientists' interest in the historical biogeography of cultured plants in the 1970s by linking collection and preservation of seeds and trees to both fears of varietal extinctions and hopes for agricultural biotechnology. The U.S. National Plant Germplasm System gradually drew together USDA and land grant scientists, botanic gardens, and fruit and flower amateurs. The new interest in collecting and utilizing all the kinds of domesticated germplasm around the globe generated significant tensions between American-dominated biotechnology corporations and governments of countries with uncollected land races, but the results have been both the preservation of ancient cultural productions and increased interest in historical varieties—whether of wheats, tomatoes, or roses.[3]

The 2004 conference volume, *Industrializing Organisms,* suggested that a new and improved history of agriculture would result from focusing attention on the interactions among evolving organisms, humanized environments, and creative improvers.[4] I advance this approach by sketching the development and propagation of an American horticultural network. This network, which extended continuously forward from the 1820s, included gentlemen amateurs, nurserymen, scientists, and government officials. They acquired Old World cultured varieties, and used selection and breeding to adapt these organisms to differing American conditions. Choices centrally involved matters of taste. It is possible to advance an argument in this area that would, in other realms, be academically foolhardy: that culture was a more important element in the history of America than in that of northwestern Europe or eastern Asia.

The implications of this history of culture reach beyond agriculture. Culture was unique in its multiple reflexivities. The idea that culture cultured the culturist and his culture was, on its simplest level, a bromide about the domestic moral uplift that resulted from gardening rather than carousing. But it also characterized the complex structures of competition, networking, and status adjustment that horticulturists developed in the nineteenth century. Finally, as horticulturists frequently reminded their audiences, it referred to the effects that the creation and propagation of new and beautiful things would have on both the culturist and on others who would encounter them, both contemporary and yet to be born.

An important consequence of the social character of culture has been that choices have seldom been arbitrary or merely individual. Sweetness was integral to European expansion from the 1500s onward. In the last three centuries standards of beauty in flowers and landscape have changed much less than in painting or in music. Choices regarding trees, perennials, and fruit varieties involve long-term planning. Finally, the tastes of a broad public, who express themselves directly as taxpayers and voters, and indirectly as selective consumers, influence design decisions. Difficulties arise because horticultural desires are loosely articulated, multiple, and conflicting. Gardeners and their audiences want vegetation that is lush but not excessive, familiar but not ordinary, new but not odd, rare but also diversified, and characteristic of the locale, the region, the continent, and the climatic zone. At any given place, only some of these concerns can be satisfied.

Resolution of these conflicts is a problem in public art. As in the more extensively examined area of urban design, the central figures are bureaucratic professionals with finite resources and limited powers. Advancing public beauty involves a balance between the articulation of desire and the awareness that choices have consequences, some of which are inherently unstable, but others are only too permanent. Such insight into the future depends on knowledge about both usable traditions and previous pitfalls. Lasting public beauty comes more frequently through open rather than closed decision-making processes, but depends on designers who respond to public interests without merely capitulating to the powerful or the vocal. This combination of intelligence and adaptability is an old evolutionary strategy and important cultural work.[5]

Notes
Acknowledgments
Index

Notes

Introduction

1. On the background of plant technology, see essays in Susan R. Schrepfer and Philip Scranton, eds., *Industrializing Organisms* (New York: Routledge, 2004).
2. Recent works on the history of gardening include Mark Laird, *The Flowering of the Landscape Garden* (Philadelphia: University of Pennsylvania Press, 1999); Charles Quest-Ritson, *The English Garden* (Boston: David R. Godine, 2003); and Rebecca Bushnell, *Green Desire* (Ithaca, NY: Cornell University Press, 2003). On botany and plant introductions, see Anna Pavord, *The Tulip* (New York: St. Martin's Press, 1999); Emma Spary, *Utopia's Garden* (Chicago: University of Chicago Press, 2000); Londa Schiebinger, *Plants and Empire* (Cambridge, MA: Harvard University Press, 2004); and Harold J. Cook, *Matters of Exchange* (New Haven, CT: Yale University Press, 2007).
3. Abigail J. Lustig, "Cultivating Knowledge in Nineteenth-Century English Gardens," *Science in Context* 13 (2000): 155–181; Brent Elliott, *The Royal Horticultural Society: A History, 1804–2004* (Chichester: Phillimore and Co. Ltd., 2004). While *agricultura* was a Latin word, *horticulture* was an English neologism, coined in the 1670s but used extensively only after 1800.
4. Alfred W. Crosby, *Ecological Imperialism* (New York: Cambridge University Press, 1986).
5. "Horticultural independence" improves on "ecological independence" through its emphasis on technology rather than nature: cf. Philip J. Pauly, "The Beauty and Menace of the Japanese Cherry Trees: Conflicting Visions of American Ecological Independence," *Isis* 87 (1996): 51–73.
6. Thomas Jefferson, "Memorandum of Services," 1800, Thomas Jefferson Papers, Library of Congress, http://www.memory.loc.gov (accessed 8 October

2006); Stephen A. Spongberg, *A Reunion of Trees* (Cambridge, MA: Harvard University Press, 1990); cf. Richard N. Mack and W. Mark Lonsdale, "Humans as Global Plant Dispersers: Getting More than We Bargained For," *Bioscience* 51 (2001): 95–102.

7. Daniel Janzen, "Gardenification of Wildland Nature and the Human Footprint," *Science* 279 (1998): 1312.
8. William Cronon, *Changes in the Land* (New York: Hill and Wang, 1983); Joyce E. Chaplin, *Subject Matter* (Cambridge, MA: Harvard University Press, 2001).
9. Philip J. Pauly, *Biologists and the Promise of American Life* (Princeton, NJ: Princeton University Press, 2000), 8–9, sketched a weaker version of this claim.
10. William McNeill, *Plagues and Peoples* (Garden City, NY: Anchor Press, 1976); Jared Diamond, *Guns, Germs, and Steel: The Fates of Human Societies* (New York: W.W. Norton, 1997); Edmund Russell, "Evolutionary History: Prospectus for a New Field," *Environmental History* 8 (2003): 204–228.
11. James D. Drake, "Appropriating a Continent," *Journal of World History* 15 (2004): 323–357.
12. William Cronon, ed., *Uncommon Ground: Toward Reinventing Nature* (New York: W.W. Norton, 1995); Ben A. Minteer and Robert E. Manning, eds., *Reconstructing Conservation: Finding Common Ground* (Washington, DC: Island Press, 2003).
13. Steven Stoll, *The Fruits of Natural Advantage: Making the Industrial Countryside in California* (Berkeley: University of California Press, 1998); Julie Guthman, *Agrarian Dreams: The Paradox of Organic Farming in California* (Berkeley: University of California Press, 2004); and Douglas Cazaux Sackman, *Orange Empire: California and the Fruits of Eden* (Berkeley: University of California Press, 2005).
14. Judith K. Major, *To Live in the New World: A.J. Downing and American Landscape Gardening* (Cambridge, MA: MIT Press, 1997); Ethan Carr, *Wilderness by Design: Landscape Architecture and the National Park Service* (Lincoln: University of Nebraska Press, 1998).
15. L.H. Bailey, *Sketch of the Evolution of Our Native Fruits* (New York: Macmillan, 1898); Peter Del Tredici, "Neocreationism and the Illusion of Ecological Restoration," *Harvard Design Magazine* 20 (2004): 87–89. For models, see Michael Pollan, *The Botany of Desire* (New York: Random House, 2001), and Mark L. Winston, *Travels in the Genetically Modified Zone* (Cambridge, MA: Harvard University Press, 2002).

1. Culture and Degeneracy

1. Thomas Jefferson, *Notes on the State of Virginia*, ed. William Peden (Chapel Hill: University of North Carolina Press, 1954; orig. pub. 1785, 1787), quotation p. 64.

2. *Thesaurus Linguae Latinae*, multivolume (Leipzig: B. G. Teubneri, 1900–); Pliny the Elder, *Natural History (Naturalis Historia)*, trans. H. Rackham and D. E. Eichholz, 10 vols. (Cambridge, MA: Harvard University Press, 1938–1962), esp. books xii–xix (vols. 4–5); Columella, *On Agriculture (De Re Rustica)*, trans. H. B. Ash et al. (Cambridge, MA: Harvard University Press, 1941–1955).
3. James Walvin, *Fruits of Empire* (New York: New York University Press, 1997).
4. Margaret T. Hodgen, *Early Anthropology in the Sixteenth and Seventeenth Centuries* (Philadelphia: University of Pennsylvania Press, 1964), 209; D. A. Brading, *The First America* (Cambridge: Cambridge University Press, 1991), 380; Londa Schiebinger, *Plants and Empire* (Cambridge, MA: Harvard University Press, 2004).
5. Karen Ordahl Kupperman, "The Puzzle of the American Climate in the Early Colonial Period," *American Historical Review* 87 (1982): 1262–1289; Joyce E. Chaplin, *Subject Matter* (Cambridge, MA: Harvard University Press, 2001), 151–154.
6. Lyman Carrier, *The Beginnings of Agriculture in America* (New York: McGraw Hill, 1923), 124; Joyce E. Chaplin, *An Anxious Pursuit* (Chapel Hill: University of North Carolina Press, 1993).
7. Chaplin, *Anxious Pursuit*; Thomas Pinney, *A History of Wine in America* (Berkeley: University of California Press, 1989), 12–54. My main source for common and scientific plant names is USDA, NRCS, The PLANTS Database (http://plants.usda.gov), National Plant Data Center, Baton Rouge, LA 70874-4490, USA; more broadly see Integrated Taxonomic Information System (ITIS) (http://www.itis.gov).
8. Compare Michael Pollan, *The Botany of Desire* (New York: Random House, 2001), 12–13; Katherine Pandora, "Knowledge Held in Common: Tales of Luther Burbank and Science in the American Vernacular," *Isis* 92 (2001): 484–516.
9. Carrier, *Beginnings of Agriculture*, 154; U. P. Hedrick, *History of Agriculture in the State of New York* (Albany: New York State Agricultural Society, 1933), 332–333.
10. Stevenson W. Fletcher, *Pennsylvania Agriculture and Country Life 1640–1840* (Harrisburg: Pennsylvania Historical and Museum Commission, 1950), 144–145; Carrier, *Beginnings of Agriculture*, 184, 254; John Canup, "Cotton Mather and 'Criolian Degeneracy,'" *Early American Literature* 24 (1989): 20–34; for an overview of concerns about degeneration, see Susan Scott Parrish, *American Curiosity* (Chapel Hill: University of North Carolina Press, 2006), 82–102.
11. Quoted in Richard Gorer, *The Growth of Gardens* (London: Faber and Faber, 1978), 51.
12. Brooke Hindle, *The Pursuit of Science in Revolutionary America, 1735–1789* (Chapel Hill: University of North Carolina Press, 1956); Raymond Stearns, *Science in the British Colonies of America* (Urbana: University of Illinois Press, 1970).

13. On Bartram see, most recently, *America's Curious Botanist,* ed. Nancy E. Hoffmann and John C. Van Horne (Philadelphia: American Philosophical Society, 2003).
14. Edmund Berkeley and Dorothy Smith Berkeley, *Dr. John Mitchell* (Chapel Hill: University of North Carolina Press, 1974); Parrish, *American Curiosity,* 93–94; Philosophical Society of Edinburgh, *Medical Essays and Observations,* 5 vols. (Edinburgh: printed by T. and W. Ruddimans, for Mr. William Monro, 1733–1744).
15. *Gentleman's Magazine* 10 (1740): 527; John Mitchell, "An Essay upon the Causes of the Different Colours of People in Different Climates," *Philosophical Transactions of the Royal Society of London* 43 (1744–1745): 102–150.
16. On the "mania" for American plants during this period, see Mark Laird, *The Flowering of the Landscape Garden* (Philadelphia: University of Pennsylvania Press, 1999), 61–98.
17. For images of Mitchell's map, and discussion, see "John Mitchell's Map: An Irony of Empire," http://www.usm.maine.edu/~maps/mitchell/ (accessed 8 October 2006).
18. "An Impartial Hand" [John Mitchell], *The Contest in America between Great Britain and France* (London: A. Millar, 1757).
19. Ray Desmond, *Kew* (London: Harvill Press, 1995), 30–39; Berkeley and Berkeley, *Mitchell,* 231–248.
20. John Mitchell, *The Present State of Great Britain and North America* (1767; reprint, New York: Research Reprints, 1970), quotation p. v.
21. Ibid., 144–145, 155–156, 166–168.
22. Ibid., 133–139, 152–155, 178–185.
23. Ibid., 170–171, 186–212, esp. 191–192.
24. Ibid., 255.
25. Ibid., 269–270.
26. Ibid., 260–261n.
27. "Proposals for Enlarging the Plan of the American Society, Held at Philadelphia, for Promoting Useful Knowledge," *Pennsylvania Gazette,* 17 March 1768; reprinted as "Preface," *Transactions of the American Philosophical Society* 1 (1789): xvii–xxiv.
28. Following complex negotiations in 1769, the American Society merged into a revived American Philosophical Society. See Hindle, *Pursuit of Science,* 121–138.
29. The original edition was Guillaume-Thomas-François Raynal, *Histoire philosophique et politique des établissemens et du commerce des Européens dans les deux Indes* (Amsterdam: [s.n.], 1770). I work primarily from Raynal, *A Philosophical and Political History of the Settlements and Trade of the Europeans in the East and West Indies,* 2nd ed., trans. J. O. Justamond, 5 vols. (London: T. Cadell, 1776). For the state of scholarship on Raynal, see Hans-Jürgen Lüsebrink and Anthony Strugnell, eds., *L'Histoire des deux Indes* (Studies on Voltaire and the Eighteenth Century, vol. 333, 1995); Gilles Bancarel and Gianluigi Goggi, eds., *Raynal, de la polémique à l'histoire* (SVEC 2000:12).

30. Raynal, *History*, 3:79–83.
31. Ibid., 5:41, 62.
32. Ibid., 5:321–323.
33. Ibid., 5:202–205, 245, 257.
34. Ibid., 5:332–333.
35. Ibid., 5:245, 337.
36. Ibid., 5:438–439.
37. G. T. F. Raynal, *The Revolution of America* (London: Lockyer Davis, 1781), 194–195.
38. Raynal, *Revolution of America*, 197–198; Raynal, *Philosophical and Political History*, trans. J. O. Justamond, 8 vols. (London: W. Strahan and T. Cadell, 1783), 7:410.
39. The dinner story is in Thomas Jefferson to Robert Walsh, 4 December 1818, *Writings of Thomas Jefferson*, ed. Andrew A. Lipscomb, 20 vols. (Washington, DC: Thomas Jefferson Memorial Association, 1907), 18:170; Thomas Paine, *Letter Addressed to the Abbe Raynal* (Philadelphia: M. Steiner, 1782).
40. Thomas Jefferson, "Garden Book, 1766–1824," 3 (3 August 1767), 5 (14 March 1769), *Thomas Jefferson Papers: An Electronic Archive* (Boston: Massachusetts Historical Society, 2003, http//:www.thomasjeffersonpapers .org (accessed 8 October 2006); Peter J. Hatch, *The Fruits and Fruit Trees of Monticello* (Charlottesville: University Press of Virginia, 1998), 6–7, 42.
41. Jefferson, "Garden Book," 13–15; Hatch, *Fruits and Fruit Trees of Monticello*, 42.
42. Hatch, *Fruits and Fruit Trees of Monticello*, 30; Silvio A. Bedini, *Thomas Jefferson: Statesman of Science* (New York: Macmillan, 1990), 74–77, 84.
43. Jefferson, "Garden Book," 15–19.
44. Ibid., 24, 26; Pinney, *History of Wine*, 79–82.
45. Fawn M. Brodie, *Thomas Jefferson* (New York: W. W. Norton, 1974), describes Martha Jefferson as an oversensitive aristocrat who manipulated her husband and held back his career; Gisela Tauber, "*Notes on the State of Virginia*: Thomas Jefferson's Unintentional Self-Portrait," *Eighteenth-Century Studies* 26 (1993): 635–648, suggests that her increasingly difficult pregnancies were due to diabetes.
46. "Marbois' Queries concerning Virginia," November 1780, from the American Philosophical Society, 7 February 1781, *Papers of Thomas Jefferson*, ed. Julian P. Boyd (Princeton, NJ: Princeton University Press, 1950), 4:166–167, 544–545.
47. Jefferson to Marbois, 20 December 1781, 24 March 1782, Marbois to Jefferson, 22 April 1782, *Papers of Thomas Jefferson*, 6:141–142, 171–172, 177–178. Due to postal inefficiency, the report took four months to travel from Charlottesville to Philadelphia; that copy soon disappeared.
48. Jefferson, "Garden Book"; Jefferson to Edmund Randolph, 16 September 1781, Jefferson to George Rogers Clark, 19 December 1781, Jefferson to Charles Thomson, 20 December 1781, *Papers of Thomas Jefferson*, 6:117–118, 139, 142–143.

49. I am grateful to Lucia Stanton, Shannon senior research historian, Monticello (personal communication, 20 May 2004), for her insights regarding the composition chronology of *Notes on the State of Virginia*.
50. G. L. L. Buffon and L. J. M. Daubenton, *Histoire naturelle, générale et particulière, avec la description du Cabinet de roi*, 15 vols. (Paris: De l'Imprimerie royale, 1749–1767), 9:97–128.
51. Jefferson, *Notes*, 43–65.
52. Ibid., 47.
53. Ibid., 73–75, 81.
54. Ibid., 75–80.
55. Ibid., 80.
56. Philip Mazzei, *Researches on the United States* (1788; reprint, Charlottesville: University Press of Virginia, 1976), 205–287.
57. Jefferson, *Notes*, 82–85; cf. Benjamin Franklin, "Observations concerning the Increase of Mankind, Peopling of Countries, &c" (1751), *Writings of Benjamin Franklin*, ed. A. H. Smyth, 10 vols. (New York: Macmillan, 1905–1907), 3:63–73.
58. Jefferson, *Notes*, 87.
59. Ibid., 138–139.
60. Ibid., 137–143, 162–163.
61. Ibid., 165–168.
62. Ibid., 169; on mythologies surrounding Arabian horses, see Margaret E. Derry, *Bred for Perfection* (Baltimore, MD: Johns Hopkins University Press, 2003), 4–5.
63. Jefferson to Chastellux, 7 June 1785, *Papers of Thomas Jefferson*, 8:184–186.
64. Jefferson to Edward Bancroft, 26 January 1788 [1789], *Papers of Thomas Jefferson*, 14:492; Edward Bancroft, *An Essay on the Natural History of Guiana, in South America* (London: T. Becket and P. A. De Hondt, 1769), esp. 367–371.
65. Brodie, *Jefferson*, 216–245; Alexander O. Boulton, "The American Paradox: Jeffersonian Equality and Racial Science," *American Quarterly* 47 (1995): 467–492; Annette Gordon-Reed, *Thomas Jefferson and Sally Hemings* (Charlottesville: University Press of Virginia, 1997); *Report of the Research Committee on Thomas Jefferson and Sally Hemings*, Thomas Jefferson Foundation, January 2000, http//:www.monticello.org (accessed 8 October 2006).

2. The United States' First Invasive Species

1. Max Savelle, *George Morgan, Colony Builder* (New York: Columbia University Press, 1932), 76–110, 183–199; Gregory Schaaf, *Wampum Belts and Peace Trees* (Golden, CO: Fulcrum Press, 1990). Morgan's farm is now the southern half of Princeton University; the faculty club occupies the site of his house.
2. Varnum Lansing Collins, " 'Prospect Near Princeton,' " *Princeton University Bulletin* 15 (1904): 164–182; Collins, ed., *A Brief Narrative of the Ravages of*

the British and Hessians at Princeton in 1776–77 (Princeton, NJ: Princeton University Library, 1906); also a note, 1786, in George Morgan, "Journal 1780–1804, 'Prospect,'" AM 12800, Special Collections, Princeton University Library, Princeton, NJ.

3. Morgan, "Journal"; Morgan, "Management of Bees," 10 March 1786, Records of the Philadelphia Society for Promoting Agriculture, Special Collections, University of Pennsylvania Library (hereafter, PSPA); Morgan, "An Essay, Exhibiting a Plan for a Farm-Yard, and Method of Conducting the Same," *Columbian Magazine* 1 (1786): 77–80; Simon Baatz, *Venerate the Plough* (Philadelphia: The Society, 1985), 7–11; Savelle, *George Morgan,* 196.

4. Morgan, "Journal."

5. This chapter is adapted from Philip J. Pauly, "Fighting the Hessian Fly: American and British Responses to Insect Invasion, 1776–1789," *Environmental History* 7 (2002): 377–400. It draws substantially on a portfolio of correspondence and notes gathered by Joseph Banks, now part of the Joseph Banks Papers in the Sutro Branch, California State Library, San Francisco, CA; a microfilm on deposit at the American Philosophical Society Library (History of Science 3.1/19) is cited hereafter as Banks-APS. The Privy Council published Banks's selection of memoranda as *Proceedings of His Majesty's Most Honourable Privy Council, and Information Received, Respecting an Insect, Supposed to Infest the Wheat of the Territories of the United States of America* (21 April 1789), reprinted in *Annals of Agriculture* 11 (1789): 406–613 (citations from the original; hereafter, *PCP*).

6. David Hackett Fischer, *Washington's Crossing* (New York: Oxford University Press, 2004), 66–81; see also Ira D. Gruber, *The Howe Brothers and the American Revolution* (New York: Atheneum, 1972).

7. Ibid., 354.

8. Ibid., 351–360; David Syrett, *Shipping and the American War, 1775–83* (London: Athlone Press, 1970), 121–139, 197; Edward E. Curtis, *The Organization of the British Army in the American Revolution* (New Haven, CT: Yale University Press, 1926), 115–117. See also R. Arthur Bowler, *Logistics and the Failure of the British Army in America, 1775–1783* (Princeton, NJ: Princeton University Press, 1975). I conform to the usage of the time and refer to the insect as a fly, not a midge.

9. Morgan's 1786 journal entries imply a two- or three-year history of observation. *Pennsylvania Mercury,* 1 April 1785, discussed a new wheat insect on Long Island.

10. "Extract of a Letter from a Gentleman, dated New-York, September 1, 1786," printed in "On the Hessian Fly," *American Museum* 1 (April 1787): 325–326.

11. George Clinton [Address to Legislature], 13 January 1787, in *New York Packet,* 16 January 1787; Samuel Bard in *New York Journal,* 18 January 1787.

12. "On the Hessian Fly," *American Museum* 1 (February 1787): 133–135; "On the Hessian Fly," *American Museum* 1 (April 1787): 324–326. See also "Agricola" in *New York Journal,* 19 April 1787.

13. George Morgan, "On the Hessian Fly," *American Museum* 1 (June 1787): 529–531, dated 20 May 1787; Morgan, "Letter Relative to the Hessian Fly," *American Museum* 2 (September 1787): 298–300, dated 25 July 1787; see also Morgan, "Journal."
14. George Morgan to Sir John Temple, 26 August 1788, copy in Thomas Jefferson Papers, Manuscripts Division, Library of Congress, Washington, DC, http://www.memory.loc.gov (accessed 8 October 2006).
15. Morgan, "Letter Relative to the Hessian Fly," 20 July 1787.
16. Morgan to George Washington, 31 July 1788, and Washington to Morgan, 25 August 1788, George Washington Papers, http://www.memory.loc.gov (accessed 8 October 2006).
17. Joanne Neel, *Phineas Bond* (Philadelphia: University of Pennsylvania Press, 1968); Doron Ben Atar, *Trade Secrets* (New Haven, CT: Yale University Press, 2004).
18. Phineas Bond to Lord Carmarthen, 22 April 1788, "Letters of Phineas Bond, British Consul at Philadelphia, to the Foreign Office of Great Britain, 1787, 1788, 1789," ed. J. Franklin Jameson, American Historical Association, *Annual Report*, 1896, 565; "Letter Relative to the Hessian Fly . . . ,"*American Museum* 2 (August 1787): 175–176; "Copy of Letter from Mr. Decius Wadsworth to Col. Jeremiah Wadsworth," *American Museum* 2 (November 1787): 458–459.
19. John Gascoigne, *Science in the Service of Empire* (Cambridge: Cambridge University Press, 1998), 65–86; Richard Drayton, *Nature's Government* (New Haven, CT: Yale University Press, 2000), 96–106; Charles R. Ritcheson, *Aftermath of Revolution* (Dallas, TX: Southern Methodist University Press, 1969), 21–24, 189–195; "A Bill for Regulating the Trade between the Subjects of His Majesty's Colonies . . . and the Countries Belonging to the United States of America," *House of Commons Sessional Papers of the Eighteenth Century*, 13 February 1788, 61:4019.
20. Gascoigne, *Science in the Service of Empire*, 77–83; "A Bill to Explain and Amend an Act for . . . Regulating the Importation and Exportation of Corn and Grain," *Sessional Papers*, 22 May 1788, 61:4069. Wheat prices are found in William Beveridge et al., *Prices and Wages in England from the Twelfth Century to the Nineteenth Century*, 2 vols. (London: Longmans, Green, 1939) 1:568; Arthur Harrison Cole, *Wholesale Commodity Prices in the United States 1700–1861: Statistical Supplement* (Cambridge, MA: Harvard University Press, 1938).
21. "Preface" [April 1789], 2 pp., Banks-APS.
22. Banks to Carmarthen, 4 June 1788, *PCP*, 3–4; Landon Carter, "Observations Concerning the Fly-Weevil, That Destroys the Wheat," *Transactions of the American Philosophical Society* 1 (1771): 274–294.
23. Carmarthen to Banks, 26 June 1788, Banks-APS, 13; Order in Council, 25 June 1788, *PCP*, 5; *New York Independent Journal*, 23 August 1788; Privy Council Register, PC 2/133, pp. 175, 191, 199, Public Record Office, Kew, United Kingdom (hereafter, PRO).

24. Samuel L. Mitchill, "An Account of the Insect; Which for Some Years Has Been Very Destructive to Wheat in Several of the United States," *American Magazine* 1 (February and March 1788): 173–176, 201–204; Banks's abstract of Mitchill's essay, 2 July 1788, Banks-APS, 173–176; "American Wheat. Minute of What Was Stated by Sir Joseph Banks," 5 July 1788, *PCP*, 7–8; "Further Account of the Hessian Fly, by Sir Joseph Banks, Baronet; and, a Statement of the Printed Accounts Published in America, 8th July 1788," *PCP*, 9–11.
25. "Report of Sir Joseph Banks, Baronet, proposing the Mode of Making Experiment on the American Wheat, 6th July 1788," *PCP*, 8–9.
26. Adam Smith, James Edgar, and David Reid to Stephen Cottrell, 21 August 1788, Banks-APS, 299; A. Onslow and E. Rigby [to Stephen Cottrell], 16 July 1788, Banks-APS, 303–305.
27. Banks [to Cottrell], 20 July 1788, Banks-APS, 306.
28. James Currie to Banks, 22 July 1788, and 23 July 1788, Banks-APS, 379–389. On Currie, see William Wallace Currie, ed., *Memoir of the Life, Writings, and Correspondence of James Currie, M.D., F.R.S., of Liverpool*, 2 vols. (London, 1831), esp. 1:10–15, 76, 152.
29. Banks to Currie, 26 July 1788, Banks-APS, 386; Banks note, n.d., Banks-APS, 306 (top).
30. "General Report of Sir Joseph Banks, Respecting the Hessian Fly, and Flying-Wevil, 24th July 1788," *PCP*, 18.
31. Ibid.
32. Ibid., 18–19.
33. Mitchill, "An Account of the Insect," 173–176, 201–204.
34. Joseph Ewart to Lord Carmarthen, 9 August 1788, *PCP*, 26; Robert Murray Keith to Carmarthen, 8 October 1788, *PCP*, 32; Mr. Walpole to Carmarthen, 23 October 1788, *PCP*, 33; Alexander Gibson to Carmarthen 29 October 1788, *PCP*, 35; Mr. Heathcote to Carmarthen, 16 February 1789, *PCP*, 57; Mr. Mathias to Carmarthen, 19 September 1788, *PCP*, 34.
35. Mr. Heathcote to Carmarthen, 16 February 1789; Lord Torrington to Carmarthen, 28 July 1788, *PCP*, 53–55.
36. Broussonet to Carmarthen, 3 September 1788, with enclosed "Extraits des Registres de la Société d'Agriculture, du 28 Août 1788," *PCP*, 23–26.
37. *New York Independent Journal*, 23 August 1788; *Pennsylvania Mercury*, 28 August 1788.
38. Temple to Carmarthen, 4 September 1788, *PCP*, 27. On Temple, see Neil R. Stout, "John Temple," *American National Biography* 21:433–435.
39. Morgan to Temple, 26 August 1788, *PCP*, 27–29; "Minute of Council [with] Colonel Morgan's Remarks," *PCP*, 30–31.
40. Morgan to Temple, 26 August 1788, *PCP*, 27–29.
41. Bond to Carmarthen, 1 October 1788, with enclosures, *PCP*, 36–52; Carmarthen to Temple, Bond, and G. Miller, 6 August 1788, FO 4/6, PRO; Bond to Carmarthen, 3 November 1788, with enclosures, *PCP*, 61–62; Bond to Carmarthen, 20 January 1789, *PCP*, 63.

42. Banks, "Preface," Banks-APS; "Report of Sir Joseph Banks, Baronet, upon the Above Correspondence and Information. Dated 2d March 1789," and "Paper Delivered in the 27th of April, by Sir Joseph Banks, Baronet, by Way of Appendix," *PCP,* 59–61.
43. Paine to Jefferson, 10 April 1789, *Papers of Thomas Jefferson,* 14:567; Jefferson to Benjamin Vaughan, 17 May 1789, *Papers of Thomas Jefferson,* 15:133–134; Vaughan to Banks, 5 May 1789, 21 May 1789, with Banks's draft reply, Banks-APS; [Duke of Grafton], "On the Hessian Fly in America," 30 June 1789, with postscript dated 1 September 1789, Banks-APS.
44. Brian M. Fagan, *The Little Ice Age* (New York: Basic Books, 2000), 162–166; William Doyle, *Origins of the French Revolution,* 3rd ed. (Oxford: Oxford University Press, 1998), 151; C.-E. Labrousse, *Esquisse du Mouvement des Prix et des Revenus de France au XVIIIe Siécle,* 2 vols. (Paris: Librairie Dalloz, 1933), 1:92, 104; Thomas Jefferson to John Jay, 29 November 1788, printed in *New York Daily Gazette,* 12 February 1789.
45. Beveridge et al., *Prices and Wages,* 568; Committee report and other papers relating to the exportation and importation of corn, 8 March 1790, PC 1/18/20, PRO.
46. William Fawkener to Duke of Leeds, 23 November 1789; Leeds to Temple, Bond, and Miller, 24 November 1789; Leeds to Temple, Bond, and Miller, 4 December 1789, all FO 4/7, PRO.
47. Statement of William Backhouse, and others, and minutes of deliberations, 27 November 1789, PC 2/134, pp. 320–323, PRO.
48. *New York Daily Gazette,* 22 February 1790; *New-York Weekly Museum,* 27 February 1790.
49. Temple to Leeds, 24 February 1790, also Bond to Leeds, 1 March 1790, FO 4/8, PRO.
50. Ritcheson, *Aftermath of Revolution,* 202–203; Bond to Leeds, 6 July 1790, FO 4/8, PRO.
51. *National Gazette,* 14 June 1792; "Jefferson's Notes on the Hessian Fly," 24 May–18 June 1791, *Papers of Thomas Jefferson,* 20:456–461; editorial comments, *Papers of Thomas Jefferson,* 20:445–449.
52. Samuel Mitchill, "Remarks on the Wheat Insect or Hessian Fly," manuscript, 4 pp., 23 June 1791, Thomas Jefferson Papers, Library of Congress; Mitchill, "Oration," *Transactions of the New York Society for the Promotion of Agriculture, Arts, and Manufactures* 1 (1792): 23; Mitchill, "Further Ravages of the Wheat Insect, or Tipula Tritici of America, and of Another Species of Tipula in Europe," *Medical Repository* 7 (1803): 97–98.
53. Asa Fitch, *The Hessian Fly, Its History, Character, Transformations, and Habits* (Albany, NY: Joel Munsell, 1846), 4–18; Balthasar Wagner, "Observations on the New Crop Gall-Gnat" (1860), translated in *Third Report of the United States Entomological Commission,* 1883, appendix 2, pp. 24–38; Hermann A. Hagen, "The Hessian Fly Not Imported from Europe," *Canadian Entomologist* 12 (1880): 197–207; Charles V. Riley, "The Hessian Fly an Imported Insect," *Canadian Entomologist* 20 (1888): 121–127.

54. Brooke Hunter, "Rage for Grain: Flour Milling in the Mid-Atlantic, 1750–1815" (Ph.D. dissertation, University of Delaware, 2001); Alan L. Olmstead and Paul W. Rhode, "Biological Innovation in American Wheat Production: Science, Policy, and Environmental Adaptation," in *Industrializing Organisms*, ed. Susan R. Schrepfer and Philip Scranton (New York: Routledge, 2004), 43–84; Steven Stoll, *Larding the Lean Earth* (New York: Hill and Wang, 2002).
55. The modern perspective on the Hessian fly is summarized in Horace Francis Barnes, *Gall Midges of Economic Importance*, vol. 7: *Gall Midges of Cereal Crops* (London: Crosby Lockwood and Son Ltd., 1956), 107; Raymond J. Gagné, *The Plant-Feeding Gall Midges of North America* (Ithaca, NY: Comstock Publishing Association, 1989), 46. The species' range and spread prior to the 1770s remains unaddressed.

3. The Development of American Culture, with Special Reference to Fruit

1. *New England Farmer* 9 (1830): 62–63, 72, 82, 187–188.
2. T. G. Fessenden, "The Course of Culture," *New England Farmer* 9 (1830): 72; Christopher Caustic [T. G. Fessenden pseudonym], *Democracy Unveiled* (Boston: David Carlisle, 1805), 104.
3. William Wilson, "A Sketch of the Different Kinds of Gardens in the United States, Particularly Those of the Middle and Eastern States," *New-York Farmer* 1 (1828): 201; P. Barry, "American Horticulture," *Horticulturist* 8 (1853): 250; Tamara Plakins Thornton, *Cultivating Gentlemen* (New Haven, CT: Yale University Press, 1989).
4. On flowers' effeminacy, see Philo Florist, "Massachusetts Horticultural Society's Exhibitions," *Horticultural Register* 2 (1836): 145–146.
5. Stephen Daniels, Susanne Seymour, and Charles Watkins, "Enlightenment, Improvement, and the Geographies of Horticulture in Later Georgian England," in *Geography and Enlightenment*, ed. David N. Livingstone and Charles W. J. Withers (Chicago: University of Chicago Press, 1999), 345–371.
6. Henry Savage Jr. and Elizabeth J. Savage, *André and Francois André Michaux* (Charlottesville: University of Virginia Press, 1986), 22, 29; Thomas Andrew Knight, letter to his daughter, 2 May 1826, in *A Selection from the Physiological and Horticultural Papers Published in the Transactions of the Royal and Horticultural Societies by Thomas Andrew Knight, to Which Is Prefaced a Sketch of His Life* (London: Longman, Orme, Brown, Green, and Longmans, 1841), 39.
7. Overviews of practice were J. C. Loudon, *Encyclopaedia of Gardening* (London: Longman, Hurst, Rees, Orme, and Brown, 1822); Patrick Neill, "Horticulture," in *American Edition of the New Edinburgh Encyclopaedia*, 18 vols. (Philadelphia : E. Parker & J. Delaplaine, 1812–1831), 11:177–315; George Lindley, *A Guide to the Orchard and Fruit Garden*, ed. John

Lindley, American ed. with notes by Michael Floy (New York: G. F. Hopkins, 1833), 321–337.

8. On English horticulture see Brent Elliott, *The Royal Horticultural Society* (London: Phillimore, 2004); Charles Quest-Ritson, *The English Garden* (Boston: David R. Godine, 2003), 154–165, 177–198; Abigail J. Lustig, "The Creation and Uses of Horticulture in Britain and France in the Nineteenth Century" (Ph.D. dissertation, University of California, Berkeley, 1997).

9. Minutes, New York Horticultural Society, 27 November and 24 December 1821, and 26 August 1823, folder 2, Records of the New York Horticultural Society, New York Botanical Garden Library, Bronx, NY (I am grateful to Peter Mickulas for this reference); David Hosack, *An Inaugural Discourse, Delivered before the New-York Horticultural Society at Their Annual Meeting on the 31st of August 1824* (New York: J. Seymour, 1824).

10. Massachusetts Horticultural Society, *History of the Massachusetts Horticultural Society, 1829–1878* (Boston: MHS, 1880), 49, 58; John Lowell, "New Present of Fruit to the Citizens of the United States, by Thomas A. Knight, Esq.," *New England Farmer* 7 (1828): 42; Thomas G. Fessenden, "Contemplated Horticultural Society," *New England Farmer* 7 (1828): 198; "Notice," *New England Farmer* 7 (1828): 302.

11. On the challenge of propagating scions, see *New England Farmer* 10 (1832): 103; William Kenrick, "New Fruits," *Horticultural Register* 1 (1835): 9–10.

12. "Sketch of the Rise and Progress of the New-York Horticultural Society," *New-York Farmer* 1 (1828): 1–3; "Correspondence between the Hon. Richard Rush, and David Hosack, M.D., Relative to the Introduction of Foreign Seeds and Plants into This Country—and the Report of a Committee of the New-York Horticultural Society, Recommending the Establishment of Three National Gardens," *New-York Farmer* 1 (1828): 37–41; Minutes, New York Horticultural Society, 29 April, 20 May, and 29 July 1828, folder 2. Minutes were suspended after 1829. More broadly, see Peter Mickulas, "Cultivating the Big Apple: The New York Horticultural Society, Nineteenth Century New York Botany, and the New York Botanical Garden," *New York History* (Winter 2002): 34–54.

13. Quoted in H. A. S. Dearborn, "Historical Sketch," *Transactions of the Massachusetts Horticultural Society* 1 (1847): 70. On Mount Auburn, see Blanche Linden-Ward, *Silent City on a Hill* (Columbus: Ohio State University Press, 1989), 178–211.

14. Albert Emerson Benson, *History of the Massachusetts Horticultural Society* (Boston: MHS, 1929), 46.

15. Massachusetts Horticultural Society, *History of the Massachusetts Horticultural Society, 1829–1878* (Boston: MHS, 1880), 101–102, 110–111.

16. [John Lowell], "The Massachusetts Horticultural Society," *New England Farmer* 13 (1835): 393; Dearborn, "Historical Sketch," 83. Boston mayor Theodore Lyman, for example, bequeathed $10,000 to the MHS in 1849.

17. Charles M. Hovey, "Some Remarks upon the Production of New Varieties of Strawberries, from Seeds," *Magazine of Horticulture* 3 (1837): 241–246;

Hovey, "An Account of the Origin, Cultivation, &c. of Hovey's Seedling Strawberry; with a Description of the Fruit, Accompanied by an Engraving," *Magazine of Horticulture* 6 (1840): 284–294. On Hovey, see B. June Hutchinson, "A Taste for Horticulture," *Arnoldia* 40 (1980): 30–48.

18. Londoniensis, "Notes and Recollections of a Visit to the Nurseries of Messrs. Hovey & Co., Cambridge," *Magazine of Horticulture* 16 (1850): 442–447.
19. Charles M. Hovey, *The Fruits of America*, 3 vols. (Boston: C. C. Little & J. Brown, and Hovey & Co., 1847–1856).
20. A. J. Downing, *The Fruits and Fruit Trees of America* (New York: Wiley and Putnam, 1845); Charles M. Hovey, "The Fruits and Fruit Trees of America" [review], *Magazine of Horticulture* 11 (1845): 297–308, esp. 305–306; S. B. Parsons, "Some Remarks on the New Work of A. J. Downing, on the Fruits and Fruit Trees of America," *Magazine of Horticulture* 11 (1845): 308–313; T. S. Humrickhouse, "Remarks on the Importance of an Uniform Nomenclature of Fruits; with a Few Preparatory Observations upon the Misconceptions Entertained by Many of Downing's Fruits and Fruit Trees of America, as Attributable, in Part, to the Author's Objectionable and Imperfect Statement of its Design," *Magazine of Horticulture* 12 (1846): 47–57 (with Hovey's addendum, 57–58).
21. *Horticulturist* 9 (1854): 293–294, 389–390, 535–536; Massachusetts Horticultural Society, Minutes, 27 May and 21 October 1854, Massachusetts Horticultural Society Library, Boston, MA. Rochester nurseryman Patrick Barry had taken over *The Horticulturist* following Downing's sudden death in 1853.
22. Hovey, "A Social Chat with Our Readers," *Magazine of Horticulture* 20 (1854): 537–542; Hovey, [Presidential Address], 18 August 1864, *Magazine of Horticulture* 30 (1864): 347.
23. Downing, *Fruits and Fruit Trees of America*, vii–viii; Hosack, *Inaugural Discourse*; William Prince, *A Short Treatise on Horticulture* (New York: T. and J. Swords, 1828), 41.
24. Humrickhouse, "Remarks on the Importance of an Uniform Nomenclature of Fruits"; Downing, "Pomological Reform," *Horticulturist* 2 (1847–1848): 175–178; Downing, "The Rules of American Pomology," *Horticulturist* 2 (1847–1848): 273–275, 480–481; Hovey, "Rules of 'American' Pomology Adopted by the Massachusetts Horticultural Society; with Remarks upon the Same," *Magazine of Horticulture* 14 (1848): 97–107.
25. "Horticultural Memoranda 1835–1838 by Charles M. Hovey," manuscript notes in Charles M. Hovey Papers, Massachusetts Horticultural Society Library; Hovey, "A Social Chat with Our Readers," 541; Downing, "The Fruits in Convention," *Horticulturist* 4 (1850): 345–351, quotations pp. 348–349.
26. A. J. Downing, "American versus British Horticulture," *Horticulturist* 7 (1852): 249–252; Hovey, "Does Our American Climate Injuriously Affect the Foreign Fruits?" *Magazine of Horticulture* 20 (1854): 105–112.

27. For an example of high culture, see Hovey, "Does Our American Climate Injuriously Affect the Foreign Fruits?" 111.
28. A. J. Downing, "Some Remarks on the Superiority of Native Varieties of Fruit," *Transactions of the Massachusetts Horticultural Society* 1 (1847): 29–32; Downing, "On the Improvement of Vegetable Races," *Horticulturist* 7 (1852): 153–156.
29. Downing, *Fruits and Fruit Trees*, 3; William Kenrick, *New American Orchardist* (Boston: Carter, Hendee, 1833), 25–34.
30. Thomas A. Knight, *A Treatise on the Culture of the Apple and Pear, and on the Manufacture of Cider & Perry* (Ludlow: H. Procter, 1797). Knight's approach was summarized in Patrick Neill, "Horticulture," *American Edition of the New Edinburgh Encyclopaedia* 11:197–198.
31. Jean Baptiste van Mons, *Arbres fruitiers*, 2 vols. (Louvain: L. Dusart et H. Vandenbroeck, 1835–1836); A. Poiteau, *Théorie Van Mons* (Paris: Madame Huzard, 1834).
32. John Lindley, *Theory of Horticulture*, American ed., with notes, etc. by A. J. Downing and A. Gray (New York: Putnam, 1841), 161, 284; Antoine Poiteau, "Considerations, on the Process, Which Nurserymen Adopt to Obtain New Ameliorated Fruits, and That Which Nature Appears to Employ to Arrive at the Same Result," trans. H. A. S. Dearborn, *New England Farmer* 8 (1830): 221; H. A. S. Dearborn, "Mr. Van Mons' Method of Raising Fruit Trees from the Seed," *Horticultural Register* 2 (1836): 203–232; Henry Ward Beecher, "Do Varieties of Fruit Run Out?" *Horticulturist* 1 (1846): 181–182.
33. Charles M. Hovey, "The Production of Plants by Hybridization," *Magazine of Horticulture* 20 (1854): 249–257; Downing, "Remarks on the Superiority of Native Varieties"; Downing, *Fruits and Fruit Trees*; Hovey, "The Fruits and Fruit Trees of America."
34. William Coxe, *A View of the Cultivation of Fruit Trees, and the Management of Orchards and Cider* (Philadelphia: M. Carey and Son, 1817), 11, 142; S. A. Beach, *Apples of New York*, 2 vols. (Albany, NY: J. B. Lyon, 1905), 1:10.
35. Downing, "Fruits in Convention."
36. Henry David Thoreau, "Wild Apples," *Harper's Monthly*, November 1862, reprinted in Thoreau, *Natural History Essays*, ed. Robert Sattelmeyer (Salt Lake City, UT: Peregrine Smith, 1988), 178–210.
37. U. P. Hedrick, *The Pears of New York* (Albany, NY: J. B. Lyon, 1921), 14–15; Tamara Plakins Thornton, "The Moral Dimensions of Horticulture in Antebellum America," *New England Quarterly* 57 (1984): 3–24.
38. On weather problems, see Floy in G. Lindley, *Guide*, 376–377; David Hosack, "Some Account of the Seckle Pear, a New Seedling Raised in the Neighbourhood of Philadelphia," *Transactions of the Horticultural Society of London* 3 (1820): 256–258; A. J. Downing, "Pomological Notices," *Magazine of Horticulture* 6 (1840): 206–207; Robert Manning, *The New England Fruit Book*, 2nd ed. (Salem, MA: W. and S. B. Ives, 1844), 43–44, 52; Hedrick, *Pears of New York*, 125, 216.

39. Downing, "Fruits in Convention"; Thoreau, *Wild Fruits*, ed. Bradley P. Dean (New York: W. W. Norton, 2000), 127.
40. Manning, *The New England Fruit Book*; Thornton, "Moral Dimensions of Horticulture."
41. Samuel Walker, "Remarks on Some of the New Pears," *Horticulturist* 2 (1847–1848): 396–398; Hovey, "Does Our American Climate"; James Thacher, *The American Orchardist*, 2nd ed. (Plymouth, MA: E. Collier, 1825), 181–183; A. H. Ernst, "On the Fire Blight in Pear Trees," *Horticulturist* 2 (1847–1848): 328–330; Henry Ward Beecher, *Plain and Pleasant Talks about Fruits, Flowers and Farming* (New York: Derby and Jackson, 1859), 341–345. In 1880 Illinois botanist T. J. Burrill established the modern view that fire blight resulted from a bacteria, later named *Erwinia amylovora*. For a historical overview, see Tom Van Der Zwet and Harry L. Keil, *Fire Blight: A Bacterial Disease of Rosaceous Plants*, USDA Agriculture Handbook #510, 1979, 1–5, 112. The species evolved in North America.
42. Charles M. Hovey, "Our Pear Culture All Quackery," *Tilton's Journal of Horticulture* 7 (1870): 8–14; Hedrick, *Pears of New York*, 38–40, 180–181.
43. On the history of the strawberry, see S. W. Fletcher, *The Strawberry in North America: History, Origin, Botany, and Breeding* (New York: Macmillan, 1917); George M. Darrow, *The Strawberry: History, Breeding and Physiology* (New York: Holt, Rinehart and Winston, 1966); Stephen Wilhelm and James E. Sagen, *A History of the Strawberry: From Ancient Gardens to Modern Markets* (Berkeley: University of California Division of Agricultural Sciences, 1974).
44. Darrow, *The Strawberry*, 47–61; Antoine Nicolas Duchesne, *L'Histoire naturelle des fraisiers* (Paris: Didot, 1766).
45. Darrow, *The Strawberry*, 76–80.
46. Charles M. Hovey, "Some Remarks upon the Production of New Varieties of Strawberries, from Seeds," *Magazine of Horticulture* 3 (1837): 241–246; Hovey, "An Account of the Origin, Cultivation, &c. of Hovey's Seedling Strawberry," *Magazine of Horticulture* 6 (1840): 284–294; Fletcher, *Strawberry in North America*, 24. According to modern breeder George Darrow (*The Strawberry*, 132), the most likely parents were Methven Scarlet (a Canadian variant of *virginiana* that was imported to England a few decades earlier and then returned to Massachusetts) and Keen's Seedling (a second- or third-generation cross between Miller's Old Pine and other unnamed English varieties).
47. Nicholas Longworth, "On the Strawberry," *Horticultural Register and Gardener's Magazine* 3 (1837): 81–83.
48. Nicholas Longworth, "Observations on Different Varieties of Strawberries; and the Means of Producing Good Crops of Fruit" [with commentary by Hovey], *Magazine of Horticulture* 8 (1842): 257–262; *Magazine of Horticulture* 9 (1843): 415–417.
49. *Magazine of Horticulture* 10 (1844): 30–33, 109–113, 187–189, 431–432; 12 (1846): 355–362; 13 (1847): 347–349; Nicholas Longworth, *A Letter from N.*

Longworth, to the Members of the Cincinnati Horticultural Society (Cincinnati: L'Hommedieu & Co., 1846); Henry Ward Beecher, "The Strawberry Controversy" [with commentary by A.J. Downing], Horticulturist 1 (1846–1847): 273–277; Cincinnati Horticultural Society, Strawberry Report (Cincinnati: Morgan & Overend, 1848); [A.J. Downing], "The Cincinnati Strawberry Report," Horticulturist 2 (1847–1848): 517–519; Horticulturist 3 (1848–1849): 19–21, 49, 50 (reference to "the great strawberry question"), 101–102, 150, 196–198, 250–251; "Strawberries and Their Culture," Horticulturist 9 (1854): 345–349; Magazine of Horticulture 22 (1856): 281–292; Hovey, "Something about Strawberries," Magazine of Horticulture 26 (1860): 337–347. Finer details are in manuscript minutes of the Cincinnati Horticultural Society, vol. 1, 1843–1848, Cincinnati Historical Society Library, Cincinnati, OH. On these discussions, see Fletcher, Strawberry in North America, 98–102; Wilhelm and Sagen, History of the Strawberry, 150–156.

50. "A Strawberry View," Gardener's Monthly 3 (1861): 208–209.
51. Thomas Pinney, A History of Wine in America: From the Beginnings to Prohibition (Berkeley: University of California Press, 1989).
52. Pinney, History of Wine in America, 117–126; Nicholas Longworth, "On the Culture of the Grape, and the Manufacture of Wine," Report of the Commissioner of Patents: Agriculture, 1847, 462–470.
53. Pinney, History of Wine in America, 161; John F. von Daacke, "Grape-Growing and Wine-Making in Cincinnati, 1800–1870," Bulletin of the Cincinnati Historical Society 25 (1967): 196–215.
54. Alexander McClurg to Robert Buchanan, 17 February 1853, Robert Buchanan Letters, Cincinnati Historical Society Library; Beecher, Plain and Pleasant Talks, 79–82, 268.
55. On Bull, see William Barrett, "Ephraim Wales Bull," Memoirs of the Members of the Social Circle of Concord, 4th series (Cambridge: Riverside Press, 1909), 145–158; Edmund A. Schofield, " 'He Sowed; Others Reaped': Ephraim Wales Bull and the Origins of the 'Concord' Grape," Arnoldia 48 (1988): 4–16.
56. Bills from William R. Prince and Co., 1847–1849, box 1, folder 11, Ephraim Wales Bull Papers; Concord Farmers Club, Records 1 (1852–1855), esp. 19 January 1854. Both in Special Collections, Concord Free Public Library, Concord, MA.
57. Madelon Bedell, The Alcotts: Biography of a Family (New York: Clarkson N. Potter, 1980), 234–235, 255–256; on Bull's philosophical views, see A Letter [from] Ephraim Wales Bull of Concord, February 12, 1843, to Henry Moore (Champlain, NY: Moorsfield Press, 1932); Barrett, "Bull," 157–158.
58. Bull in Hovey, "Description," 66; Liberty Hyde Bailey, Sketch of the Evolution of Our Native Fruits, 2nd ed. (New York: Macmillan, 1906), 72; William D. Brown, "Grapes," Report of the Massachusetts State Board of Agriculture 3 (1855): 274.
59. Schofield, "He Sowed," 10; Henry Thoreau journal entry, 28 August 1853, quoted in Thoreau, Wild Fruits: Thoreau's Rediscovered Last Manuscript, ed. Bradley P. Dean (New York: W. W. Norton, 2000), 152.

60. Charles M. Hovey, "Description and Engraving of the Concord Grape, a New Seedling, Raised by E. W. Bull, Concord, Mass.," *Magazine of Horticulture* 20 (1854): 62–67; Pinney, *History of Wine,* 212; a contract between Bull and Hovey, probably a revision of an earlier one, is in box 2, folder 23, Bull Papers; discussions included *Magazine of Horticulture* 20 (1854): 360–361, 428–432, 553–559; *Horticulturist* 9 (1854): 124–125, 188, 236, 399–400, 515.
61. E. W. Bull, "Process of Wine-Making," *Report of the Massachusetts State Board of Agriculture* 9 (1861): 223–227; William D. Brown, "Grapes," *Report of the Massachusetts State Board of Agriculture* 3 (1855): 272; "Statement by E. W. Bull," *Report of the Massachusetts State Board of Agriculture* 5 (1857): 308.
62. Greeley, quoted in Schofield, "He Sowed," 12.
63. Barrett, "Bull," 155–156; Margaret M. Lothrop, *The Wayside* (New York: American Book Company, 1940), 157–163.
64. See Frank Lamson-Scribner et al., *Report on the Fungus Diseases of the Grape Vine,* U.S. Department of Agriculture, Botanical Division, bulletin 2, 1886.
65. Pinney, *History of Wine,* 391.
66. Ibid., 385–390.
67. "Tradition," http://www.schapiro-wine.com; "About Us," http://www.manischewitzwine.com (both accessed 8 October 2006).

4. Fixing the Accidents of American Natural History

1. John Opie, *Ogallala,* 2nd ed. (Lincoln: University of Nebraska Press, 2000), 345–382; Joel Jason Orth, "The Conservation Landscape: Trees and Nature on the Great Plains" (Ph.D. dissertation, Iowa State University, 2004).
2. John Evelyn, *Sylva, or a Discourse of Forest-Trees, and the Propagation of Timber in His Majesties Dominions* (London: Jo. Martyn and Ja. Allestry, 1664); Duhamel de Monceau, *Des Semis et Plantations des Arbres, et de leur Culture* (Paris: H. L. Guerin and L. F. Delatour, 1760).
3. Nicolai Cikovsky Jr., "'The Ravages of the Axe': The Meaning of the Tree Stump in Nineteenth-Century Art," *Art Bulletin* 61 (1979): 611–627; David Lowenthal, *George Perkins Marsh* (Seattle: University of Washington Press, 2000), 4–5.
4. Carolyn Merchant, *Ecological Revolutions* (Chapel Hill: University of North Carolina Press, 1989); Michael Williams, *Americans and Their Forests* (Cambridge: Cambridge University Press, 1989); Char Miller, ed., *American Forests* (Lawrence: University Press of Kansas, 1997); Miller, *Gifford Pinchot and the Making of Modern Environmentalism* (Washington, DC: Island Press, 2001).
5. Henry Thoreau, "The Succession of Forest Trees," in Thoreau, *Faith in a Seed,* ed. Bradley P. Dean (Washington, DC: Island Press, 1993).
6. Charles S. Sargent, "A Few Suggestions on Tree-Planting in Massachusetts," Massachusetts Board of Agriculture, *Agriculture of Massachusetts,* 1875–1876, 259.

7. George B. Emerson, *Report on the Trees and Shrubs Growing Naturally in the Forests of Massachusetts* (Boston: Dutton and Wentworth, 1846), 23.
8. Wilson Flagg, *The Woods and By-Ways of New England* (Boston: J. R. Osgood, 1872), 426–433.
9. J. L. Russell, "Introduction of Native Trees and Shrubs into Artificial Planting," *Magazine of Horticulture* 20 (1854): 219–228; Thomas J. Campanella, *Republic of Shade* (New Haven, CT: Yale University Press, 2003).
10. In the twentieth century, American taxonomists acknowledged that the giant sequoia was not a *Sequoia*. But because *Wellingtonia* had by then been used elsewhere, the genus name *Sequoiadendron* was created: Harold St. John and Robert W. Krauss, "The Taxonomic Position and the Scientific Name of the Big Tree Known as *Sequoia Gigantea*," *Pacific Science* 8 (1954): 341–358; Richard J. Hartesvedt et al., *The Giant Sequoia of the Sierra Nevada* (Washington, DC: National Park Service, 1975), 22–25.
11. A. J. Downing to Millard Fillmore, 3 March 1851, in John W. Reps, "Downing and the Washington Mall," *Landscape* 16, no. 3 (Spring 1967): 6–11.
12. J. G. Cooper, "The Forests and Trees of Northern America, as Connected with Climate and Agriculture," *Report of the Commissioner of Patents: Agriculture*, 1860, 445.
13. Andrew S. Fuller, *The Forest Tree Culturist* (New York: Geo E. & F. W. Woodward, 1866), 158–161, 170; William Saunders to J. Sterling Morton, 20 August 1893, RG 16, Letters received by the Secretary of Agriculture 1893–1906, National Archives II, College Park, MD.
14. Russell, "Introduction of Native Trees," 228.
15. Stephen Spongberg, *A Reunion of Trees* (Cambridge, MA: Harvard University Press, 1990), 88; A. J. Downing, "On the Employment of Ornamental Trees and Shrubs in North America," *American Gardeners Magazine* 1 (1835): 449; "Shade Trees in Cities," *Horticulturist* 8 (1852): 345–349. For a different perspective on the poplar and ailanthus, see Flagg, *Woods and By-Ways*, 329, 346.
16. "The Japanese in New-York," *New York Times*, 22 June 1860, 8; Parsons & Co., "Japanese Plants," *Horticulturist* 17 (1862): 186–187; Clay Lancaster, *The Japanese Influence in America* (New York: Walton H. Rawls, 1963), 66–69, 189–215.
17. Reid's Nurseries, *Catalogue for 1863 and 1864, of Fruit and Ornamental Trees* (Elizabeth, NJ: Drake, Drake, and Drake, 1863); Spongberg, *Reunion of Trees*.
18. Sargent, "Suggestions," 260–262. On the history of tree introductions more generally, see Spongberg, *Reunion of Trees*.
19. George B. Emerson, "Restoration of the Forests," Massachusetts Board of Agriculture, *Agriculture of Massachusetts*, 1875–1876, 140–151, and audience comments, 151–164; Charles S. Sargent, "A Few Suggestions on Tree-Planting in Massachusetts," ibid., 250–282, esp. 266–282; Franklin B. Hough, *Report on Forestry* (Washington, DC: Government Printing Office, 1878), 1:403–433; Ida Hay, *Science in the Pleasure Ground* (Boston: Northeastern University Press, 1995).

20. John Lindley, *The Theory of Horticulture,* 1st American ed. with notes, etc., by A. J. Downing and A. Gray (New York: Wiley and Putnam, 1841); Philip J. Pauly, *Biologists and the Promise of American Life* (Princeton, NJ: Princeton University Press, 2000), 31–33.
21. Asa Gray, "Diagnostic Characters of New Species of Phaenogamous Plants, Collected in Japan by Charles Wright, Botanist of the U.S. North Pacific Exploring Expedition," American Academy of Arts and Sciences, *Memoirs,* n.s., 6 (1859): 437–449.
22. Asa Gray, "Sequoia and Its History" (1872), in Asa Gray, *Scientific Papers of Asa Gray,* ed. C. S. Sargent, 2 vols. (Boston: Houghton, Mifflin, 1889), 2:142–171; see also Gray, "Forest Geography and Archaeology," *American Journal of Science* (hereafter, *AJS*), 3rd ser., 16 (1878): 85–94.
23. *New York Tribune,* 27 August 1872.
24. Wilmon H. Droze, *Trees, Prairies, and People* (Denton: Texas Woman's University, 1977).
25. Caleb Atwater, "On the Prairies and Barrens of the West," *AJS* 1 (1818): 116–125.
26. R. W. Wells, "On the Origin of Prairies," *AJS* 1 (1819): 331–337.
27. Leo Lesquereux, "On the Origin and Formation of Prairies," *AJS* 39 (1865): 317–327, and 40 (1866): 23–31; Josiah Whitney, *Report of the Geological Survey of the State of Wisconsin* 1 (1862): 114; James D. Dana, "On the Origins of Prairies," *AJS* 40 (1865): 293–304; J. S. Newberry, "Physical Geography of Ohio," *Report of the Geological Survey of Ohio* 1 (1873): 26–30; J. G. Cooper, "On the Distribution of the Forests and Trees of North America," *Annual Report of the Smithsonian Institution,* 1858, 275–280. For a review of that literature, see B. Shimek, "The Prairies," *Bulletin of the Laboratories of Natural History of the State University of Iowa* 6 (1911): 171–240; see also James C. Malin, *The Grassland of North America, Prolegomena to Its History, with Addenda* (Lawrence, KS: James C. Malin, 1956).
28. J. A. Allen, "The Flora of the Prairies," *American Naturalist* 4 (1870): 584; Lorin Blodget, "Forest Cultivation on the Plains," *USDA Annual Report,* 1872, 316–332; Henry Engelmann, "Remarks upon the Causes Producing the Different Characters of Vegetation Known as Prairies, Flats, and Barrens in Southern Illinois, with Special Reference to Observations Made in Perry and Jackson Counties," *AJS* 32 (1863): 384–396; On fire as the cause of the prairies, see Stephen J. Pyne, *Fire in America* (Princeton, NJ: Princeton University Press, 1982), 84–99; Gerald Williams, "Aboriginal Use of Fire: Are There Any 'Natural' Plant Communities?" in Charles Kay and Randy T. Simmons, eds., *Wilderness and Political Ecology* (Salt Lake City: University of Utah Press, 2002).
29. Asa Gray, "Forest Geography and Archaeology," *AJS,* 3rd ser., 16 (1878): 93–94; Charles S. Sargent, *Report on the Forests of North America (Exclusive of Mexico),* Department of Interior, Census Office, vol. 9, 1884, 5; J. W. Powell, *Report on the Arid Regions of the United States* (Washington, DC:

Government Printing Office, 1879), http://memory.loc.gov (accessed 8 October 2006).

30. T. J. Burrill and G. W. McCluer, "The Forest Tree Plantation," *University of Illinois Agricultural Experiment Station Bulletin* 26 (1893).

31. Blodget, "Forest Cultivation," 318–319; see discussion in *Transactions of the Illinois State Horticultural Society* 6 (1872): 98–100; W. C. Flagg, "Conditions of Tree Growth," *Transactions of the Illinois State Horticultural Society* 7 (1873): 36–42; Robert Douglas, "Tree-Growing upon the Prairies," ibid., 198–208.

32. "On Trees and Tree Planting: Report of Committee [of the State Horticultural Society]," *Transactions of the Illinois State Agricultural Society* 5 (1861–1864): 841.

33. Williams, *Americans and Their Forests*, 384; Burton J. Williams, "Trees but No Timber: The Nebraska Prelude to the Timber Culture Act," *Nebraska History* 53 (1972): 77–86; C. Barron McIntosh, "Use and Abuse of the Timber Culture Act," *Annals of the Association of American Geographers* 65 (1975): 347–362.

34. Robert W. Furnas, *Arbor Day* (Lincoln, NE: State Journal Company, 1888); N. H. Egleston, *Arbor Day: Its History and Observance*, U.S. Department of Agriculture, Report 56, 1896; L. E. Schmidt, "From Arbor Day to the Environmental Sabbath: Nature, Liturgy, and American Protestantism," *Harvard Theological Review* 84 (1991): 299–323.

35. O. B. Galusha, "Cultivation of Forest Trees on the Prairies," *Transactions of the Illinois State Agricultural Society* 5 (1861–1864): 808–814.

36. William M. Baker, "An Essay on Climatology," *Transactions of the Illinois State Agricultural Society* 7 (1867–1868): 575–582; J. H. Tice, "Meteorological Effects of Forests," *Transactions of the Illinois State Horticultural Society* 4 (1870): 162–177; W. C. Flagg, "The Climatic Effects of Forests," *Transactions of the Illinois State Horticultural Society* 5 (1871): 215–224.

37. Jonathan Periam, "Forest Tree Planting as a Means of Wealth," *Transactions of the Illinois State Horticultural Society* 5 (1871): 31–42. For typical exchanges on this subject, see A. G. Humphrey, "The Destruction of Forests, and the Necessity of Government Aid in Forest Culture," *Transactions of the Illinois State Horticultural Society* 7 (1873): 46–50, and discussion following, pp. 50–61; E. Gale, "Forest Tree Culture," *Transactions of the Kansas State Horticultural Society* 2 (1872): 13–18, and discussion, pp. 19–20, where Joseph Stayman (originator of the eponymous apple variety) "could not be satisfied that timber encouraged the fall of rain"; H. H. McAfee, "Report of the Committee on Forestry," *Transactions of the Iowa State Horticultural Society* 9 (1874): 135–143, and discussion, pp. 143–144. See also I. A. Lapham, J. G. Knapp, and H. Crocker, *Report on the Disastrous Effects of the Destruction of Forest Trees, Now Going on so Rapidly in the State of Wisconsin* (Madison, WI: Atwood & Rublee, 1867), 14–18.

38. Henry Nash Smith, "Rain Follows the Plough: The Notion of Increased Rainfall for the Great Plains, 1840–1880," *Huntington Library Quarterly*

10 (1946–1947): 169–193; Smith, *Virgin Land* (Cambridge, MA: Harvard University Press, 1950); David M. Emmons, "American Myth, Desert to Eden: Theories of Increased Rainfall and the Timber Culture Act of 1873," *Forest History* 15, no. 3 (1971): 6–13; Charles R. Kutzleb, "American Myth, Desert to Eden: Can Forests Bring Rains to the Plains?" *Forest History* 15, no. 3 (1971): 14–21; Walter Kollmorgen and Johanna Kollmorgen, "Landscape Meteorology in the Plains Area," *Annals of the Association of American Geographers* 63 (1973): 424–441; Everett N. Dick, *Conquering the Great American Desert: Nebraska* (Lincoln: Nebraska Historical Society, 1975), 118–128; Williams, *Americans and Their Forests,* 379–386.

39. Thomas Meehan, "Forests the Result and Not the Cause of Climate," *Prairie Farmer* 44 (1873): 370; Arthur Bryant, "Report on Timber-Planting," *Transactions of the Illinois State Horticultural Society* 7 (1873): 301–306; F. W. Hart, "Report on Forestry—Needs of Our Prairie State," *Transactions of the Iowa State Horticultural Society* 14 (1879): 274–279.
40. O. Ordway, "On the Culture of Forest and Ornamental Trees on the Prairies," *Transactions of the Illinois State Agricultural Society* 3 (1857–1858): 478.
41. A. N. Godfrey, "Kansas Trees for Kansas Forests," *Kansas State Horticultural Society Report* 11 (1881): 167–173. Botanist J. C. Arthur advanced the opposite view that the region's flora was not diverse: "Native Shrubs for Cultivation," *Transactions of the Iowa State Horticultural Society* 15 (1880): 223.
42. Lapham et al., *Report on the Disastrous Effects,* 57–88, emphasized trees from all over the continent. See also Minier, "Report of the Committee on Trees," *Transactions of the Illinois State Horticultural Society* 5 (1871): 320–321; Willard C. Flagg, "Report upon Ornamental and Timber Trees for Southern Illinois," *Transactions of the Illinois State Horticultural Society* 6 (1872): 91.
43. Disdain for black locusts grew: see Lapham et al., *Report on the Disastrous Effects,* 66. On failure of the chestnut, see *Transactions of the Illinois State Horticultural Society* 5 (1871): 86; Burrill and McCluer, "Forest Tree Plantation."
44. *Transactions of the Illinois State Horticultural Society* 5 (1871): 84–86. On openness to all types and the need to correct "nature's faulty distributions," see J. L. Budd, "Not Well Known Trees and Shrubs Promising to Do Well in Iowa," *Transactions of the Iowa State Horticultural Society* 17 (1882): 139–140. Biographical information on Douglas provided by the Waukegan (Illinois) Historical Society.
45. Frederick Starr, "American Forests: Their Destruction and Preservation," U.S. Department of Agriculture, *Report,* 1865, 219; Willard C. Flagg, "Report of Committee on Ornamental and Forest Trees," *Transactions of the Illinois State Horticultural Society* 6 (1872): 91; *Evergreen and Forest Tree Grower* 2 (1872): 24; E. Y. Teas, "Some Points of Excellence Possessed by the European Larch," *Transactions of the Illinois State Horticultural Society* 7 (1873): 326–327 (followed by H. H. McAfee's wager, p. 328).

46. Douglas, "Evergreens for the Prairies," *Transactions of the Illinois State Horticultural Society* 4 (1870): 51; Douglas, "Tree-Growing upon the Prairies," *Transactions of the Illinois State Horticultural Society* 7 (1873): 198–208; Douglas, "Report on New, Rare and Valuable Conifers Adapted to the Northwest," *Transactions of the Iowa State Horticultural Society* 14 (1879): 391.

47. An early Illinois list included 109 deciduous and evergreen trees: *Transactions of the Illinois State Horticultural Society* 1 (1867): 276–278; for a description of the number of species and their qualities, see "Report upon Ornamental and Timber Trees for Southern Illinois," *Transactions of the Illinois State Horticultural Society* 6 (1872): 85–97; W. L. Brockman, "Timber Culture," *Transactions of the Iowa State Horticultural Society* 8 (1873): 132; Suel Foster, "Forestry," *Transactions of the Iowa State Horticultural Society* 14 (1879): 63–64; Foster, "Catalpa Speciosa: Some Account of Its Introduction and Planting in Iowa," *Transactions of the Iowa State Horticultural Society* 18 (1883): 187–188; H. W. Lathrop, "Forestry Manual," *Transactions of the Iowa State Horticultural Society* 15 (1880): 309–330.

48. *Transactions of the Illinois State Horticultural Society* 1 (1867): 276–278; "Observations Accompanying Annual Report of 1868 of the Commissioner of the General Land Office on Forest Culture," U.S. Department of the Interior, *Report*, 1868–1869, 196–197.

49. *Kansas State Horticultural Society Report* 11 (1881): 238–242, with supporting letters from Hough and Sargent; F. P. Baker and R. W. Furnas, *Preliminary Report on the Forestry of the Mississippi Valley, and Tree Planting on the Plains*, USDA Report 28, 1883, 39–42; see also, for example, J. T. Allan, *Prize Essay on Forest Growing* (Brownville, NE: Advertiser Book and Job Office, 1873); J. B. Schlichter, "Forest Culture," *Kansas State Horticultural Society Report* 11 (1881): 200–204; Nathaniel Egleston. "Tree Planting in the Prairie States, as Shown by the Reports," U.S. Department of Agriculture, *Report on Forestry* 4 (1884): 100–104. On this point, see also Hough, *Report on Forestry* 1 (1878): 19; B. E. Fernow, *What Is Forestry?* U.S. Department of Agriculture, Forestry Division, Bulletin 5, 1891, 40–41.

50. Gilbert C. Fite, *The Farmer's Frontier, 1865–1900* (New York: Holt, Rinehart, and Winston, 1966).

51. Raymond J. Pool, "Fifty Years on the Nebraska National Forest," *Nebraska History* 34 (1953): 139–179, esp. 157–160; Richard Overfield, *Science with Practice* (Ames: Iowa State University Press, 1993).

52. Pool, "Fifty Years," 173–178; Clyde M. Brundy, "Trees for the Prairies," *American Forests* 39 (1933): 553–555, 574; Droze, *Trees, Prairies, and People*; R. Douglas Hurt, *The Dust Bowl* (Chicago: Nelson-Hall, 1981), 121–138.

53. Robert H. Mohlenbrock, *Guide to the Vascular Flora of Illinois*, rev. ed. (Carbondale: Southern Illinois University Press, 1986).

5. Immigrant Aid

1. David Fairchild, *The World Was My Garden* (New York: Charles Scribner's Sons, 1938), 109–110.
2. L. H. Bailey, *Sketch of the Evolution of Our Native Fruits* (New York: Macmillan, 1898), 21–41; Thomas Pinney, *A History of Wine in America: From the Beginnings to Prohibition* (Berkeley: University of California Press, 1989), 117–126, 136–139; House Committee on Public Lands, *Report... on the Petition of Peter S. Chazotte and Others, in Behalf of the American Coffee Land Association, February 20, 1822,* 12 December 1822, H. Rpt. 17(1)-14; Canter Brown Jr., "The East Florida Coffee Land Expedition of 1821: Plantations or a Bonapartist Kingdom of the Indies?" *Tequesta* 51 (1991): 7–28.
3. Nelson Klose, *America's Crop Heritage* (Ames: Iowa State College Press, 1950), 24–37; Wayne D. Rasmussen, "Diplomats and Plant Collectors: The South American Commission, 1817–1818," *Agricultural History* 29 (1956): 22–31.
4. U.S. Office of Experiment Stations, *The Cotton Plant,* Bulletin 33, 1896.
5. Thomas Spalding, "Cotton—Its Introduction and Progress of Its Culture, in the United States," *Southern Agriculturist and Register of Rural Affairs* 8 (1835): 35–46, 81–88; "The Beginning of Cotton Cultivation in Georgia," *Georgia Historical Quarterly* 1 (1917): 39–45; S. G. Stephens, "The Origin of Sea Island Cotton," *Agricultural History* 50 (1976): 391–399; figures from *Report of the Commissioner of Patents: Agriculture,* 1853, 181.
6. L. P. Brockett, *The Silk Industry in America* (New York: Silk Association of America, 1876), 38–40; Bailey, *Sketch of the Evolution of Our Native Fruits,* 127–157; Nelson Klose, "Sericulture in the United States," *Agricultural History* 37 (1963): 225–234; Alma Burner Creek, "*Bombyx Mori* and Americans: Or, Here We Go Round the Mulberry Bush," *University of Rochester Library Bulletin* 39 (1986): 36–52.
7. E. C. Genet, "Notes on the Growth and Manufacture of Silk in the United States," *New England Farmer* 4 (1826): 380, 387; House Committee on Agriculture, *Mulberry—Silk Worm,* 2 May 1826, H. Rpt. 19(1)-182; Richard Rush, "Circular Letter to the Several Governors of States and Territories," 29 July 1826, in *New England Farmer* 5 (1827): 69–70; House, *Silk-Worms. Letter from James Mease Transmitting a Treatise on the Rearing of Silk-Worms, by Mr. De Hazzi, of Munich,* 2 February 1828, H. Doc. 20(1)-226; Felix Pascalis, *Practical Instructions for the Culture of Silk and the Mulberry Tree,* vol. 1 (New York: William Gilley, 1829), 13.
8. "Practical Instructions for the Culture of Silk" [review], *American Journal of Science* 18 (1830): 278–288; Felix Pascalis, "Culture of Silk," *American Journal of Science* 19 (1831): 175–176; "The Chinese Mulberry. Morus Multicaulis," *Naturalist* 1 (1831): 315–316.
9. *The Silk Question Settled* (New York: Saxton and Miles, 1844); Brockett, *Silk Industry in America,* 41–43.

10. *Dictionary of American Biography*, 3:110–111; Kenneth W. Dobyns, *The Patent Office Pony* (Fredericksburg, VA: Sergeant Kirkland's, 1994).
11. House, *Report from the Commissioner of Patents, 1837*, 17 January 1838, H. Doc. 25-112, 5–6.
12. H. L. Ellsworth to J. K. Polk, 22 February 1838, in House Committee on Agriculture, *Agriculture and Useful Arts*, 1838, H. Rpt. 25(2)-655; House, *Memorial of Charles Lewis Fleischmann, on the Subject of Improving the Agriculture of this Country*, 16 April 1838, H. Doc. 25(2)-334.
13. Senate, *Letter from the Commissioner of Patents to the Chairman of the Committee on Patents and the Patent Office, in Relation to the Collection and Distribution of Seeds and Plants*, 28 January 1839, S. Doc. 25(3)-151.
14. *Dictionary of American Biography*, 6:357–358; Charles Mason diary, January 1857, *Life and Letters of Charles Mason Chief Justice of Iowa 1804–1882*, ed. Charles Mason Remey, 1939 (typescript, 12 vols., on deposit at the New York Public Library), 2:162–163; Paul W. Gates, *The Farmer's Age: Agriculture 1815–1860* (New York: Holt, Rinehart and Winston, 1960), 333–336.
15. Charles Mason, "Preliminary Remarks," *Report of the Commissioner of Patents: Agriculture*, 1853, v–vi; Charles Mason, "Preliminary Remarks. Experiments with Seeds," *Report of the Commissioner of Patents: Agriculture*, 1854, v–ix; D. J. Browne, "Report on the Seeds and Cuttings Recently Introduced into the United States," *Report of the Commissioner of Patents: Agriculture*, 1854, x–xxxv; D. J. Brown[e], "An Agricultural Bureau" (Letter to the House Agriculture Committee, 31 March 1856), *The Plough, the Loom and the Anvil* 8 (1856): 674–676; Senate, *Report of the Commissioner of Patents . . . Respecting the Purchase of Seeds by That Department*, 26 February 1857, Sen. Ex. Doc. 34(3)-61; Browne, "Sorghum Canes: Report of the United States Agricultural Society," *Report of the Commissioner of Patents: Agriculture*, 1857, 181–226; Lorin Blodget, "Agricultural Climatology of the United States Compared with That of Other Parts of the Globe," *Report of the Commissioner of Patents: Agriculture*, 1853, 328–381; Blodget, *Climatology of the United States* (Philadelphia: J. B. Lippincott, 1857); James R. Fleming, *Meteorology in America* (Baltimore: Johns Hopkins University Press, 1990), 110–115.
16. D. J. Browne, "The Cotton Manufactures of the United States," *Report of the Commissioner of Patents: Agriculture*, 1857, 305–418; Browne, "Report on the Seeds and Cuttings Recently Obtained by the Patent Office, with Suggestions as to the Expediency of Introducing Others," *Report of the Commissioner of Patents: Agriculture*, 1856, xlii–xlviii; Browne, "Tea Culture: On the Practicability of the Tea-Culture in the United States," *Report of the Commissioner of Patents: Agriculture* 1857, 166–181; Thomas G. Clemson, "Preliminary Remarks," *Report of the Commissioner of Patents: Agriculture*, 1860, 15.
17. D. J. Browne, "Preparation for a Government Propagating Garden at Washington," *Report of the Commissioner of Patents: Agriculture*, 1858, 280–282;

"Government Experimental and Propagating Garden," *Report of the Commissioner of Patents: Agriculture,* 1859, 1–22. For southern reactions, see "Tea, Its Culture and Commerce," *De Bow's Review* 28 (1860): 125–132.

18. "Agricultural Humbug at Washington," *American Agriculturist. Designed to Improve All Classes Interested in Soil Culture* 17 (1858): 40, 72, 104, 198–199, 230–231; *Letter from David Landreth, of Philadelphia, to the Commissioner of Patents* (n.s., 1858); "Vindication of the Agricultural Division of the Patent Office," 4 June 1858, in "Patent Office—Agricultural Department," *American Farmer's Magazine* 12 (1858): 402–408; "Public Gardens—the Patent Office," *Gardener's Monthly* 3 (1861): 240; "The Patent Office Report," *Gardener's Monthly* 3 (1861): 317; "Patent Office Report for 1859," *Horticulturist,* May 1861, 17.

19. John Lewis Russell, "The Distribution of Seeds by the United States Patent Office," *Transactions of the Massachusetts Horticultural Society,* 1858, 97–104; reprinted as "Introduction of New Seeds," *Magazine of Horticulture* 24 (1858): 542–552.

20. Russell, "Distribution of Seeds."

21. Edward Rand Jr., "The Culture of Our Native Plants," *Transactions of the Massachusetts Horticultural Society,* 1858, 44–54. Asa Gray, *Manual of the Botany of the Northern United States,* 5th ed. (New York: Ivison, Blakeman, Taylor, & Co., 1867), 375–376, considered both species European; in current classifications, *Convolvulus arvensis* (field bindweed) is European, while *Calystegia sepium* (hedge false bindweed) is divided into one European and two eastern North American subspecies, all of which grow in New England: see http://www.plants.usda.gov (accessed 8 October 2006).

22. Asa Gray, "Notices of New Native Plants, Introduced to Our Gardens from the Cambridge Botanic Garden," *Transactions of the Massachusetts Horticultural Society,* 1859, 29–35.

23. Charles J. Sprague, "Introduced Plants," *Transactions of the Massachusetts Horticultural Society,* 1860, 18–20.

24. Russell, "Distribution of Seeds."

25. Susan Warren Lanman, " 'For Profit and Pleasure': Peter Henderson and the Commercialization of Horticulture in Nineteenth-Century America," in *Industrializing Organisms,* ed. Susan R. Schrepfer and Philip Scranton (New York: Routledge, 2004), 19–42; Parsons & Co., "Japanese Plants," *Horticulturist* 17 (1862): 190; *Gardener's Monthly* 5 (1863): 245; after 1872, plant packages of up to four pounds could be sent as "printed matter" (eight cents per pound): *United States Domestic Postage Rates 1789–1956,* Post Office Publication 15, 1956, 33.

26. Edward S. Rand Jr., *The Rhododendron and "American Plants"* (Boston: Little, Brown, 1871).

27. James Wilson et al., "The Russian Thistle," *Iowa Agricultural Experiment Station Bulletin* 26 (1894); John J. Winberry and David M. Jones, "Rise and Decline of the 'Miracle Vine': Kudzu in the Southern Landscape," *Southeastern Geographer* 13 (1973): 63.

28. "Operations at the Government Experimental Garden," *Report of the Commissioner of Patents: Agriculture*, 1860, 28.
29. U.S. Department of Agriculture, *Report*, 1874, 7; 1875, 8; 1877, 9; 1878, 5.
30. Jack R. Harlan, "Human Interference with Grass Systematics," in *Grasses and Grasslands*, ed. James R. Estes et al. (Norman: University of Oklahoma Press, 1982), 37–50.
31. B. T. Galloway to W. T. Swingle, 29 April 1892, box 19, folder "Correspondence 1885–1892," Walter T. Swingle Papers, Department of Special Collections, University of Miami, Coral Gables, FL.
32. George Vasey, *Grasses for the South*, U.S. Department of Agriculture, Division of Botany, Bulletin 3, 1887, 5.
33. Henry Muhlenberg, "On the Cultivation of the Tall-Meadow-Oats (*avena elatior*) for Pasture and Hay—and on Gypsum and Stone-Coal as a Manure," *Transactions of the Society Promoting Agriculture, Arts, and Manufactures, New York* 1 (1793): 180–182; Charles L. Flint, *Culture of the Grasses*, Agricultural Tract No. 1 (Boston: W. White, 1860); Charles E. Bessey, "A Dozen Grasses and Clovers for Nebraska," *Annual Report of the Nebraska State Board of Agriculture*, 1890, 100–108; George Vasey, *The Agricultural Grasses of the United States*, U.S. Department of Agriculture, Report 32, 1884, 94–95.
34. John Stanton Gould, "The Grasses and Their Culture," *Transactions of the New York State Agricultural Society* 29 (1869): 191–402.
35. C. W. Howard, "Grasses for the South," *Report of the Commissioner of Patents: Agriculture*, 1860, 224–239.
36. Henry Muhlenberg, *Descriptio Uberior Graminum et Plantarum Calamariarum Americae Septentrionalis* (Philadelphia: Solomon W. Conrad, 1817); John Torrey, *A Flora of the State of New-York*, 2 vols., Natural History of New York, Division 2 (Albany, NY: Carroll and Cook, 1843), 2:413–479.
37. John J. Thomas, "On the Introduction and Culture of New Agricultural Products," *Transactions of the New York State Agricultural Society* 3 (1843): 175–185; quotation p. 182.
38. Carleton Ball, *Johnson Grass: Report of Investigations Made During the Season of 1901*, U.S. Bureau of Plant Industry, Bulletin 11, 1902, 7–9.
39. Benjamin Willard, "Choice Stock and Seed of a New Grass," *Maine Farmer*, 2 May 1850; Charles L. Flint, *Practical Treatise on Grasses and Forage Plants* (New York: G. P. Putnam, 1857), 102–106; Gould, "The Grasses and Their Culture," 203–204.
40. William James Beal, *Grasses of North America*, 2 vols. (New York: H. Holt, 1887–1896); Ray Stannard Baker, *An American Pioneer in Science* (Amherst, MA: privately printed, 1925), 18; Gray, *Manual*, 1867, 637; *Transactions of the New Hampshire Agricultural Society* (1853): 142–143.
41. "Grasses for the South," U.S. Department of Agriculture, *Report*, 1874, 157; George Vasey, *Grasses of the South*, U.S. Department of Agriculture, Division of Botany, Bulletin 3, 1887, 15–17.
42. Ball, *Johnson Grass*, 12–13.

43. On Vasey, see William M. Canby and J. N. Rose, "George Vasey: A Biographical Sketch," *Botanical Gazette* 18 (1893): 170–183; "Grasses of the Plains and Eastern Slope of the Rocky Mountains," U.S. Department of Agriculture, *Report,* 1870, 220.
44. "Report of the Botanist," U.S. Department of Agriculture, *Report,* 1889, 377–378; on Gilded Age federal science, see Philip J. Pauly, *Biologists and the Promise of American Life* (Princeton, NJ: Princeton University Press, 2000), 44–70; on Hitchcock and Chase, see Pamela Henson, " 'What Holds the Earth Together': Agnes Chase and American Agrostology," *Journal of the History of Biology* 36 (2003): 437–460.
45. "Report of the Botanist and Chemist on Grasses and Forage Plants," U.S. Department of Agriculture, *Report,* 1878, 157–158; "Report of the Botanist on Grasses," U.S. Department of Agriculture, *Report,* 1879, 349–350; "Report of the Botanist," U.S. Department of Agriculture, *Report,* 1881, 231–255; "Report of the Commissioner," U.S. Department of Agriculture, *Report,* 1888, 31.
46. S. M. Tracy, "Mississippi Experiment Station," U.S. Department of Agriculture, *Report,* 1890, 378–383; Tracy, "Cooperative Branch Stations in the South," U.S. Department of Agriculture, *Report,* 1891, 344–367.
47. George Vasey, *Report of an Investigation of the Grasses of the Arid Districts of Kansas, Nebraska, and Colorado.* U.S. Department of Agriculture, Division of Botany, Bulletin 1, 1886, 9.
48. George Vasey and Beverly T. Galloway, *A Record of Some of the Work of the Division, Including Extracts from Correspondence and Other Communications,* U.S. Department of Agriculture, Division of Botany, Bulletin 8, 1889, 9–17; "Report of the Botanist," U.S. Department of Agriculture, *Report,* 1889, 380; J. A. Sewall, "Grass and Forage Experiment Station at Garden City Kansas," U.S. Department of Agriculture, *Report,* 1891, 342.
49. Samuel Aughey, "Improvement of Western Pasture-Land," *Science* 1 (1883): 335; N. S. Shaler, "Improvement of the Native Pasture-Lands of the Far West," *Science* 1 (1883): 186–187; Charles E. Bessey, "The Grasses and Forage Plants of Nebraska," *Annual Report of the Nebraska State Board of Agriculture,* 1889, 159; W. J. Beal, *Grasses of North America,* 2 vols. (New York: H. Holt, 1887–1896), 1:305.
50. "Report of the Division of Botany," U.S. Department of Agriculture, *Report,* 1893, 240.
51. E. S. Morse, *Japanese Homes and Their Surroundings* (New York: Harper, 1885), 273–274; William P. Brooks, "Fruits and Flowers of Northern Japan," *Transactions of the Massachusetts Horticultural Society,* 1890, 39–60.
52. C. C. Georgeson, "The Economic Plants of Japan," *American Garden* 12 (1891): 7ff.
53. C. S. Sargent, *Forest Flora of Japan* (Boston: Houghton, Mifflin, 1894).
54. U.S. Department of Agriculture, *Report,* 1894, 58. When Congress required continuation of the program, Morton published an eighty-page list of seed

packet recipients, broken down by legislator: U.S. Department of Agriculture, *Report*, 1896, 155–237; Peter Dreyer, *A Gardener Touched with Genius*, rev. ed. (Berkeley: University of California Press, 1985); *Bulletin of the New York Botanical Garden* 1 (1898): 141.

55. Earley Vernon Wilcox, *Tama Jim* (Boston: The Stratford Co., 1930); Russell Lord, *The Wallaces of Iowa* (Boston: Houghton Mifflin, 1947); Earle D. Ross, *A History of the Iowa State College of Agriculture and Mechanic Arts* (Ames: Iowa State College Press, 1942).
56. T.H. Hoskins, "Tested Russian Apples," *American Garden* 11 (1890): 524–525.
57. James Wilson et al., "The Russian Thistle," *Iowa Agricultural Experiment Station Bulletin* 26 (1894).
58. Rose S. Taylor, *To Plant the Prairies and the Plains* (Mount Vernon, IA: Bios, 1941), 23–25.
59. Fairchild, *World Was My Garden*.
60. U.S. Department of Agriculture, Division of Botany, *Foreign Seeds and Plants: Imported by the Section of Seed and Plant Introduction*, Inventories 1– [1898–].
61. Isabel Cunningham, *Frank N. Meyer, Plant Hunter in Asia* (Ames: Iowa State University Press, 1984); Roy W. Briggs, *"Chinese" Wilson* (London: HMSO, 1993).
62. M.A. Carleton, *The Basis for the Improvement of American Wheats*, U.S. Department of Agriculture, Division of Vegetable Physiology and Pathology, Bulletin 24, 1900.
63. Joseph C. Bailey, *Seaman A. Knapp* (New York: Columbia University Press, 1945).
64. C.V. Piper and W.J. Morse, *The Soybean* (New York: McGraw Hill, 1923); *Soybeans: Improvement, Production, Uses*, ed. B.E. Caldwell, 1973 (Agronomy 16), 11–12; "Soybeans: National Statistics," http://www.nass.usda.gov (accessed 8 October 2006); see also publications of Theodore Hymowitz, such as "Introduction of the Soybean to Illinois," *Economic Botany* 41 (1987): 28–32, and T. Hymowitz and W.R. Shurteff, "Debunking Soybean Myths and Legends in the Historical and Popular Literature," *Crop Science* 45 (2005): 473–476.
65. Bailey, *Knapp*, 220–221, emphasized that agricultural extension began as Knapp's application of Methodist organizational and exhortational techniques to farming.
66. Fairchild, *World Was My Garden*; see, for example, U.S. Bureau of Plant Industry, Office of Foreign Seed and Plant Introduction, *The Dasheen: A New Vegetable of Great Value from the South*, 1918.
67. Fairchild, *World Was My Garden*, 411–413.

6. Mixed Borders

1. On the broader dimensions of the pest concept, see Sarah Jansen, *"Schaedlinge"* (Frankfurt: Campus Verlag, 2003); on federal activities, see

Vivian D. Wiser, *Protecting American Agriculture: Inspection and Quarantine of Imported Plants and Animals*, Agricultural Economic Report No. 266 (Washington, DC: Economic Research Service, U.S. Department of Agriculture, 1974); on plant diseases, see C. Lee Campbell et al., *The Formative Years of Plant Pathology in the United States* (St. Paul: APS Press, 1999). I learned of Peter Coates, *American Perceptions of Immigrant and Invasive Species: Strangers on the Land* (Berkeley: University of California Press, 2007) too late for consideration here.

2. E. Michener, *A Manual of Weeds, or the Weed Exterminator* (Philadelphia: King & Baird, 1872); Thaddeus William Harris, *A Treatise on Some of the Insects of New England: Which Are Injurious to Vegetation* (Cambridge, MA: John Owen, 1842); Asa Fitch, *First and Second Report on the Noxious, Beneficial and Other Insects of the State of New-York* (Albany, NY: C. Van Benthuysen, 1856).

3. C. V. Riley, "Imported Insects and Native American Insects," *American Entomologist* 2 (1870): 110–112; Asa Gray, "The Pertinacity and Predominance of Weeds," *American Journal of Science*, n.s., 18 (1879): 161–167; L. H. Bailey, "Why Have Our Enemies Increased?" (1892), in Bailey, *The Survival of the Unlike* (New York: Macmillan, 1896), 180–192.

4. C. V. Riley, *The Colorado Beetle* (London: Routledge, 1877), 16–17, 104–111; Jansen, *"Schaedlinge,"* 230–237.

5. Michener, *Manual of Weeds*; William Saunders, *Insects Injurious to Fruits* (Philadelphia: Lippincott, 1883).

6. Edmund H. Fulling, "Plant Life and the Law of Man, IV: Barberry, Currant and Gooseberry, and Cedar Control," *Botanical Review* 9 (1943): 488.

7. "An Act to Prevent the Growth of Canada Thistles," 3 November 1791, *Statutes of the State of Vermont . . . 1791*, 314 (*Early American Imprints*, 1st ser., 23939).

8. "An Act to Amend an Act, Entitled, 'An Act Relative to the Duties and Privileges of Towns,'" *Laws of New-York*, 35th Session, chap. 91, 8 June 1812; Lyster H. Dewey, *Legislation against Weeds*, U.S. Department of Agriculture, Division of Botany, Bulletin 17, 1896.

9. L. O. Howard, *Legislation against Injurious Insects*, U.S. Department of Agriculture, Division of Entomology, Bulletin 33, 1895; Erwin F. Smith, *Legal Enactments for the Restriction of Plant Diseases*, U.S. Department of Agriculture, Division of Vegetable Physiology and Pathology, Bulletin 11, 1896, 21–23; Lyster H. Dewey, *The Russian Thistle*, U.S. Department of Agriculture, Division of Botany, Bulletin 15, 1894, 23–26.

10. Daniel C. Sanders, "Observations on the Canada Thistle," *American Medical and Philosophical Repository* 1 (1810): 209–211; L. H. Bailey, "Coxey's Army and the Philosophy of Weediness" (1894), in Bailey, *Survival of the Unlike*, 193–201. For early skepticism about stigmatization of the thistle, see "The Canada Thistle," *New England Farmer*, 5 August 1825.

11. Kevin Starr, *Americans and the California Dream, 1850–1915* (New York: Oxford University Press, 1973); Steven Stoll, *The Fruits of Natural Advantage* (Berkeley: University of California Press, 1998).

12. Frederick M. Maskew, *A Sketch of the Origin and Evolution of Quarantine Regulations* (Sacramento: California State Association of County Horticultural Commissioners, 1925), 13–34; Thomas Pinney, *A History of Wine in America: From the Beginnings to Prohibition* (Berkeley: University of California Press, 1989), 342–347; Ian Tyrrell, *True Gardens of the Gods* (Berkeley: University of California Press, 1999).
13. C. L. Marlatt, "Insect Control in California," U.S. Department of Agriculture, *Yearbook*, 1896, 217–235; B. F. Lelong, "The Inspection of Trees, Plants, Fruit, etc., as Conducted under the Laws in California," *Proceedings of the National Convention for the Suppression of Insect Pests and Plant Diseases by Legislation,* ed. B. T. Galloway (Washington, DC: Government Printing Office, 1897), 12–19.
14. Robert J. Spear, *The Great Gypsy Moth War* (Amherst: University of Massachusetts Press, 2005), 18–47.
15. Ibid., 118–120, 126–127.
16. *Gibbons v. Ogden*, 22 U.S. 1 (1824); *Railroad Co. v. Husen*, 95 U.S. 465 (1878); U.S. Department of State, *Quarantine Laws of the United States (State and National)* (Washington, DC: Government Printing Office, 1887); Felix Frankfurter, *The Commerce Clause under Marshall, Taney and Waite* (Chapel Hill: University of North Carolina Press, 1937).
17. John B. Smith, "Report of Investigations on the San Jose Scale," New Jersey State Board of Agriculture, *Annual Report*, 1896, 128.
18. L. O. Howard, "Injurious Insects and Commerce," *Insect Life* 7 (1895): 335. California confirmed these values in 1899, mandating that infested plants "shall be immediately sent out of the state or destroyed at the option of the owner" (California State Commission of Horticulture, *Quarantine Laws and Orders,* Bulletin 1, 1911, 2).
19. Smith, "Report of Investigations on the San Jose Scale," 132.
20. L. O. Howard, "Eastern Occurrences of the San José Scale," *Insect Life* 7 (1894): 153–154; Howard, "Further Notes on the San Jose Scale," *Insect Life* 7 (1895): 283–295.
21. "Legislation against Plant Diseases and Injurious Insects," *Garden and Forest* 6 (27 September 1893): 401–402.
22. "The San José or Pernicious Scale," *Insect Life* 6, no. 5 (September 1894); L. O. Howard and C. L. Marlatt, *San Jose Scale*, U.S. Department of Agriculture, Division of Entomology, Bulletin 3, n.s., 1896; see citations in note 9 above.
23. L. O. Howard and A. F. Burgess, *The Laws in Force against Injurious Insects and Foul Brood in the United States,* U.S. Bureau of Entomology, Bulletin 61, 1906, 201; *National Nurseryman* 3 (1895): 74–75; *National Nurseryman* 4 (1896): 70–71.
24. C. H. Tyler Townsend, "Report on the Mexican Cotton-Boll Weevil," *Insect Life* 7 (1895): 295–309, esp. 309; *The Mexican Cotton-Boll Weevil,* U.S. Department of Agriculture, Division of Entomology, Circular 6, 2nd ser., 1895; Douglas Helms, "Technological Methods for Boll Weevil Control," *Agricultural History* 53 (1979): 287.

25. F. M. Webster, "Report on the Introduction and Dissemination of Injurious Insects," Ohio State Horticultural Society, *Annual Report,* 1895, 195, 207; Howard and Burgess, *The Laws in Force against Injurious Insects.*
26. B. T. Galloway, ed., *Proceedings of the National Convention for the Suppression of Insect Pests and Plant Diseases by Legislation* (Washington, DC: Government Printing Office, 1897); "National Fruit Growers' Convention," *Garden and Forest* 10 (17 March 1897): 108.
27. See commentaries and letters in *National Nurseryman* 5 (1897): 36–37, 38, 45, 48–50, 52–56, 84–86.
28. *National Nurseryman* 6 (1898): 3; *Congressional Record,* 55(2), 18 January 1898, 743; House Committee on Agriculture, *Rules and Regulations Governing Importation of Trees, Etc.,* 16 February 1898, H. Rpt. 55(2)-456, 1898.
29. "Germany Bars Our Fruit," *New York Times,* 3 February 1898, 1; "Bugs as Weapons of War," *New York Times,* 6 February 1898, 18; John B. Smith, "Germany's Exclusion of American Fruits," *North American Review* 166 (April 1898): 460–468; L. O. Howard, "International Relations Disturbed by an Insect," *Forum* 25 (1898): 569–573.
30. John B. Smith, "The San Jose Scale Scare," *American Agriculturist* 60 (1897): 414, 435; Smith, "Quarantine against Foreign Insects: How Far Can It Be Effective?" *Proceedings of the Society for the Promotion of Agricultural Science* 19 (1898): 90–100, esp. 99–100. For a nursery industry assessment of the "buggers," see N. H. Albaugh, "Is the Insect Agitation of the Day a Good or Bad Thing for Nurserymen?" *National Nurseryman* 6 (1898): 71–72.
31. Herbert Osborn, "The Duty of Economic Entomology," *Proceedings of the Association of Economic Entomologists,* 1898, 9.
32. On Marlatt, see E. Ralph Sasser et al., "Charles Lester Marlatt 1863–1954," *Proceedings of the Entomological Society of Washington* 57 (1955): 37–43.
33. Charles L. Marlatt, "The Laisser-Faire Philosophy Applied to the Insect Problem," *Proceedings of the Association of Economic Entomologists,* 1899, 5–23.
34. Marlatt, "Laisser-Faire Philosophy" [discussion], 19–23. Orlando Harrison, *National Nurseryman* 7 (1899): 118, claimed that the entomologists severely edited these proceedings to eliminate the more controversial and personal comments. Smith supported Marlatt in "Truth about the San Jose Scale," *American Agriculturist* 65 (1900): 2, 8.
35. Massachusetts, House, *Gypsy Moth Inquiry,* Doc. 1138, 1900; Spear, *Great Gypsy Moth War,* 227–236.
36. House Committee on Agriculture, *Transportation of Insect Pests, etc.,* 15 February 1905, H. Rpt. 58(2)-4617; A. Eckert, "The Need of a National Inspection Law, and How to Secure It," California State Commission of Horticulture, *Biennial Report,* 1905–1906, 409–416; *National Nurseryman* 16 (1908): 21, 54–59, 230. Marlatt reviewed the legislative history up to 1911 in "Need of National Control of Imported Nursery Stock," *Journal of Economic Entomology* 4 (1911): 107–124.

37. Charles Lester Marlatt, *An Entomologist's Quest* (Washington, DC: Monumental Printing, 1953); author's interview with Florence Marlatt (daughter of Charles Marlatt and Helen McKey-Smith Marlatt), 11 July 1995.
38. C. L. Marlatt, *The San Jose or Chinese Scale,* U.S. Bureau of Entomology Bulletin 62, 1906; Florence Marlatt interview; Sasser et al., "Marlatt," 40.
39. House Committee on Agriculture, *Inspection of Nursery Stock at Ports of Entry of the United States,* 12 February 1909, H. Rpt. 60(2)-2138; *Congressional Record* 60(2), 16 February 1909, 2492–2495; Senate Committee on Agriculture and Forestry, *Inspection of Nursery Stock at Ports of Entry of the United States,* 19 February 1909, S. Rpt. 60(2)-1042; William Pitkin testimony, House Committee on Agriculture, *Inspection of Nursery Stock* Hearing, 28 April 1910, 3 vols., 3:558.
40. *Journal of Economic Entomology* 3 (1910): 68.
41. Roland M. Jefferson and Alan E. Fusonie, *The Japanese Flowering Cherry Trees of Washington, D.C.: A Living Symbol of Friendship,* U.S. Department of Agriculture, Agricultural Research Service, National Arboretum Contribution No. 4, 1977, 9. Memos are reprinted on pp. 49–54.
42. Roger Daniels, *The Politics of Prejudice: The Anti-Japanese Movement in California and the Struggle for Japanese Exclusion* (Berkeley: University of California Press, 1962), 31–45.
43. Walter LeFeber, *The Clash* (New York: W. W. Norton, 1997), 87–92.
44. G. W. F., "Cherry Trees of Japan," *New York Times,* 31 August 1909, 6. On this theme, see Emiko Ohnuki-Tierney, *Kamikaze, Cherry Blossoms, and Nationalisms* (Chicago: University of Chicago Press, 2002).
45. LeFeber, *The Clash,* 88.
46. "Destroy Tokyo Gift Trees," *New York Times,* 29 January 1910, 1; "Wounding to Japanese Sensibilities" [editorial], *New York Times,* 31 January 1910, 6.
47. LeFeber, *The Clash,* 97; Jefferson and Fusonie, *Cherry Trees,* 16–21.
48. C. L. Marlatt testimony, *Inspection of Nursery Stock* Hearing, 3:496, 499, 505.
49. *Inspection of Nursery Stock* Hearing, 3:543, 559.
50. *Hawaiian Forester and Agriculturist* 7 (1910): 336; O. E. Bremner, *A Fruit Fly Menace,* California State Commission of Horticulture, 1911.
51. A. L. Quaintance, *The Mediterranean Fruit-Fly,* U.S. Bureau of Entomology Circular 160, 1912, 2–8.
52. California State Commission of Horticulture, *Quarantine Laws and Orders Bulletin* 1, July 1911, 14–15.
53. House Committee on Agriculture, "Quarantine against Importation of Diseased Nursery Stock," 6 January 1911, H. Rpt. 61(3)-1858; *Congressional Record* 61(3), 2 March 1911, 3964–3968.
54. California, *Horticultural Statutes with Court Decisions and Legal Opinions Relating Thereto* (Sacramento: F. W. Richardson, 1912).
55. California State Commission of Horticulture, *Quarantine Laws and Orders Bulletin* 1, 12–13.

56. Ibid., 16.
57. "Attitude of the Nurseryman toward a National Inspection and Quarantine Law," *Journal of Economic Entomology* 4 (1911): 278–281.
58. C. L. Marlatt, "Pests and Parasites: Why We Need a National Law to Prevent the Importation of Insect-Infested and Diseased Plants," *National Geographic Magazine* 22 (1911): 321–346.
59. William Pitkin, "The Federal Inspection Bill," *National Nurseryman* 20 (1912): 14.
60. *Journal of Economic Entomology* 5 (1912): 63–64; House Committee on Agriculture, *A Bill to Regulate the Importation and Interstate Transportation of Nursery Stock* Hearing, 19 February 1912, 51–107; Senate Committee on Agriculture, *Importation and Interstate Transportation of Nursery Stock,* 23 July 1912, S. Rpt. 62-961.
61. Author's interview with Florence Marlatt; U.S. Department of Agriculture, *Report,* annually.
62. Gustavus A. Weber, *The Plant Quarantine and Control Administration* (Washington, DC: Brookings Institution, 1930).
63. *Bulletin of the New York Botanical Garden* 1 (1898): 141; I. C. Williams, "The New Chestnut Bark Disease," *Science* 34 (1911): 397–400; D. G. Fairchild, "The Discovery of the Chestnut Bark Disease in China," *Science* 38 (1913): 297–299; Pennsylvania Chestnut Tree Blight Commission, *The Publications of the Pennsylvania Chestnut Tree Blight Commission, 1911–13* (Harrisburg, PA: W. S. Ray, 1915).
64. Perley Spalding, *The Blister Rust of White Pine,* U.S. Bureau of Plant Industry *Bulletin* 206, 1911, 7–9; "The Lesson of the Pine Blister Canker," *American Forestry* 22 (1916): 752–753; "Shall the United States Continue to Be a Dumping Ground for Pests?" Massachusetts Forestry Association *Bulletin* 121, 1917; *The Massachusetts Forestry Association: Its First Twenty-Five Years 1898–1922* (Boston: n.p., 1922), 21.
65. C. L. Marlatt, "Losses Caused by Imported Tree and Plant Pests," *American Forestry* 23 (1917): 75–80; W. D. Hunter, *The Pink Bollworm,* U.S. Department of Agriculture Departmental Bulletin 723, 1918.
66. Marlatt, "Losses," 79; "Resolution," *American Forestry* 23 (1917): 75; *Congressional Record* 65(2), 4 January 1918, 4336; D. F. Houston to M. L. Dean, 31 January 1918, in U.S. Federal Horticultural Board, *Service and Regulatory Announcements,* January 1918, 5–6; March 1918, 30; October–November 1918, 100–110; Charles Lathrop Pack, "Excluding Enemy Aliens with Appetites De Luxe," *American Forestry* 25 (1919): 1053.
67. Meetings for interested commercial groups were held in Washington on 28 May 1918 (reported in *Florists' Exchange,* 1 June 1918, 1094, 1112), and 18 October 1918 (*Florists' Exchange,* 26 October 1918, 649). See also C. L. Marlatt, "Plant Quarantine No. 37," *American Florist,* 13 September 1919, 411–420; "Protest against the Horticultural Import Prohibition," *Gardeners' Chronicle of America* 23 (1919): 35; John Scheepers, "Condemnation of Plant Import Prohibition," *Gardeners' Chronicle of America* 23

(1919): 51; "Memorandum concerning Quarantine No. 37," U.S. Federal Horticultural Board, *Service and Regulatory Announcements,* January 1919, 4–6.
68. "Great Forestry Conference and Annual Meeting," *American Forestry* 22 (1916): 747; "Ravages of Pests in Forests of U.S.," *Washington Evening Star,* 19 January 1917, 15.
69. David G. Fairchild, "The Independence of American Nurseries," *American Forestry* 23 (1917): 213–216; quotation p. 216.
70. The risks of opposing quarantine within the bureaucracy briefly became public at the FHB's stakeholders' meeting in May 1918. When Fairchild spoke about the problems that the proposed rules would pose, plant pathologist R. K. Beattie condemned his office for having introducing bamboo smut into government greenhouses: *Florists' Exchange,* 1 June 1918, 1112.
71. B. T. Galloway, "Protecting American Crop Plants against Alien Enemies," *Transactions of the Massachusetts Horticultural Society,* 1919, 75–84, discussion pp. 84–87; the typescript copy in the B. T. Galloway Papers, National Fungus Collections, USDA-ARS, Beltsville, MD, had a handwritten note on the title page that "B. T. G. was heckled & finally went off & left others talking."
72. "Mr. Wilson on Quarantine No. 37," *Horticulture* 29 (1919): 466; "Horticultural Conference on Quarantine No. 37 Held in New York, June 15, 1920," *Bulletin of the Massachusetts Horticultural Society,* No. 4, 1920; A. C. Burrage and Charles S. Sargent, "The Case against Quarantine No. 37," *Garden Magazine,* September 1920, 45–46; "United Horticultural Interests Confer on Quarantine 37," *Florists' Exchange,* 19 June 1920, 1373, 1393.
73. Ernest Morrison, *J. Horace McFarland: A Thorn for Beauty* (Harrisburg: Pennsylvania Historical and Museum Commission, 1995), 215–232; American Joint Committee on Horticultural Nomenclature, *Standardized Plant Names* (Salem, MA: American Joint Committee on Horticultural Nomenclature, 1923).
74. Morrison, *McFarland,* 106–122, 153–193.
75. "To Every Friend of American Horticulture," *Garden Magazine,* January 1921, 266–267.
76. C. L. Marlatt, "Restrictions on Entry of Foreign Plants Widely Misunderstood," 25 February 1921, U.S. Federal Horticultural Board, *Service and Regulatory Announcements,* January–June 1921, 44–51; C. L. Marlatt, "Memorandum for the Secretary," 4 pp., 2 February 1921, box 842, RG 16, Correspondence of the Secretary of Agriculture, National Archives II, College Park, MD (hereafter, USDA Records).
77. "American Rose Society, Washington, D.C., Pilgrimage, June 1–2," *Florists' Exchange* 51 (1921): 134. Promoted heavily by the American Rose Society, "Mary Wallace" was one of the most popular varieties of the 1920s.
78. "Statement to the Secretary of Agriculture by the Committee on Horticultural Quarantine," 11 pp., with supporting documentation, 39 pp., 30 January 1922, box 920, USDA Records.

79. *Report of Conference to Consider the Advisability of Any Modifications—Additions to or Deductions from—of the Classes of Plants Permitted Entry under Permit for Immediate Sale under Regulation 3 of Quarantine 37, Held by the Federal Horticultural Board, May 15, 1922,* typescript of stenographic record, copy at National Agricultural Library, Beltsville, MD; substantial files on the subject are in box 842 (1921) and box 920 (1922), USDA Records.
80. *Report of Conference,* 2, 24, 29–47, 97–98, 103ff.
81. U.S. Federal Horticultural Board, *Service and Regulatory Announcements,* January–June 1922, 17–22; C. L. Marlatt, "Memorandum for the Secretary," 24 June 1922; Henry C. Wallace to J. Horace McFarland, [5] August 1922; McFarland to Wallace, 8 August 1922, box 920, USDA Records.
82. J. Horace McFarland, "What about Plant Quarantine 37?" *Landscape Architecture* 13 (1922–1923): 55; Marlatt's response was in *Landscape Architecture* 13 (1922–1923): 147.
83. Alfred C. Kinsey, "Foreign Pests and Our Gardens," *Farm and Garden* 11, no. 2 (August 1923): 6–10; J. Horace McFarland, "What Is America, Anyhow?" *Farm and Garden* 11, no. 3 (September 1923): 18–19.
84. *A Constructive Criticism of the Policies Governing the Establishment and Administration of Quarantines against Horticultural Products* ([New York?]: n.p., 1925); Neil Van Aken, "Worms and Dutch Bulbs," *Nation* 121 (1925): 701–702; W. M. Jardine, "The Purpose of Plant Quarantine," *Nation* 122 (1926): 33–34; "Bugs and Bureaucracy," *Outlook,* 3 February 1926, 170–172.
85. *Oregon-Washington R. & Nav. Co. v. State of Washington,* 270 U.S. 87 (1926); C. L. Marlatt, "The Effect of the Supreme Court Decision of March 1, 1926 in the Case of the *Oregon-Washington Railroad and Navigation Company vs. The State of Washington,* on the Basic Quarantine Laws of the Various States," *Journal of Economic Entomology* 20 (1927): 447–453; Weber, *Plant Quarantine and Control Administration,* 14–15.
86. For concerns about warrantless searches, see Henry Alan Johnston, *What Rights Are Left?* (New York: Macmillan, 1930), 91–92, 102–107; House Committee on Agriculture, *Plant Quarantine Act Amendment* Hearing, 17, 18 January 1927, 6, 14; Weber, *Plant Quarantine and Control Administration,* 15–16; *United States Statutes at Large,* 70th Congress, 45:468.
87. Weber, *Plant Quarantine and Control Administration,* 93–128.

7. Gardening American Landscapes

1. Norman T. Newton, *Design on the Land: The Development of Landscape Architecture* (Cambridge, MA: Belknap Press of Harvard University Press, 1971); Elizabeth Barlow Rogers, *Landscape Design: A Cultural and Architectural History* (New York: Harry N. Abrams, 2001).
2. Raymond J. O'Brien, *American Sublime* (New York: Columbia University Press, 1981); Kenneth Myers, *The Catskills: Painters, Writers, and Tourists*

in the Mountains 1820–1895 (Yonkers, NY: Hudson River Museum, 1987), 21–36; Tom Lewis, *The Hudson: A History* (New Haven, CT: Yale University Press, 2005).

3. Charles Quest-Ritson, *The English Garden: A Social History* (Boston: David Godine, 2003), 120–144; Edward Hyams, *Capability Brown and Humphry Repton* (New York: Scribner, 1971).

4. See illustrations in *Hudson River Portfolio* (1825) and in Jacques Milbert, *Itinéraire pittoresque du fleuve Hudson et des parties latérales de l'Amérique du Nord* (Paris: H. Gaugain et cie, 1828–1829), http://www.digitalgallery.nypl.org (accessed 8 October 2006); N. P. Willis, *American Scenery*, 2 vols. (London: George Virtue, 1840), 1:4.

5. Christine Chapman Robbins, *David Hosack: Citizen of New York* (Memoirs of the American Philosophical Society, vol. 62, 1964), 177–179; J. E. Spingarn, "Henry Winthrop Sargent and the Early History of Landscape Gardening and Ornamental Horticulture in Dutchess County, New York," *Dutchess County Historical Society Yearbook* 22 (1937): 36–70; Patricia M. O'Donnell and Cynthia Zaitzevsky, *Cultural Landscape Report for Vanderbilt Mansion National Historic Site*, Cultural Landscape Publication No. 1 (Boston: National Park Service, 1992), 29–55. Hosack's property is now the Vanderbilt Mansion National Historic Site; the Franklin Roosevelt home is two miles to the south.

6. André Parmentier, "Landscapes and Picturesque Gardens," in T. G. Fessenden, *New American Gardener* (Boston: J. B. Russell, 1828), 184–185.

7. David Schuyler, *The New Urban Landscape* (Baltimore: Johns Hopkins University Press, 1986); Schuyler, *Apostle of Taste: Andrew Jackson Downing, 1815–1852* (Baltimore: Johns Hopkins University Press, 1996); Judith K. Major, *To Live in the New World: A. J. Downing and American Landscape Gardening* (Cambridge, MA: MIT Press, 1997).

8. A. J. Downing, "Notices on the State and Progress of Horticulture in the United States," *Magazine of Horticulture* 3 (1837): 5; A. J. Downing, *A Treatise on the Theory and Practice of Landscape Gardening, Adapted to North America* (New York: Wiley and Putnam, 1841).

9. Downing, *Treatise* (1841), 33–35.

10. Ibid., 34–35, 175, 127.

11. A. J. Downing, *A Treatise on the Theory and Practice of Landscape Gardening, Adapted to North America*, 2nd ed. (New York: Wiley and Putnam, 1844), 59, 493; Spingarn, "Henry Winthrop Sargent," 48–49.

12. "The Neglected American Plants," *Horticulturist* 5 (1851): 201–203; Therese O'Malley, " 'A Public Museum of Trees': Mid-Nineteenth Century Plans for the Mall," in *The Mall in Washington, 1791–1991*, ed. Richard Longstreth (Hanover, NH: University Press of New England, 1991), 66–70.

13. "On the Improvement of Vegetable Races," *Horticulturist* 7 (1852): 154–156.

14. "American versus British Horticulture," *Horticulturist* 7 (1852): 249–251.

15. "Shade Trees in Cities," *Horticulturist* 7 (1852): 345–349; Behula Shah, "The Checkered Career of *Ailanthus altissima*," *Arnoldia* 57, no. 3 (1997): 20–27.
16. A. J. Downing, *A Treatise on the Theory and Practice of Landscape Gardening, Adapted to North America*, 6th ed. (New York: A. O. Moore, 1859); Spingarn, "Henry Winthrop Sargent," 58–59.
17. N. P. Willis, *Out-doors at Idlewild; or, the Shaping of a Home on the Banks of the Hudson* (New York: Charles Scribner, 1855); O'Brien, *American Sublime*.
18. Willis, *Out-doors at Idlewild*.
19. Francis R. Kowsky, *Country, Park, & City* (New York: Oxford University Press, 1998), 54–65; Calvert Vaux, *Villas and Cottages* (New York: Harper and Bros., 1857), 39–40, 248.
20. The Greensward Plan and associated illustrations in edited form are in C. C. McLaughlin et al., eds., *The Papers of Frederick Law Olmsted*, multivolume (Baltimore: Johns Hopkins University Press, 1977), 3:117–188. See also Kowsky, *Country, Park, & City*, 91–98; Roy Rosenzweig and Elizabeth Blackmar, *The Park and the People* (Ithaca, NY: Cornell University Press, 1992), 121–149.
21. Commissioners of the Central Park, *First Annual Report on the Improvement of Central Park* (New York City, Board of Aldermen, Document 5, 19 January 1857), 52, 54, 79.
22. "Advertising for Plans," *New York Times*, 29 August 1857, 4; Commissioners of the Central Park, *First Annual Report*, 37.
23. Charles Rawolle and I. A. Pilat, *Catalogue of Plants. Gathered in August and September 1857, in the Ground of the Central Park* (New York: M. W. Siebert, 1857); Board of Commissioners of Central Park (hereafter, BCCP), *Minutes of Proceedings*, August–October 1857; F. L. Olmsted to the Board of Commissioners, 16 October 1857, in *Papers of Frederick Law Olmsted*, 3:106–111.
24. *Papers of Frederick Law Olmsted*, 3:133, 162–168, 176–177.
25. BCCP, *Annual Report* 2 (1858/1859): 68.
26. BCCP, *Annual Report* 3 (1859/1860): 42, and accompanying map dated 1 January 1860.
27. BCCP, *Minutes of Proceedings*, 7 November 1861; BCCP, *Annual Report* 6 (1862/1863): 15; BCCP, *Annual Report* 7 (1863/1864): 32. A drawing of the conservatory is in BCCP, *Annual Report* 5 (1861/1862), facing p. 46.
28. BCCP, *Minutes of Proceedings*, 9 July 1863. The arboretum was completely eliminated from the map (dated 1 January 1863) printed in BCCP, *Annual Report* 6 (1862/1863).
29. BCCP, *Annual Report* 8 (1864/1865): 26. The report noted an additional virtue of vegetation by comparing it with sculpture. The latter could often be controversial, whereas everyone, whether cultivated or ignorant, accepted and appreciated plants. Referring to commission president and Democratic politician Henry Stebbins's efforts to place his sister Emma's *Angel of the Waters* on the Terrace (the Bethesda Fountain), the report recommended "leaving out the

petrified emblems of exploded faiths." Submission to the "predominance of nature" would "illustrate the purer faith of our age."

30. Information on Pilat is compiled in Franziska Kirchner, *Der Central Park in New York und den Einfluss der deutschen Gartentheorie und—praxis auf seine Gestaltung* (Worms, Germany: Wernersche Verlagsgesellschaft, 2002), 203–213; see also M. M. Graff, *Central Park, Prospect Park* (New York: Greensward Foundation, 1985), 46–55.
31. *Papers of Frederick Law Olmsted*, 3:175; Olmsted to Pilat, 26 September 1863, *Papers of Frederick Law Olmsted*, 5:85–92.
32. Clarence Cook, *A Description of the New York Central Park* (New York: F. J. Huntington, 1869), 111–114.
33. Ibid., 106–107, 128.
34. Rawolle and Pilat, *Catalogue of Plants . . . 1857;* "Catalogue of Trees, Shrubs and Herbaceous Plants on the Central Park, December 31, 1863," BCCP, *Annual Report* 7 (1863/1864): 91–123; "Supplemental Catalogue of Plants Cultivated on the Central Park, 1865," BCCP, *Annual Report* 9 (1865/1866): 103–106; "A List of Certain Classes of Plants on the Central Park in 1873, Prepared by Robert Demcker, Landscape Architect," Appendix L, Part X, Board of Commissioners of the Department of Public Parks, *General Report* 3 (1872–1873 [1875]): 352–462.
35. Ida Hay, *Science in the Pleasure Ground* (Boston: Northeastern University Press, 1995); S. B. Sutton, *Charles Sprague Sargent and the Arnold Arboretum* (Cambridge, MA: Harvard University Press, 1970).
36. George Bentham and Joseph Hooker, *Genera Plantarum Plantarum ad Exemplaria Imprimis in Herbariis Kewensibus Servata Definita*, 3 vols. (London: Reeve, Williams & Norgate, 1862–1883); Stephen A. Spongberg, "Establishing Traditions at the Arnold Arboretum," *Arnoldia* 49, no. 1 (1989): 11–20.
37. Hay, *Science in the Pleasure Ground*, 79–91.
38. Cynthia Zaitzevsky, *Frederick Law Olmsted and the Boston Park System* (Cambridge, MA: Harvard University Press, 1982), 57; F. L. Olmsted, "Report of the Landscape Architect," Boston, Board of Commissioners of the Department of Parks, *Annual Report*, 1880, 6–16; Boston, Board of Commissioners of the Department of Parks, *Annual Report*, 1883, 16; F. L. Olmsted, "Report on Back Bay," Boston, Board of Commissioners of the Department of Parks, *Annual Report*, 1884, 13–16.
39. F. L. Olmsted, "Report of the Landscape Architect," Boston, Board of Commissioners of the Department of Parks, *Annual Report*, 1880, 12.
40. Zaitzevsky, *Olmsted and the Boston Park System*, 188–189.
41. Ibid., 189–190; on Fischer, see Kirchner, *Der Central Park in New York*, 213–217. On problems of maintaining this system see Nancy S. Seasholes, *Gaining Ground: A History of Landmaking in Boston* (Cambridge, MA: MIT Press, 2003), 220–222.
42. *Garden and Forest* 1 (1888): 266. On Stiles see Phyllis Andersen, " 'Master of a Felicitous English Style': William Augustus Stiles, Editor of *Garden and Forest*," *Arnoldia* 60, no. 2 (2000): 39–43.

43. Frederick Law Olmsted, "Foreign Plants and American Scenery," *Garden and Forest* 1 (1888): 418–419. Stiles conceded most of Olmsted's claims, but drew a line at apple trees in pinewoods. The problem was not the species itself, but the fact that the presence of such a tree showed that "the land had once been tilled." Such "startling" evidence that seemingly natural areas were actually abandoned farms was unwelcome because it destroyed the "mental associations" that made nature restful.
44. Olmsted to Fischer, 6 August 1889, and Fischer to Olmsted, 9 August 1889, quoted in Zaitzevsky, *Olmsted and the Boston Park System*, 193.
45. Zaitzevsky, *Olmsted and the Boston Park System*, 196; the plant lists are reproduced on pp. 216–220.
46. Ibid., 198.
47. Charles A. Platt, *Italian Gardens* (New York: Harper & Brothers, 1894); Keith N. Morgan, *Charles A. Platt, the Artist as Architect* (New York: Architectural History Foundation; Cambridge, MA: MIT Press, 1985), 49–53; Mac Griswold and Eleanor Weller, *The Golden Age of American Gardens* (New York: Abrams, 1991); Guy Lowell, ed., *American Gardens* (Boston: Bates and Guild Company, 1902).
48. Thomas S. Hines, "The Imperial Mall: The City Beautiful Movement and the Washington Plan of 1901–1902," in *The Mall in Washington*, 79–99.
49. L. H. Bailey, Wilhelm Miller, and C. E. Hunn, *The 1895 Chrysanthemums* (Ithaca, NY: Cornell University, 1896).
50. E. H. Wilson, *Plantae Wilsonianae: An Enumeration of the Woody Plants Collected in Western China for the Arnold Arboretum of Harvard University during the Years 1907, 1908, and 1910*, ed. Charles Sprague Sargent, 3 vols. (Cambridge, MA: Harvard University Press, 1913–1917).
51. George F. Pentecost Jr., "The Formal and the Natural Style," *Architectural Record* 12 (1902): 174–194; Glenn Brown, ed., *European and Japanese Gardens: Papers Read before the American Institute of Architects* (Philadelphia: Henry T. Coates, 1902); David C. Streatfield, "The Olmsteds and the Landscape of the Mall," in *The Mall in Washington*, 117–141; Brooklyn Botanic Garden, *Record* 1 (1911): 25–60; 3 (1914): 112–113.
52. Frank A. Waugh, *The Natural Style in Landscape Gardening* (Boston: Richard G. Badger, 1917), 16–19, 48–52, 63–73.
53. Olmsted, Vaux & Co., *Report Accompanying Plan for Laying out the South Park* (Chicago: Evening Journal, 1871), 9–14.
54. H. W. S. Cleveland, *Landscape Architecture, as Applied to the Wants of the West; with an Essay on Forest Planting on the Great Plains* (Chicago: Jansen, McClurg & Co., 1873); O. C. Simonds, "A Plan for Rural Grounds," *Michigan Horticulturist* 1 (1885): 27–28; Simonds, "Twelve Good Native Shrubs," *Michigan Horticulturist* 1 (1885): 171–172; Barbara Geiger, "'Nature as the Great Teacher': The Life and Work of Landscape Designer O. C. Simonds" (MLA thesis, University of Wisconsin, Madison, 1997).

55. Daniel Bluestone, *Constructing Chicago* (New Haven, CT: Yale University Press, 1991); Henry Regnery, *The Cliff Dwellers* (Evanston: Chicago Historical Bookworks, 1990).
56. James [Jens] Jensen, "Parks and Politics," *American Park and Outdoor Art Association Annual Meeting* 6 (1902): 11–14; also Bryan Lathrop, "Parks and Landscape Gardening," ibid., 7–10.
57. Jens Jensen, "Landscape Art—An Inspiration from the Western Plains," *The Sketch Book* 6 (1906): 21–28; Thomas McAdam, "Landscape Gardening under Glass," *Country Life in America* 21 (15 December 1911): 11–13, 51–52; Jens Jensen, Plan for Dewey Ranch, 1909, copy in box 1, Friends of Konza Historical Records, Department of Special Collections, Hale Library, Kansas State University, Manhattan, KS.
58. Wilhelm Miller, *What England Can Teach Us about Gardening* (Garden City, NY: Doubleday, Page, 1911), esp. 5–7; see also Miller's assessment of the field: "Have We Progressed in Gardening?" *Country Life in America* 20 (15 April 1912): 25–26, 68. On Miller, see Christopher Vernon, "Wilhelm Miller and *The Prairie Spirit in Landscape Gardening*," in *Regional Garden Design in the United States*, ed. Therese O'Malley and Marc Treib (Washington, DC: Dumbarton Oaks, 1995), 271–276.
59. Wilhelm Miller, *The "Illinois Way" of Beautifying the Farm*, University of Illinois Agricultural Experiment Station Circular 174, 1914; Miller, "First Draft of 'Planting Motive' Scheme, June 20, 1914; Miller and Franz A. Aust, "The Illinois Planting Motive: A Combination of Seven Trees, Shrubs, and Perennial Flowers Which Epitomizes the Scenery of the 'Prairie State,' " 31 July 1914, frames 123–140, reel 3, Franz A. Aust Papers, Wisconsin Historical Society, Madison; Miller, *The Prairie Spirit in Landscape Gardening*, University of Illinois Agricultural Experiment Station Circular 184, 1915; reprinted in *Regional Garden Design in the United States.*
60. Miller, *Prairie Spirit*, 33–34; [Miller], "Comments on the Work of the Division of Landscape Extension," typescript, 16 pp., 27 March 1915, frames 217–233, reel 3, Aust Papers; John H. Small Jr., "The Prairie Spirit in Landscape Architecture" [review], *Architectural Record* 40 (1916): 590; Christopher Vernon, "Wilhelm Miller and *The Prairie Spirit in Landscape Gardening*," in *Regional Garden Design in the United States*, 271–276.
61. John Nolen, *Madison: A Model City* (Boston: n.p., 1911), 69–74. On the history of the arboretum, see Nancy D. Sachse, *A Thousand Ages* (Madison: University of Wisconsin Press, 1963); William R. Jordan III, ed., *Our First 50 Years: The University of Wisconsin-Madison Arboretum 1934–1984* (Madison: University of Wisconsin-Madison Arboretum, 1984), http://www.digital.library.wisc.edu (accessed 8 October 2006).
62. J. W. Jackson to Franz Aust, 23 November 1933, general files, G. William Longenecker, box 1, 38/7/2, University of Wisconsin Arboretum Records, University Archives, Steenbock Library, University of Wisconsin, Madison (hereafter, UW Archives); Sachse, *A Thousand Ages,* 17; "Memorandum,

Proposed Chair of Conservation," 1933, general correspondence, general files, J. W. Jackson 1928–1934, 38/3/2, UW Archives; Arthur Hasler, interview by William R Jordan, 1979, University of Wisconsin Oral History 049, UW Archives; Thomas J. Blewett and Grant Cottam, "History of the University of Wisconsin Arboretum Prairies," *Transactions of the Wisconsin Academy of Sciences, Arts and Letters* 72 (1984): 130–144.

63. Aldo Leopold, "The Arboretum and the University" (1934), in Leopold, *The River of the Mother of God, and Other Essays*, ed. Susan L. Flader and J. Baird Callicott (Madison: University of Wisconsin Press, 1991), 209–211; Leopold, "What Is the Arboretum?" address to Nakoma Women's Club, 20 September 1934, 5, 38/7/2, UW Archives.
64. Sachse, *A Thousand Ages*, 45–48.
65. Ibid., 30–33; figure estimated from statement in "Memorandum for President Dykstra on a Research Program for the University of Wisconsin Arboretum," 10 July 1938, General Files, G. William Longenecker, box 12, folder "Research Arboretum," 38/7/3, UW Archives.
66. Aldo Leopold to C. L. Christensen, 23 May 1934, and Leopold to E. M. Gilbert, 21 July 1937, both in General Correspondence, General File 1932–1950 (J. W. Jackson), Folder Leopold, A., 38/3/1, UW Archives.
67. Theodore S. Cochrane and Hugh H. Iltis, *Atlas of the Wisconsin Prairie and Savanna Flora*, Wisconsin Department of Natural Resources Technical Bulletin 191, 2000, http://www.digicoll.library.wisc.edu (accessed 8 October 2006); "Plantings of Fassett and Thomson 1935," typescript, 8 pp., General Files, G. William Longenecker, box 10, folder "Prairie," 38/7/3, UW Archives; Theodore Sperry, "Prairie Restoration on the University of Wisconsin Arboretum," spring 1939, typescript, 6 pp., General Files, G. William Longenecker, box 1, 38/7/2, UW Archives.
68. Walter C. L. Muenscher, *Poisonous Plants of the United States* (New York: Macmillan, 1939).
69. Theodore Sperry, "Prairie Restoration on the University of Wisconsin Arboretum," spring 1939, typescript, 6 pp., General Files, G. William Longenecker, box 1, 38/7/2; and "Prairie Plantings by Sperry, University of Wisconsin Arboretum 1936–1940," typescript, 11 pp., General Files, G. William Longenecker, box 10, folder "Prairie," 38/7/3, UW Archives.

8. The Horticultural Construction of Florida

1. T. Stanton Dietrich, *The Urbanization of Florida's Population* (Gainesville: University of Florida Bureau of Economic and Business Research, 1978), 14–23.
2. Robert W. Long, "Origin of the Vascular Flora of Southern Florida," in *Environments of South Florida*, vol. 2, ed. Patrick Gleason (Coral Gables, FL: Miami Geological Society, 1984), 118–126; Walter S. Judd, personal communication, 4 March 2003; James J. Miller, *An Environmental History of Northeast Florida* (Gainesville: University Press of Florida, 1998).

3. The 36,000 square kilometers included in the south Florida region contain 1,574 flowering plant species, about 142 of which are endemic (to the state). An early-twentieth-century flora of the Bahamas (10,000 square kilometers) counted 995 species with 133 endemics; a more recent Jamaica (10,800 square kilometers) compilation lists 3,247 species and 784 endemics. See R. W. Long and O. Lakela, *A Flora of Tropical Florida* (Miami: Banyan Books, 1976), 15, 24; N. L. Britton and C. F. Millspaugh, *The Bahama Flora* (New York: Britton and Millspaugh, 1920), vii; C. D. Adams, *Flowering Plants of Jamaica* (Mona, Jamaica: University of the West Indies, 1972), 22.
4. Jerald T. Milanich, *Florida Indians and the Invasion from Europe* (Gainesville: University Press of Florida, 1995), 1–2, 213–230, estimates the pre-Columbian population within the present state boundaries at 350,000.
5. George R. Fairbanks, "Is the Orange Indigenous in Florida?" *Semi-Tropical* 3 (1877): 154–156.
6. Jane G. Landers, ed., *Colonial Plantations and Economy in Florida* (Gainesville: University Press of Florida, 2000).
7. *American National Biography*, 17:358–360; Nelson Klose, "Dr. Henry Perrine, Tropical Plant Enthusiast," *Florida Historical Quarterly* 27 (1948): 189–201; Jerry Wilkinson, "Dr. Henry Edward Perrine," http://www.keyshistory.org (accessed 8 October 2006).
8. Henry Perrine, "Cases of Periodical Disease Treated with Ergot, in Mississippi, 1825," *American Journal of the Medical Sciences* 13 (1833): 278–279.
9. House Committee on Agriculture, "Dr. Henry Perrine—Tropical Plants," 17 February 1838, H. Rpt. 25(2)-564, 17.
10. *New York Farmer and Horticultural Register* 2 (1829): 124; Perrine to Secretary of the Treasury, 26 November 1831, in "Foreign Trees and Plants," 6 April 1832, H. Doc. 22(1)-198, 12; Perrine to Congress, 29 December 1834, in H. Rpt. 25(2)-564, 29.
11. Perrine to Secretary of the Treasury, 8 November 1831, H. Doc. 22(1)-198, 8–11.
12. Perrine to *New York Farmer*, 7 June 1834; Perrine to Secretary of the Treasury, 1 February 1834, H. Rpt. 25(2)-564, 50, 54–59.
13. Perrine to Congress, 29 December 1834; Perrine to Secretary of the Treasury, 1 February 1834; Perrine to Secretary of State, 20 February 1834, in H. Rpt. 25(2)-564, 27–32, 47–54. Perrine was a polygenist—a supporter of the scientific position that not all humans belonged to one species.
14. John K. Mahon, *History of the Second Seminole War 1835–1842*, rev. ed. (Gainesville: University Press of Florida, 1991).
15. Perrine in *Farmers' Register* 8 (1840): 29.
16. Mahon, *History of the Second Seminole War*, 226–236; *Congressional Globe*, 24 January 1838, 126–127; Thomas Jesup to Joel Poinsett, 11 February 1838, printed in John T. Sprague, *The Origin, Progress, and Conclusion of the Florida War* (New York: D. Appleton, 1848), 199–201.
17. Perrine to Senate Agriculture Committee, 9 January 1838, 3 February 1838, 12 March 1838, S. Doc. 25(2)-300, 8, 11.

18. Perrine in *Farmers' Register* 7 (1839): 354.
19. Charles Howe and Perrine to Secretary of the Navy, 22 April 1839, in *Farmer's Register* 7 (1839): 472.
20. Perrine in *Farmers' Register* 8 (31 January 1840): 29–31; *House Journal* 26, no. 1 (16 March 1840): 612; T. R. Robinson, "Henry Perrine, Pioneer Horticulturist of Florida," *Proceedings of the Florida State Horticultural Society* 50 (1937): 78–82; Robinson, "Further Notes on the Perrine Episode," *Proceedings of the Florida State Horticultural Society* 51 (1938): 83–84; John Viele, *The Florida Keys: A History of the Pioneers* (Sarasota, FL: Pineapple Press, 1996), 41–68.
21. John T. Foster Jr. and Sarah Whitmer Foster, *Beechers, Stowes, and Yankee Strangers* (Gainesville: University Press of Florida, 1999); Florida Commissioner of Lands and Immigration [J. S. Adams], *Florida: Its Climate, Soil, and Productions, with a Sketch of Its History, Natural Features and Social Condition* (New York: The Florida Improvement Co., 1869), 7–8.
22. Lawrence N. Powell, *New Masters* (New Haven, CT: Yale University Press, 1980); *Florida: Its Climate, Soil, and Productions*, 9.
23. Mary B. Graff, *Mandarin on the St. John's* (Gainesville: University Press of Florida, 1953), 14; Foster and Foster, *Beechers, Stowes, and Yankee Strangers*, 54–75.
24. Harriet Beecher Stowe, *Oldtown Folks* (Boston: Fields, Osgood, 1869); Stowe, *Palmetto-Leaves* (Boston: J. R. Osgood, 1873), 188–195.
25. Catherine Beecher and Harriet Beecher Stowe, *The American Woman's Home* (New York: J. B. Ford & Co., 1869); Stowe, *Palmetto-Leaves*, 116–130; her campaign was seconded, belatedly, by Henry's wife, Eunice: E. Beecher, *Letters from Florida* (New York: Appleton, 1879).
26. Joseph A. Fry, *Henry S. Sanford* (Reno: University of Nevada Press, 1982), 1–11.
27. Henry Sanford to Frederick Seward, 10 January 1868, Seward Papers, University of Rochester Library, Rochester, NY; *Some Account of Belair, Also of the City of Sanford, with a Brief Sketch of Their Founder* (Sanford, FL: n.p.,1889), 44.
28. Jerry Woods Weeks, "Florida Gold: The Emergence of the Florida Citrus Industry, 1865–1895," Ph.D. dissertation, University of North Carolina, Chapel Hill, 1977, 124–178. During Reconstruction USDA curator of gardens and grounds William Saunders worked to introduce new citrus varieties, but this effort faded after 1876. See George W. Atwood, "Fruits of Florida," U.S. Department of Agriculture, *Report*, 1867, 140–147; "Southern Fruit-Growing for Market," U.S. Department of Agriculture, *Report*, 1871, 154; *Some Account of Belair*, 51; Fry, *Sanford*, 98–101.
29. In *Semi-Tropical* 3 (1877), see, for example, "Law and Order vs. Anarchy and Violence," 217–220, "California vs. Florida," 513–517, and Charles Beecher, "Florida a Hundred Years Hence," 389–391.
30. *Some Account of Belair*, 45, 51; Fry, *Sanford*, 98–101.
31. Fry, *Sanford*.

32. *Some Account of Belair;* Weeks, "Florida Gold"; Charlie C. Carlson, *When Celery Was King,* Sanford Historical Series No. 3 (Sanford, FL: Sanford Historical Society, 2000).
33. On Flagler see David Leon Chandler, *Henry Flagler* (New York: Macmillan, 1986); Edward N. Akin, *Flagler, Rockefeller Partner and Florida Baron* (Kent, OH: Kent State University Press, 1988).
34. Susan R. Braden, *The Architecture of Leisure* (Gainesville: University Press of Florida, 2002); *Florida, the American Riviera: St. Augustine, the Winter Newport, the Ponce de Leon, the Alcazar, the Casa Monica* (New York: Gilliss Brothers & Turnure, 1887).
35. Roberta Smith Favis, *Martin Johnson Heade in Florida* (Gainesville: University Press of Florida, 2003).
36. Akin, *Flagler,* 134–142, 188–197.
37. J. Wadsworth Travers, *History of Beautiful Palm Beach* (Palm Beach: n.p., 1929), 5; Charles W. Pierce, *Pioneer Life in Southeast Florida,* ed. D. W. Curl (Coral Gables, FL: University of Miami Press, 1970), 115–117; *Souvenir of the Royal Poinciana, Palm Beach (Lake Worth) Florida* (New York: U. Grant Duffield, 1894).
38. Braden, *Architecture of Leisure,* 219.
39. Chandler, *Henry Flagler,* 166–172.
40. Norman J. Pinardi, *The Plant Pioneers* (Torrington, CT: Rainbow Press, 1980), 33; Elizabeth Ogren Rothra, *Florida's Pioneer Naturalist* (Gainesville: University Press of Florida, 1995), 36–41.
41. Pinardi, *Plant Pioneers,* 65–71; Rothra, *Florida's Pioneer Naturalist,* 33–34; Janet Snyder Matthews, *Edge of Wilderness* (Sarasota, FL: Coastal Press, 1983), 315–384; *Report on the Condition of Tropical and Semi-Tropical Fruits in the United States in 1887. Prepared under the Direction of the Commissioner of Agriculture,* U.S. Department of Agriculture, Division of Pomology, Bulletin 1, 1888, 7–110.
42. B. T. Galloway to W. T. Swingle and E. F. Smith, 13 June 1892, Walter T. Swingle Papers, University of Miami Library, box 19; Swingle to H. J. Webber, 12 August 1892, Swingle Papers, box 17; Swingle and Webber, "The Sub-Tropical Laboratory and Its Objects," *Proceedings of the Florida State Horticultural Society* 6 (1893): 59–66; Swingle, "Some Citrus Fruits That Should Be Introduced into Florida," *Proceedings of the Florida State Horticultural Society* 6 (1893): 111–121; Swingle to Galloway, 28 April 1893, Swingle Papers, box 17. On Swingle, see Frank D. Venning, "Walter Tennyson Swingle, 1871–1952," *The Carrell: Journal of the Friends of the University of Miami Library* 18 (1977): 1–32.
43. Leah La Plante, "The Sage of Biscayne Bay," *Tequesta* 45 (1995): 60–82.
44. David Fairchild, *The World Grows Round My Door* (New York: Scribner's, 1947), 18; John C. Gifford, *On Preserving Tropical Florida,* ed. Elizabeth Ogren Rothra (Coral Gables, FL: University of Miami Press, 1972); Charles T. Simpson, *Ornamental Gardening in Florida* (Little River, FL: Charles T. Simpson, 1916), 17, 52.

45. David Fairchild, Trip Diary, 1921, 115 and later note on facing page, David Fairchild Papers, Fairchild Tropical Gardens Research Center, Miami, FL.
46. Simpson, *Ornamental Gardening.*
47. Ibid., xiii.
48. David Fairchild, "Report of Florida Trip 1912," Fairchild Papers; Fairchild, *World Grows Round My Door,* esp. 23–28, 122–124.
49. Fairchild to Deering, 13 December 1921, Fairchild Papers; David Fairchild, *The Plant Introduction Garden of the Federal Department of Agriculture, Chapman Field, Miami, Florida,* pamphlet, about 1934, Fairchild Papers; Marjorie Stoneman Douglas, *An Argument for a Botanical Garden in South Florida to Be Called the Fairchild Tropical Garden* (Coral Gables, FL: Kells Press, 1937).
50. Linda Vance, "May Mann Jennings and Royal Palm State Park," *Florida Historical Quarterly* 55 (1976): 1–17; John C. Paige, *Historic Resource Study for Everglades National Park* ([Washington, DC?]: National Park Service, 1986); David McCally, *The Everglades: An Environmental History* (Gainesville: University Press of Florida, 1999).
51. Charles T. Simpson, "Catalogue of Plants in the Collection of Charles Deering," typescript, 8 pp., Charles Deering Collection, Historical Society of South Florida, Miami; a series of articles by Eleanor Bisbee in the *Miami Daily Metropolis,* 1922, on the plants and animals of Buena Vista, clippings in Deering Collection, especially "Deering Estate Has Shrubs for Hedges and Windbreaks; Others Can Grow Them Also" (26 June 1922), "Charles Deering Attempted to Introduce Various Rare Birds in the Whole County but Success Was Confined to His Estate" (29 July 1922), "It Is Extremely Simple to Plant Cactus and the Deering Plantation Gives Proof That Western Varieties Will Grow Here" (25 September 1922).
52. Janet Snyder Matthews, *Historical Documentation: The Charles Deering Estate at Cutler* (Miami: Metro Dade County Parks and Recreation Department, 1992), 51–64.
53. John K. Small, "Exploration in 1915," *Journal of the New York Botanical Garden* 17 (1916): 42; Daniel F. Austin, ed., *The Florida of John Kunkel Small* (Contributions from the New York Botanical Garden, vol. 18, 1987).
54. Rothra, *Florida's Pioneer Naturalist,* 125–143; Matthews, *Historical Documentation,* 98–157, provides a synopsis of the Deering–Small correspondence.
55. Fairchild Trip Diary 1922, 76, Fairchild Papers; Matthews, *Historical Documentation;* John K. Small, *From Eden to Sahara: Florida's Tragedy* (Lancaster, PA: Science Press, 1929), esp. 7, 114; also Small, "The Everglades," *Scientific Monthly* 28 (1929): 80–87.
56. House Committee on Public Lands, *Establishment of Everglades National Park* Hearing, 71(3), 15 December 1930, 11, 27–28, 45, 49–52, 71.
57. George Buker, "Engineers vs. Florida's Green Menace," *Florida Historical Quarterly* 60 (1982): 413–427; Don C. Schmitz et al., "Exotic Aquatic Plants in Florida: A Historical Perspective and Review of the Present Aquatic Plant Management Program," *Proceedings of the Symposium on Exotic Pest*

Plants, 1988, NPS/NREVER/NRTR Technical Report 91/06, 304–305; Herbert J. Webber, *The Water Hyacinth, and Its Relation to Navigation in Florida,* U.S. Department of Agriculture, Division of Botany, Bulletin 18, 1897, 11–12; W. H. H. Benyaurd and James B. Quinn to John M. Wilson, 30 November 1898, in House, *Water Hyacinth Obstructions,* 17 December 1898, H. Doc. 55(3)-91, 5–6.

58. Royal Palm Nurseries, *Annual Catalogue and Price List of the Royal Palm Nurseries, Manatee, Florida* (Harrisburg, PA: J. Horace McFarland, 1888), 39; Brij Gopal, *Water Hyacinth* (Amsterdam: Elsevier, 1987), 21; John K. Small, *Manual of the Southeastern Flora,* 3rd ed. (New York: John K. Small, 1933), 267.

59. Karen M. O'Neill, *Rivers by Design: State Power and the Origins of U.S. Flood Control* (Durham, NC: Duke University Press, 2006).

60. E. S. Crill to C. M. Cooper, 9 February 1895, "Petition of Citizens of Palatka, Fla., and Vicinity," 12 August 1896, and W. H. H. Benyaurd to W. P. Craighill, 18 and 21 September 1896, in Senate, *Obstruction of Navigable Waters of Florida, and Other South Atlantic and Gulf States by the Aquatic Plant Known as the Water Hyacinth,* 5 January 1897, S. Doc. 54(2)-36, 2–3, 6–8.

61. J. R. Parrott (Vice President, Florida East Coast Railway), to Craighill, 3 November 1896; Benyaurd to Craighill, 9 November 1896; Craighill to D. S. Lamont (Secretary of War), 12 November 1896; *Obstruction of Navigable Waters,* S. Doc. 54(2)-36, 10–13; "Clogged by Hyacinths," *New York Sun,* 20 September 1896; "The Water Hyacinth in Florida," *New York Times,* 26 October 1896; "Hyacinths Closing a Florida River," *New York Times,* 19 December 1896.

62. U.S. Army Corps of Engineers, *Report of the Chief Engineer,* 1903, 279.

63. James Whorton, *Before Silent Spring* (Princeton, NJ: Princeton University Press, 1974), 20–22; U.S. Army Corps of Engineers, *Report of the Chief Engineer,* 1901, 2:1746–1748; 1903, 1:279, 2:1184–1186; 1906, 2:1236. The engineers estimated that their solutions were 1 percent arsenic trioxide, probably by volume.

64. U.S. Army Corps of Engineers, *Report of the Chief Engineer,* 1903, 2:1185–1186; 1903–1904, 2:1713; *Congressional Record,* 58(3), 13 February 1905, 2450, 23 February 1905, 3195.

65. U.S. Army Corps of Engineers, *Report of the Chief Engineer,* 1906, 2:1235–1239.

66. U.S. Army Corps of Engineers, *Report of the Chief Engineer,* 1911–1912, 566.

67. Herbert J. Webber, *The Citrus Industry* (Berkeley: University of California Press, 1948), 123.

68. J. Eliot Coit, *Citrus Fruits* (New York: Macmillan, 1915), 106–113, 344–354; Douglas Cazaux Sackman, *Orange Empire* (Berkeley: University of California Press, 2005).

69. H. Harold Hume, *Citrus Fruits and Their Culture* (Jacksonville: H. & W. B. Drew Co., 1904); L. B. Skinner, "The Florida Growers' and Shippers' League," *Proceedings of the Florida State Horticultural Society* 27 (1914): 91.

70. E. W. Berger, "Citrus Canker: Its Origin, Distribution and Spread," *Proceedings of the Florida State Horticultural Society* 28 (1915): 71–80; H. E. Stevens, *Citrus Canker—III,* University of Florida Agricultural Experiment Station Bulletin 128, November 1915; K. W. Loucks, "Citrus Canker and Its Eradication in Florida," unpublished manuscript, 1934, copy provided by the Division of Plant Industry, Florida Department of Agriculture, Gainesville; Tim R. Gottwald, James H. Graham, and Tim S. Schubert, "Citrus Canker: The Pathogen and Its Impact," *Plant Health Progress,* DOI:10.1094/PHP-2002-0812-01-RV.

71. Lloyd Tenny, "Address," *Proceedings of the Florida State Horticultural Society* 27 (1914): 92–110; E. W. Berger, "Citrus Canker in the Gulf Coast Country," ibid., 120–132.

72. Clara Hasse, "*Pseudomonas Citri:* The Cause of Citrus Canker," *Journal of Agricultural Research* 4 (1915): 97–100; Florida State Plant Board, *Quarterly Bulletin* 1 (1916): 11; "Report of the Citrus Canker Committee," *Proceedings of the Florida State Horticultural Society* 30 (1917): 51–59; Wilmon Newell, "The Citrus Canker Situation," *Florida Grower,* 5 April 1919, 6–7. On Newell, see "Wilmon Newell," 1998, http://www.flaentsoc.org (accessed 8 October 2006). The bacterium reappeared around 1986 and continues to be a problem: "Citrus Canker," http://www.aphis.usda.gov (accessed 8 October 2006).

73. Charles Marlatt to David Fairchild, 10 September 1929, Fairchild Papers.

74. "Notes Made by Arthur C. Brown at Orlando, Florida, during the First Few Days after Arrival with D. B. Mackie," and Brown, "Activities in Connection with Early Discovery of Medfly in Florida," typescript, 6 pp., n.d., both in Mediterranean Fruit Fly files #24/1, Florida Department of Plant Industry Library, Gainesville; Wilmon Newell, "The Mediterranean Fruit Fly Is in Florida," *Florida State Plant Board Monthly Bulletin* 13, 1929, 121–126. The official account of the campaign was Newell, "Mediterranean Fruit Fly Report," Florida State Plant Board, *Biennial Report* 8 (1930): 10–87.

75. House, *Eradication of Mediterranean Fruit Fly,* 24 April 1929, H. Doc. 71(1)-7; *Congressional Record,* 26 April 1929, 71(1), 608–609; "House Votes Fund to Fight Fruit Fly," *New York Times,* 27 April 1929, 4.

76. House Committee on Appropriations, *Mediterranean Fruit Fly Hearing,* February 1930, 71(2), 84–111, 971 (hereafter, *Fruit Fly Hearing*); "To Search Baggage in Florida for Fruit Spreading Fly Pest," *New York Times,* 2 June 1929, 24; Burkley Wheeler to Gov. Doyle Carleton, 17 June 1929, and Capt. William C. Price to Maj. R. B. Lyle, 30 June 1929, Records of the State Plant Board, Florida State Archives, Tallahassee. Wheeler, who was jailed, was African American.

77. "Eradication of the Mediterranean Fruit Fly," *Science* 70 (9 August 1929): 146–147; Newell, "Mediterranean Fruit Fly Report."

78. *Florida State Plant Board Monthly Bulletin* 13, 1929, 141; *Florida Grower,* August 1929; *Florida Clearing House News,* 1 August 1929; "Hyde and Marlatt Explain New Plans," *Seald-Sweet Chronicle,* 1 August 1929, 2, 6–7;

H. Harold Hume, "The Mediterranean Fruit Fly Situation," *Florida State Plant Board Monthly Bulletin* 14, no. 2 (August 1929): 29–42.
79. "Will Ask $26,000,000 to Wipe Out Fruit Fly," *New York Times,* 27 August 1929, 19; Paul de Kruif, "Death to the Med-Fly," *Country Gentleman* 95 (February 1930): 12–14, 76–81; Wilmon Newell, "The Mediterranean Fruit Fly Situation," *Journal of Economic Entomology* 23 (1930): 512–523, esp. 522.
80. *Fruit Fly Hearing,* 18, 1112–1124, 1134–1182, 1258–1297; "Political Fight Delays Fruit Fly Program," *Florida Grower,* February 1930, 10; Asa Allan Adams, "Fruit Fly Hearing Features Tragedy and Comedy," *Florida Grower,* April 1930, 7–8, 28–30; Newell, "What of the Mediterranean Fruit Fly?" *Proceedings of the Florida State Horticultural Society* 43 (1930): 149–158.
81. Marvin H. Walker, "Has the Fruit Fly Been Eradicated?" *Florida Grower,* August 1930, 12; Walker, "Remove Medfly Restrictions," *Florida Grower,* September 1930, 16.
82. Gordon F. Ferris, "Concerning the Mediterranean Fruit Fly," *Science* 70 (8 November 1929): 451–453; Ferris, "Plant Quarantines Run Wild," *New Republic* 6 (August 1930): 335–338; Glenn W. Herrick, "The Procession of Foreign Insect Pests," *Scientific Monthly* 29 (1929): 269–274; Herrick, "Concerning Quarantines against the Japanese Beetle" (testimony to Plant Quarantine and Control Administration, 31 October 1929), box 3, Glenn W. Herrick Papers, 21/23/844, Rare and Manuscript Collections, Kroch Library, Cornell University, Ithaca, NY; Herrick, "Some Economic Aspects of Plant Quarantines," *Florists Exchange and Horticultural Trade World* 74, no. 12 (December 1930): 9, 32, 37.
83. *Report of the Chief of the Bureau of Plant Quarantine,* 1933 and 1934; R. Kent Beattie, "The Threat of the Dutch Elm Disease," *American Forests,* 1931, 489; Curtis May and G. F. Gravatt, *The Dutch Elm Disease,* U.S. Department of Agriculture, Circular 170, 1931; May, *Outbreaks of the Dutch Elm Disease,* U.S. Department of Agriculture, Circular 322, 1934; "Bureaus Can Never Die," *New York Times,* 8 March 1932, 22; "Bureaus Wiped Out, 'Deadwood' Cut Off by Roosevelt Axe," *New York Times,* 7 April 1933, 1.
84. Lee A. Strong, "The Biological Aspects of Quarantine No. 37," 26 March 1935, Florida Department of Plant Industry library copy; Thomas J. Headlee, "The Application of Horse Sense to Plant Quarantine," *Journal of Economic Entomology* 26 (1933): 606–609; "Another Noble Experiment Doomed," *Horticulture* 13 (1935): 50; Senate Committee on Agriculture and Forestry, *Amending the Plant Quarantine Act* Hearing, 74(2), 17 March 1936, 48.

9. Culturing Nature in the Twentieth Century

1. L. H. Bailey, ed., *Cyclopedia of American Horticulture,* 4 vols. (New York: Macmillan, 1900–1902); L. H. Bailey, "What Is Horticulture?" *Proceedings of the Society for Horticultural Science* 1 (1904): 53–60.

2. Charles Lathrop Pack, *The War Garden Victorious* (Philadelphia: J. P. Lippincott, 1919); Alexandra Eyle, *Charles Lathrop Pack* (Syracuse, NY: ESP College Foundation, 1992); American Joint Committee on Horticultural Nomenclature, *Standardized Plant Names: A Catalogue of Approved Scientific and Common Names of Plants in American Commerce. Prepared by Frederick Law Olmsted, Frederick V. Colville, Harlan P. Kelsey, Subcommittee* (Salem, MA: The Committee, 1923).
3. Including my father, Vincent Pauly. One of seven children of a machinist, he was born in 1911 in the Ohio River floodplain town of Covington, Kentucky, and raised in a two-bedroom house on a street that dead-ended into a railroad trestle. In 1926 the family moved across the river to Pleasant Ridge, one of Cincinnati's many hilltop suburbs. He learned about plants from newspapers and library books, and built a rock and water garden while unemployed during the Depression. On marrying in 1940, he and my mother, Edyth Geile, settled on a half-acre lot in an exurban radial suburb, and ten years later built a home on one and a half acres subdivided from a nursery. He planted vegetables, fruit and nut trees, and a large number of mail-order conifers, including the newly introduced dawn redwood *(Metasequoia glyptostroboides)*. J. I. Rodale's *Organic Gardening and Farming* was one of the household's few magazine subscriptions in the 1950s. I learned to operate a three-hundred-pound self-propelled Whirlwind walk-behind lawn mower at about age nine. On broader trends see Virginia Scott Jenkins, *The Lawn* (Washington, DC: Smithsonian Institution Press, 1994); Ted Steinberg, *American Green* (New York: Norton, 2006).
4. *Horticultural Organizations of the United States and Canada* (Washington, DC: Bureau of Plant Industry, 1935).
5. House Committee on Agriculture, *National Arboretum* Hearing, 13 January 1925; Senate Committee on Agriculture and Forestry, *National Arboretum* Hearing, 20 January 1926; "Advisory Council, National Arboretum, Report of First Meeting, June 17, 1927, Washington," typescript, 46 pp., box 1245, RG 16, Correspondence of the Secretary of Agriculture, National Archives II, College Park, MD; see also F. V. Coville, "The Proposed National Arboretum at Washington," *Science* 62 (1925): 579–581; Coville, "The National Arboretum," *Science* 71 (1930): 176–178; U.S. National Capital Park and Planning Commission, *Plans and Studies, Washington and Vicinity* (Washington, DC: Government Printing Office, 1929).
6. Historic American Engineering Record, Bronx River Parkway Reservation HAER No. NY-327, http://www.westchesterarchives.com (accessed 8 October 2006); William Barnaby Faherty, *A Gift to Glory In: The First Hundred Years of the Missouri Botanical Garden (1859–1959)* (Ocean Park, WA: Harris & Friedrich, 1989); Henry N. Andrews Jr., "The Missouri Botanical Garden," *Parks and Recreation* 29 (1946): 297–302.
7. N. L. Britton to H. A. Gleason, 14 January 1919, 16 January 1920, and Gleason to Britton, 6 December 1922, 23 December 1923, in N. L. Britton correspondence, administrative: H. A. Gleason, New York Botanical Garden

Archives; Edgar Anderson, *Plants, Man and Life* (Boston: Little, Brown, 1952).
8. "Open to the Public," *New York Times*, 18 April 1954; author's personal experience from the early 1970s.
9. L. H. Bailey, "Current Tendencies in Horticulture," *Proceedings of the American Society for Horticultural Science* (1926): 16–18. For the range of concerns among horticultural science leaders, see *Presidential Addresses: American Society for Horticultural Science*, compiled by Jules Janick (West Lafayette, IN: Purdue University, Department of Horticulture, 1994).
10. E. Gorton Davis, "The Landscape Department," *Cornell Alumni News*, 13 November 1924; "Landscape Architecture, Cornell University, History of the School, 1903–1917," box 2, folder 1, Department of Landscape Architecture Records 15/2/1622, Kroch Library, Cornell University, Ithaca, NY.
11. "A Project of Far-Reaching Importance," *Horticulture* 24 (1 June 1946), 303–304; American Horticultural Council, *United Horticulture for the United States*, pamphlet, 1946, New York Botanical Garden Library vertical files; *United Horticulture 1946–1947–1948*, esp. Robert Pyle, "Opening Address October 1947," 23; William J. Robbins, "Horticulture in the Life of the Nation," 25; Robert Pyle, "How We May Advance" (1948), 66–71; Walter Gould, "How to Build Up Organizations," *Proceedings of the American Horticultural Congress* 9 (1954): 23–27; A. J. Irving, "President's Report," *Proceedings of the American Horticultural Congress* 14 (1959): 4–6.
12. For anxiety about the classification of gardening as a hobby, see John C. Wister, "What's Ahead?" *Parks and Recreation* 28 (1945): 145; M. Truman Fossum, "Measurement of American Horticulture," *United Horticulture 1946–1947–1948*, 61.
13. Ernestine Abercrombie Goodman, *The Garden Club of America: History 1913–1938* (Philadelphia: Edward Stern, 1938), 52; Anne F. Scott, *Natural Allies: Women's Associations in American History* (Urbana: University of Illinois Press, 1991).
14. Katherine S. White, *Onward and Upward in the Garden*, ed. E. B. White (New York: Farrar, Strauss, and Giroux, 1979). White's literary skills were not perfect: she did not discern that White Flower Farm's opinionated old-Yankee spokesman, Amos Pettingill, was a persona created by the business's cosmopolitan Manhattan-based owners, *Fortune* editor William Harris and *New York Times* writer Jane Grant (formerly the wife and partner of *New Yorker* founder Harold Ross, White's initial employer).
15. White, *Onward and Upward*, 221; Harold W. Rickett, *Wild Flowers of the United States*, 6 vols. in 14 (New York: McGraw Hill, 1966–1973).
16. *Proceedings, Western Weed Control Conference*, 1938–1967; *Proceedings, North Central States Weed Control Conference*, 1944, 9–19; Walter C. L. Muenscher, *Weeds* (New York: Macmillan, 1935); Clinton L. Evans, *The War on Weeds in the Prairie West* (Calgary: University of Calgary Press, 2002).

17. James Whorton, *Before Silent Spring* (Princeton, NJ: Princeton University Press, 1974); Thomas R. Dunlap, *DDT: Scientists, Citizens, and Public Policy* (Princeton, NJ: Princeton University Press, 1981); Linda Lear, *Rachel Carson* (New York: H. Holt, 1997). Along with uncounted other children in the 1950s I frolicked in the cool mist of the DDT trucks sent through our neighborhood by local government.
18. L. W. Kephart and S. W. Griffin, "Chemical Weed Killers after the War," *Proceedings, North Central States Weed Control Conference,* 1944, 78–82; *Proceedings, North Central States Weed Control Conference,* 1945, 16–75, esp. 16–18, 68–75; Harold Gunderson and E. P. Sylwester, "Spraying—Babyhood to Manhood in 5 Years," typescript, 7 pp., 1950, box 5, folder 13, E. P. Sylwester Papers, 9/18/51, Iowa State University Library, Ames, IA; C. J. Willard, "Weed Control: Past, Present, Prospects," *Agronomy Journal* 46 (1954): 481–484. *Weeds* began in 1951, and the Weed Society of America first met in early 1956. In 1968 the names changed to *Weed Science* and the Weed Science Society of America.
19. Rachel Carson, *Silent Spring* (Boston: Houghton Mifflin, 1962), 68–78; Arthur W. Galston, "Changing the Environment: Herbicides in Vietnam II," *Scientist and Citizen* 9 (1967): 122–129; U.S. Department of Commerce National Technical Information Service, *Evaluation of Carcinogenic, Teratogenic, and Mutagenic Activities of Selected Pesticides and Industrial Chemicals,* 2 vols., by Bionetics Research Labs., Inc., PB-223 160, 1968, 2:18–21, table A-22.
20. "U.S. Curbs Use of Weed Killer That Produces Rat Deformities," *New York Times,* 30 October 1969, 80; Thomas Whiteside, *The Withering Rain* (New York: E. P. Dutton, 1971); "Herbicides in Vietnam: AAAS Study Finds Widespread Devastation," *Science* 171 (1971): 43–47; Fred Wilcox, *Waiting for an Army to Die* (New York: Random House, 1983); Joe Conley, "More and Better Science: Dioxin and the Strategic Management of Scientific Doubt, 1965–1995," unpublished manuscript, Princeton University, 2005.
21. U.S. Animal and Plant Health Inspection Service, *A 25-Year Retrospective of the Animal and Plant Health Inspection Service, 1972–1997,* http://www.aphis.gov (accessed 8 October 2006); House Committee on Agriculture, *Federal Noxious Weed Act* Hearing, 13 September 1973 (Serial 93-Z); Senate Committee on Agriculture and Forestry, *Noxious Weed Act* Hearing, 3 October 1974; Senate Committee on Agriculture and Forestry, *Control of Noxious Weeds,* S. Rpt. 93-1313, 9 December 1974, esp. pp. 8–9.
22. Author's telephone interview with Robert Eplee (USDA-APHIS, retired), 24 March 2004; *Exotic Organisms,* Executive Order 11987, 24 May 1977; Marc L. Miller, "The Paradox of U.S. Alien Species Law," in *Harmful Invasive Species: Legal Responses,* ed. Marc L. Miller and Robert N. Fabian (Washington, DC: Environmental Law Institute, 2004), 146–148.
23. Michael E. Soulé and Bruce A. Wilcox, eds., *Conservation Biology* (Sunderland, MA: Sinauer Associates, 1980); Daniel S. Simberloff, "Community Effects of Introduced Species," in *Biotic Crises in Ecological and Evolutionary Time,* ed. Matthew H. Nitecki (New York: Academic Press, 1981), 53–82.

24. R. H. Grove and J. J. Burdon, eds., *Ecology of Biological Invasions* (Cambridge: Cambridge University Press, 1986); Harold A. Mooney and James A. Drake, eds., *Ecology of Biological Invasions of North America and Hawaii* (New York: Springer-Verlag, 1986); see esp. Richard N. Mack, "Alien Plant Invasion into the Intermountain West: A Case History," 191–213; David Pimentel, "Biological Invasions of Plants and Animals in Agriculture and Forestry," 149–162; and Daniel Simberloff, "Introduced Insects: A Biogeographic and Systematic Perspective," 3–26. The review was Peter Kareiva, "The Ecology of Invasions: Theory or Anecdotes?" *Ecology* 68 (1987): 1556; see also National Research Council, *Ecological Knowledge and Environmental Problem-Solving* (Washington, DC: National Academy Press, 1986), 57–58.

25. For background see J. J. Ewel, "Invasibility: Lessons from South Florida," in *Ecology of Biological Invasions of North America*, 214–230; David McCally, *The Everglades: An Environmental History* (Gainesville: University Press of Florida, 1999).

26. House, *Water-Hyacinth Obstructions in the Waters of the Gulf and South Atlantic States*, H. Doc. 85-37, 1957, esp. 99–108; A. E. Hitchcock et al., "Water Hyacinth: Its Growth, Reproduction, and Practical Control by 2, 4-D," *Contributions of the Boyce Thompson Institute for Plant Research* 15 (1949): 363–401; Jeffrey D. Schardt, "Maintenance Control," in *Strangers in Paradise: Impact and Management of Nonindigenous Species in Florida*, ed. Daniel Simberloff, Don C. Schmitz, and Tom C. Brown (Washington, DC: Island Press, 1997), 229–243.

27. Don C. Schmitz et al., "Exotic Aquatic Plants in Florida: A Historical Perspective and Review of the Present Aquatic Plant Regulation Program," *Proceedings of the Symposium on Exotic Pest Plants, November 2–4, 1988, University of Miami, Miami, Florida*, Technical Report NPS/NREVER/NRTR-91/06, 303–326; 1979 7 *Code of Federal Regulations* 360:100.

28. R. W. Klukas, "The Exotic Plant Problem in Everglades National Park," *The Anhinga*, April 1969.

29. Michael J. Bodle and Robert F. Doren, "The Exotic Pest Plant Councils," *Castanea* 61 (1996): 252–254.

30. Don C. Schmitz to the author, 5 July 2005; Melaleuca Control Act of 1991, H.R. 3753, 102(1), 12 November 1991.

31. P. D. Hebert et al., "Ecological and Genetic Studies on *Dreissena Polymorpha* (Pallas): A New Mollusc in the Great Lakes," *Canadian Journal of Fisheries and Aquatic Sciences* 45 (1989): 1587–1591; Wilfred Laurier LePage, "The Impact of *Dreissena Polymorpha* on Waterworks Operations at Monroe, Michigan: A Case History," in *Zebra Mussels: Biology, Impacts, and Control*, ed. Thomas F. Nalepa and Donald W. Schloesser (Boca Raton, FL: Lewis Publishers, 1993), 333–358; William Ashworth, *The Late, Great Lakes* (New York: Knopf, 1986); Phil Weller, *Fresh Water Seas* (Toronto: Between the Lines, 1990).

32. *Great Lakes Exotic Species Prevention Act* (Introduced in House), 20 March 1989; House Committee on Merchant Marine and Fisheries, *Zebra Mussels*

and Exotic Species Hearing, 14 June 1990; Constance Harriman, prepared statement, Senate Committee on Environment and Public Works, *The Nonindigenous Aquatic Nuisance Act of 1990* Hearing, 19 June 1990, 27–30.

33. U.S. Congress, Office of Technology Assessment (OTA), *Harmful Non-Indigenous Species in the United States,* OTA-F-565 (Washington, DC: Government Printing Office, 1993), 52.

34. Ibid., 7, 15–50; author's telephone interview with Phyllis Windle, 26 March 2004; House Committee on Merchant Marine and Fisheries, *Introduction of Harmful Non-Indigenous Species into the United States* Hearing, 5 October 1993, ser. 103-164.

35. OTA, *Harmful Non-Indigenous Species,* iv–vi, 311–318; Stanley A. Temple, "The Nasty Necessity: Eradicating Exotics," *Conservation Biology* 4 (1990): 113–115.

36. Don C. Schmitz to the author, 5 July 2005; Florida Department of Environmental Protection, *An Assessment of Invasive Non-indigenous Species in Florida's Public Lands,* ed. Don C. Schmitz and Tom C. Brown (Tech. Rpt. TSS-94-100), 1994, appendix 2; Daniel Simberloff, "Impacts of Introduced Species in the United States," *Consequences* 2 (1996): 13–22; Don C. Schmitz and Daniel Simberloff, "Biological Invasions: A Growing Threat," *Issues in Science and Technology* 13, no. 4 (1997): 33–40; Don C. Schmitz and Daniel Simberloff, "Needed: A National Center for Biological Invasions," *Issues in Science and Technology* 17, no. 4 (2001): 57–62; draft letter of twelve ecologists to Vice President Albert Gore, March 1997, copy at http://www.aquat1.ifas.ufl.edu/schlet2.html (accessed 8 October 2006). One of Simberloff's initiatives was a critique of my first publication in this area: "Letter to Editor," *Isis* 87 (1996): 676–677.

37. Faith Campbell, "How Can Ecologists Encourage Environmental Organizations to Become Active in Fighting This Threat?" Florida Department of Environmental Protection, *An Assessment of Invasive Non-indigenous Species,* 5–6.

38. See, for example, Chris Bright, *Life Out of Bounds* (New York: Norton, 1998); Robert S. Devine, *Alien Invasion* (Washington, DC: National Geographic Society, 1998); Jason Van Driesche and Roy Van Driesche, *Nature Out of Place* (Washington, DC: Island Press, 2000); Harold A. Mooney and Richard J. Hobbs, eds., *Invasive Species in a Changing World* (Washington, DC: Island Press, 2000); Yvonne Baskin, *A Plague of Rats and Rubbervines* (Washington, DC: Island Press, 2002).

39. Daniel B. Botkin, *Discordant Harmonies* (New York: Oxford University Press, 1990); Mark Sagoff, "Why Invasive Species Are Not as Bad as We Fear," *Chronicle of Higher Education,* 23 June 2000, B7; Kristin Shrader-Frechette, "Non-Indigenous Species and Ecological Explanation," *Biology and Philosophy* 16 (2001): 507–519; Daniel B. Botkin, "The Naturalness of Biological Invasions," *Western North American Naturalist* 61 (2001): 261–266; David M. Lodge and Kristin Shrader-Frechette, "Nonindigenous Species: Ecological Explanation, Environmental Ethics, and Public Policy," *Conservation Biology* 17 (2003): 31–37. The most energetic criticism was

from a seedsman: David I. Theodoropoulos, *Invasion Biology: Critique of a Pseudoscience* (Blythe, CA: Avvar Books, 2003).

40. OTA, *Harmful Non-Indigenous Species,* 306.
41. For more than a decade, anti-alien ecologist David Pimentel of Cornell was a national director of the Carrying Capacity Network, a lobbying organization that identified human immigration as the most important environmental problem facing Americans: http://www.carryingcapacity.org (accessed 8 October 2006).
42. *Invasive Species,* Executive Order 13112, 3 February 1999; Miller, "Paradox of U.S. Alien Species Law."
43. On the history of ecological restoration see Marcus Hall, *Earth Repair: A Transatlantic History of Environmental Restoration* (Charlottesville: University of Virginia Press, 2005).
44. H. Nicholas Jarchow, *Forest Planting* (New York: Orange Judd, 1893), 14–15, 47–53; James W. Toumey, *Seeding and Planting* (New York: John Wiley, 1916); Toumey, *Foundations of Silviculture upon an Ecological Basis* (New York: John Wiley, 1928), esp. 296–321; Ralph C. Hawley, *The Practice of Silviculture* (New York: John Wiley, 1921), 19–23; R. T. Fisher, "The Harvard Forest as a Demonstration Tract," *Quarterly Journal of Forestry* 25 (1931): 130–139.
45. David. R. Foster and John F. O'Keefe, *New England Forests through Time* (Cambridge, MA: Harvard University Press, 2000); Hugh M. Raup and Reynold E. Carlson, *The History of Land Use in the Harvard Forest,* Harvard Forest Bulletin 20, 1941; David Foster and John D. Aber, eds., *Forests in Time* (New Haven, CT: Yale University Press, 2004).
46. H. H. Bennett and W. R. Chapline, *Soil Erosion a National Menace,* U.S. Department of Agriculture Circular 33, 1928; Bennett, *Soil Conservation* (New York: McGraw-Hill, 1939); Thomas D. Isern, "The Erosion Expeditions," *Agricultural History* 59 (1985): 181–190; Allan J. Soffar, "The Forest Shelterbelt Project, 1934–1944," *Journal of the West* 14 (1975): 95–107; Stanley W. Trimble, "Perspectives on the History of Soil Erosion Control in the Eastern United States," *Agricultural History* 59 (1985): 162–180.
47. Bennett, *Soil Conservation,* 378–418; John J. Winberry and David M. Jones, "Rise and Decline of the 'Miracle Vine': Kudzu in the Southern Landscape," *Southeastern Geographer* 13 (1973): 61–70; Mart Allen Stewart, "Cultivating Kudzu: The Soil Conservation Service and the Kudzu Distribution Program," *Georgia Historical Quarterly* 81 (1997): 151–167; Albert E. Cowdrey, *This Land, This South: An Environmental History* (Lexington: University Press of Kentucky, 1983), 149–194.
48. "A Research Program for the University of Wisconsin Arboretum," general files, G. William Longenecker, box 12, folder "Research-Arboretum," 38/7/3, UW Archives; Arthur Hasler interview by William R. Jordan, 1979, University of Wisconsin Oral Histories 049, UW Archives.
49. J. T. C.[urtis], "Prairie Burning Experiments—U. W. Arboretum," late 1942, typescript, 3 pp., general files, G. William Longenecker, box 10, folder

"Prairie," 38/7/3, UW Archives; John T. Curtis and Max L. Partch, "Effect of Fire on the Competition between Blue Grass and Certain Prairie Plants," *American Midland Naturalist* 39 (1948): 437–443.

50. David Archbald, "Man-Environment Collision," 7–11, Peter Schramm, "Preface," 3, and Schramm, "A Practical Restoration Method for Tall-Grass Prairie," 63–65, in Schramm, ed., *Proceedings of a Symposium on Prairie and Prairie Restoration . . . September 14 and 15, 1968* (Galesburg, IL: Knox College, Biological Field Station, 1970), http://www.digital.library.wisc.edu (accessed 8 October 2006).

51. William K. Stevens, *Miracle under the Oaks* (New York: Pocket Books, 1995), 110–117; on Fermilab prairies, see "Prairies," http://www.fnal.gov (accessed 8 October 2006).

52. Stevens, *Miracle under the Oaks*.

53. Jon Mendelson, Stephen P. Aultz, and Judith Dolan Mendelson, "Carving Up the Woods: Savanna Restoration in Northeastern Illinois," *Restoration and Management Notes* 10 (1992): 127–131; Henry F. Howe, "Managing Species Diversities in Tallgrass Prairies: Assumptions and Implications," *Conservation Biology* 8 (1994): 691–704; Steve Packard, "Restoring Oak Ecosystems," *Restoration and Management Notes* 11 (1993): 16; Laurel M. Ross, "The Chicago Wilderness: A Coalition for Urban Conservation," *Restoration and Management Notes* 15 (Summer 1997): 17–24; Debra Shore, "Controversy Erupts over Restoration in the Chicago Area," ibid., 25–31; Paul Gobster, "The Other Side: A Survey of the Arguments," ibid., 32–37; Stephen Packard and Cornelia F. Mutel, eds., *Tallgrass Restoration Handbook* (Washington, DC: Island Press, 1997); Paul Gobster and R. Bruce Hull, eds., *Restoring Nature* (Washington, DC: Island Press, 2000).

54. James C. Malin, "An Introduction to the History of Bluestem-Pasture Region of Kansas," *Kansas Historical Quarterly* 11 (1942): 3–28. Further south the problem was mesquite: B. W. Allred, "Distribution and Control of Several Woody Plants in Texas and Oklahoma," *Journal of Range Management* 2 (1949): 17–34; C. C. Wright, "The Mesquite Tree: From Nature's Boon to Aggressive Invader," *Southwestern Historical Quarterly* 69 (1965): 38–43; Steve Archer, "Tree-Grass Dynamics in a *Prosopis*-thornscrub Savanna Parkland: Reconstructing the Past and Predicting the Future," *Ecoscience* 2 (1995): 83–99.

55. A. E. Aldous, "Management of Kansas Permanent Pastures," *Kansas State Agricultural Experiment Station Bulletin* 272, 1935; Aldous, "Effect of Burning on Kansas Bluestem Prairies," *Kansas State Agricultural Experiment Station Technical Bulletin* 38, 1934; Thomas B. Bragg and Lloyd C. Hulbert, "Woody Plant Invasion of Unburned Kansas Bluestem Prairie," *Journal of Range Management* 29 (1976): 19–24.

56. Victor Cahalane, "A Proposed Great Plains National Monument," 10 November 1938, files of the Committee on the Ecology of Grasslands, National Academy of Sciences Archives, Washington, DC; U.S. National Park Service, *A Proposed Prairie National Park* ([Washington, DC?]: National Park Service,

1961); Hal K. Rothman and Daniel J. Holder, *Tallgrass Prairie National Preserve Historic Resource Study* (Omaha, NE: National Park Service Midwest Regional Office, 2000), and Rebecca Conard and Susan Hess, *Tallgrass Prairie National Preserve Legislative History, 1920–1996* (Omaha, NE: National Park Service Midwest Support Office, 1998), both at http://www.cr.nps.gov (accessed 8 October 2006).

57. Conard and Hess, *Tallgrass Prairie National Preserve Legislative History;* Senate Committee on Interior and Insular Affairs, *Establishment of Prairie National Park* Hearing, 8 July 1963 ([88] SIni-T.25), and 8 August 1963 ([88] SIni-T.21).

58. "The Need and Use of a Flint Hills Natural Area," attachment, 6 pp., Lloyd C. Hulbert to Glenn H. Brett, 14 March 1958, Konza Prairie Records 18/21, University Archives, Hale Library, Kansas State University, Manhattan, KS (hereafter, Konza Records).

59. E. Raymond Hall testimony, *Establishment of Prairie National Park* Hearing, 8 July 1963, 34–53; Lloyd Hulbert to Walter Boardman (Nature Conservancy), 4 January 1963, Konza Records 8/37; Lloyd C. Hulbert to Senator Alan Bible, 31 July 1963, with attached "Statement in support of S 986 to establish a Prairie National Park," Konza Records 10/5.

60. "A Proposal for the Establishment of a Prairie Research Area Near Kansas State University, Prepared by the Division of Biology, July, 1968," brochure, 32 pp., Konza Records 16/11.

61. See correspondence in Konza Records 9/3 and 9/4; William D. Blair Jr., *Katherine Ordway* (Washington, DC: Nature Conservancy, 1989), 28–34, quotation p. 49; on private versus public funding, compare "A Plan for Natural Areas in Kansas," *Transactions of the Kansas Academy of Science* 69 (1966): 9.

62. Kling L. Anderson, "Burning Flint Hills Bluestem Ranges," *Tall Timbers Conference* No. 3, 1964, 89–96; Lloyd C. Hulbert, "Management of Konza Prairie to Approximate Pre-White-Man Fire Influences," *Third Midwest Prairie Conference Proceedings: Kansas State University, Manhattan, September 22–23, 1972* (1973), 14–17, http://www.digital.library.wisc.edu (accessed 8 October 2006); "Konza Prairie Research Natural Area, Report for 1972 to 1974," typescript, 12 pp., Konza Records 9/5.

63. Blair, *Ordway*, 50–64; "Jens Jensen's Landscape Design for Konza," box 1, Friends of Konza Historical Records, University Archives, Hale Library, Kansas State University, Manhattan, KS.

64. Lloyd C. Hulbert, "History and Use of Konza Prairie Research Natural Area," *Prairie Scout* 5 (1985): 63–93.

65. M. D. Smith and A. K. Knapp, "Physiological and Morphological Traits of Exotic, Invasive Exotic, and Native Plant Species in Tallgrass Prairie," *International Journal of Plant Sciences* 162 (2001): 785–792; John M. Briggs, Greg A. Hoch, and Loretta C. Johnson, "Assessing the Rate, Mechanisms, and Consequences of the Conversion of Tallgrass Prairie to *Juniperus virginiana* Forest," *Ecosystems* 5 (2002): 578–586; John M. Briggs, A. K.

Knapp, and B.L. Brock, "Expansion of Woody Plants in Tallgrass Prairie: A Fifteen-Year Study of Fire and Fire-Grazing Interactions," *American Midland Naturalist* 147 (2002): 287–294; John M. Briggs et al., "An Ecosystem in Transition: Causes and Consequences of the Conversion of Mesic Grassland to Shrubland," *Bioscience* 55 (2005): 245–254.

66. Carl O. Sauer, "Grassland Climax, Fire, and Man," *Journal of Range Management* 3 (1950): 16–20; Ashley Schiff, *Fire and Water: Scientific Heresy in the Forest Service* (Cambridge, MA: Harvard University Press, 1962); regarding possible suppression, see the introduction to Omer Stewart, *Forgotten Fires* (Norman: University of Oklahoma Press, 2002).

67. Stephen J. Pyne, " 'These Conflagrated Prairies': A Cultural Fire History of the Grasslands," 131–137, and Lloyd C. Hulbert, "Fire Effects on Tallgrass Prairie," 138–142, in *Proceedings of the Ninth North American Prairie Conference*, 1984 (Fargo, ND: Tri-College University Center for Environmental Studies, 1986); Pyne, *Fire in America* (Princeton, NJ: Princeton University Press, 1982); Pyne, *Tending Fire* (Washington, DC: Island Press, 2004); William Cronon, *Changes in the Land* (New York: Hill and Wang, 1983); William R. Jordan III, "Festival of the Prairie: A Vision," *Restoration and Management Notes* 10, no. 2 (Winter 1992): 179.

68. Paul S. Martin, *Twilight of the Mammoths* (Berkeley: University of California Press, 2005); Daniel H. Janzen and Paul S. Martin, "Neotropical Anachronisms: The Fruit the Gomphotheres Ate," *Science* 215 (1982): 19–27; Connie Barlow, *The Ghosts of Evolution* (New York: Basic Books, 2000).

69. Paul S. Martin and David A. Burney, "Bring Back the Elephants!" *Wild Earth*, Spring 1999, 57–64; Martin, *Twilight of the Mammoths*, 200–211; C. Josh Donlan et al., "Re-wilding North America," *Nature* 436 (18 August 2005): 913–914.

10. America the Beautiful

1. Katharine Lee Bates, *The College Beautiful, and Other Poems* (Cambridge, MA: H.O. Houghton, 1887), 1–7; Bates, "America the Beautiful," in *Selected Poems of Katharine Lee Bates* (Boston: Houghton Mifflin, 1930), 92–93; Jean Glasscock, ed., *Wellesley College 1875–1975: A Century of Women* (Wellesley, MA: Wellesley College, 1975), 265–267, 285; Dorothy Whittemore Bates Burgess, *Dream and Deed: The Story of Katharine Lee Bates* (Norman: University of Oklahoma Press, 1952), 30–32, 101–102.

2. Bruno Latour, *Politics of Nature: How to Bring the Sciences into Democracy*, trans. Catherine Porter (Cambridge, MA: Harvard University Press, 2004).

3. J.R. Harlan, "Our Vanishing Genetic Resources," *Science* 188 (1975): 618–621; U.S. Congress, Office of Technology Assessment, *Grassroots Conservation of Biological Diversity in the United States*, OTA-BP-F-38 (Washington, DC: Government Printing Office, 1985); D.O. Bramwell et al., eds., *Botanic Gardens and the World Conservation Strategy* (Orlando, FL: Academic Press, 1987); Cary Fowler and Pat Mooney, *Shattering: Food, Politics, and the Loss*

of Genetic Diversity (Tucson: University of Arizona Press, 1990); National Research Council, *Managing Global Genetic Resources* (Washington, DC: National Academy Press, 1991); Brent C. Dickerson, *The Old Rose Adviser* (Portland, OR: Timber Press, 1992).

4. Susan R. Schrepfer and Philip Scranton, eds., *Industrializing Organisms* (New York: Routledge, 2004).

5. Michele H. Bogart, *The Politics of Urban Beauty: New York and Its Art Commission* (Chicago: University of Chicago Press, 2006).

Acknowledgments

If someone had told me in 1968, when I was in Washington, D.C., making the transition from buttoned-down Ohio high school student to collegiate seeker, that thirty-five years later I would be waiting expectantly at the book delivery window of the New York Public Library for nineteenth-century copies of the *Annual Report of the Nebraska State Horticultural Society*, I would have been incredulous or worse. I came to this project by steps and serendipity—childhood interests in natural history and in invention, graduate assistantships at the Smithsonian Institution, and historical research and teaching on biotechnology, exploration, and science in the United States. My initial work on tensions between plant introduction and pest exclusion, focused on the Department of Agriculture in the early 1900s, came out of research on the activities of federal biologists in the Gilded Age. I incorporated that into *Biologists and the Promise of American Life,* but saw from the start that those scientific bureaucrats were participants in an important and understudied historical process. This book is the result.

For the last seven years the New York Public Library has in fact been my chief entry point into the past. Research in its extensive nineteenth-century collections was doubly pleasant in 2003–2004 because I held a fellowship at the Dorothy and Lewis B. Cullman Center for Scholars and Writers. Both this work setting and the opportunity to share discoveries daily with the center's diverse group of fellows were invaluable; its importance as a center for intellectual hybridization was epitomized for me one day in a casual but intense conversation with its director, Jean Strouse, about collie fancying at the end of the nineteenth century.

I plundered three other libraries in the New York–New Jersey area. With guidance from the Rutgers University Libraries' professional staff—most notably, Dean Meister, Christopher Lee, Mary Fetzer, and Stacy DeMatteo—I was able to utilize the system's large but unwieldy collection of historic agricultural materials. The LuEsther T. Mertz Library of the New York Botanical Garden held many unique printed and manuscript sources. I was glad to have regular access to New York University's Bobst Library.

The archives I consulted are listed in the notes. I want to express gratitude, however, to the Fairchild Tropical Gardens Research Center's Bert Zuckerman and Nancy Korber, whose volunteer activities made David Fairchild's papers usable. Andy Reasoner, proprietor of the Royal Palm Nurseries in Bradenton, Florida, generously allowed me to examine historical materials in his possession.

Research and writing were made possible by a grant (0233922) from the National Science Foundation's Science and Technology Studies Program. I appreciate the strong support for faculty research at Rutgers, provided structurally through its sabbatical program and individually through the policies of Deans Holly Smith and Barry Qualls, and of Deborah White and Ziva Galili as chairs of the History Department.

Scientists and historians responded to my queries and provided documents, guidance, and encouragement. Current and former Agriculture Department scientists were uniformly helpful with information and insights; these included John B. Welch, Roger Ratcliffe, Andrew Liebhold, James Hatchett, and Robert Eplee. Phyllis Windle, Don C. Schmitz, Thomas McOwiti, Karen Madsen, Richard Mack, and Kathy Burks answered questions and sent me copies of correspondence, draft reports, and gray literature. Mark L. Winston donated time at an early stage.

In my own scholarly community, I was helped by John Van Sickle, Lucia Stanton, Robert J. Spear, Philip Sloan, David Schuyler, Ben Minteer, Sara Cedar Miller, Marc Miller, Peter Mickulas, Jane Maienschein, Barbara Kimmelman, Pamela Henson, Eric Grimm, Angela Creager, Joe Conley, John Burnham, Graham Burnett, Michele Aldrich, and numerous others who appeared to encourage my excited monologues at conventions. Abigail Lustig taught me the nineteenth-century meaning of horticulture. Gregg Mitman provided good advice and displayed good humor at an important juncture. Conversations with Matt Chew have been particularly stimulating. I presented preliminary versions of different chapters, and received valuable criticism, at Yale, University of Pennsylvania, Minnesota–Twin Cities, California-Berkeley, Brown, Arizona State, and the Rutgers Department of Human Ecology.

Rutgers colleagues who provided information, suggestions, and support include Mark Wasserman, Andrew Vayda, Philip Scranton, Julie Livingston, James Livingston, Jan Lewis, Gerald Grob, Paul Clemens, Jack Cargill, Herman Bennett, and Mia Bay. Deborah Popper educated me about the historical geography of the Midwest when we were fellows at the Rutgers Center for Historical Analysis in 2002–2003. Much of my knowledge of environmental history has come from my longtime colleague, Susan Schrepfer, who generously read and commented on the manuscript.

Completion of this project, like my previous book, was possible due to the timely interventions of medical professionals. The staffs of the St. Luke's–Roosevelt Hospital Center Department of Hematology and Oncology, and of the Mt. Sinai Hospital Bone Marrow Transplantation Program, provided humane, professional care. Paul Israel and Kelly Enright kindly volunteered in March 2006 to take over my classes, and the staff of the Rutgers University History Department provided essential support. My brother, Mark Pauly, offered practical guidance at a critical moment, as well as continuing encouragement. My sister, Rita Verderber, most generously shared with me her immune system. I now experience life as a bisexual chimera, and wonder if this will increase my sympathy toward some of the organisms discussed in this book.

I remember with pleasure the continuing encouragement of Leo Bogart (1921–2005), a principled investigator and wonderful man, and am pleased to continue to share ideas and plans with Agnes Bogart. My son Nick's athleticism, musicianship, and good sense form a welcome counterexample to vulgar notions of biological determinism. My incredible good fortune in having the loving companionship of my wife, Michele Bogart, continues day by day. She educates me about the built environment and demonstrates how to be an engaged scholar. And she holds our world together.

Brooklyn, New York
June 16, 2007

Index

Note: Page numbers in *italics* refer to illustrations and maps.

Adams, John S., 204
Adlum, John, 74
Agent Orange, 239
Agricultural Research Service, 233
Agriculture Department, 115; in Florida, 212–214, 223–227; and grass, 119–123; grounds in Washington, D.C., 113, *114;* and pest control, 142, 146–147, 151–153
Ailanthus, 93, 96–97, 169, 171, 182
Alcott, Bronson, 75
Aldous, A. E., 250, 252, 255
America and eastern Asia compared, 19–20, 87, 107, 124
America characterized, 7, 23
"America the Beautiful," 259
American Agriculturist, 107
American Association of Horticultural Inspectors, 146
American Association of Nurserymen, 143–147, 151–152
American Coffee Land Association, 101
American Forestry Association, 155
American Horticultural Council, 235; organizational chart, *236*
American Institute (New York), 103
American Museum, 38, 41
American Philosophical Society, 19, 27
American Plant Pest Committee, 155

"American plants," 16, 112–113
American Pomological Society, 63
American Rose Society, 160–161, 231
American Society for the Promotion of Useful Knowledge, 19
Anderson, Edgar, 233
Apple, 13, 67, 125–126, 187, 307n43; Baldwin, 68; Newtown Pippin, 67
Arbor Day, 92–93
Arboriculture, 80–81, 91–97
Argyll, Duke of, 16
Army Corps of Engineers, 220–222, 241
Arnold Arboretum, 86, 179–180, 232; planting scheme, *180*
Arsenic: as herbicide, 221, 222; as insecticide, 137, 225
Artichoke, 111
Association of Economic Entomologists, 144–145
Astor, William B., 169
Atwater, Caleb, 89
Aughey, Samuel, 123

Back Bay Fens (Boston), *181, 183,* 184
Bailey, Liberty Hyde, 133, 135, 231, 234
Baltimore-Washington Parkway, 233
Bancroft, Edward, 31
Banks, Joseph, 40–46
Bartram, John, 14

Bates, Katharine Lee, 259
Beal, W. J., 123
Beecher, Henry Ward, 59, 66, 74
Bell, Marian, 127
Bessey, Charles, 123
Boll weevil, 142
Bond, Phineas, 39, 45, 48
Botanic gardens, decline of, 232–233
Bourdeaux mixture, 78
Breadfruit, 40
Britton, Nathaniel, 233
Bronx River Parkway, 232
Brooklyn Botanic Garden, 186
Browne, D. J., 106–107, 110
Budd, Joseph L., 125, 126
Buena Vista (Deering estate), 215
Buffon, Comte de, 9–10, 27–28
Bull, Ephraim Wales, 74–76, 77
Burbank, Luther, 125
Bute, Earl of, 16–17

California, pest control in, 135–136, 141, 152
California State Board (Commission) of Horticulture, 131, 136; Quarantine Division seal, *132*
Canada, degeneration in, 18, 21
Canada thistle, 135
Carleton, Mark, 127–128
Carrying Capacity Network, 322n41
Catalpa, 84, 93, 96, 97, 177, 183
Catesby, Mark, 16
Central Park (New York), 173, 174; Arboretum, 174, *175, 176*, 178; Ramble, *178*
Chastellux, Marquis de, 31
Chestnut: American, 94–95; Chinese, 125, 154
Chestnut blight, 154
Chicago, landscape gardening in, 187–190
Chicago Wilderness, 251
Citrus, 12, 198, 204–207, 211, 222–227
Citrus canker, 223–224
Civilian Conservation Corps, 192
Cleveland, Grover, administration (1893–1897), 123, 125, 142
Clifford, George, 70
Climate, 11, 13, 17, 22; American, 11, 18, 21; and arboriculture, 94; of Virginia, 24–29
Clover, 111–112

Coe, Ernest, 217
Collinson, Peter, 14, 16
Colorado potato beetle, 134
Committee on Horticultural Plant Quarantine, 159
Concord, Massachusetts, 75
Congressional seed distribution, 105, 113, 295–296n54
Cornell University, landscape architecture at, 234
Cosmopolitanism, 86
Cotton, 101–102
Coxe, William, 67
Creoles, 11, 21
Crosby, Alfred, 2, 3
Culture, 6, 11, 66, 263–266
Currie, James, 41
Curtis, William B., 249–250
Curtis Prairie, Wisconsin, 192–193, 249; burning, *250;* planting, *192*
Cutler (Deering estate), 215–216, *217*
Cyclopedia of American Horticulture, 189, 230

Davis, Robert W., 221
DDT and children, 319n17
De Kruif, Paul, 226
Dearborn, H. A. S., 56, 66
Death camas, 193
Deering, Charles, 215–217
Degeneration, 11, 14, 21–22, 29, 32, 65, 67
Dewey Ranch, Kansas, 189, 255
Dogwood, 84, 256
Douglas, David, 84
Douglas, Robert, 95
Downing, Andrew Jackson, 59–60, 64–67, 85, 168–171
Dufour, Jean, 73–74, 101
Duschesne, Antoine Nicholas, 70
Dutch elm disease, 228

Ecologists and pest controllers, 241, 244, 245
Edinburgh, University of, 15
Ehlers, Hans Jacob, 169
Ellsworth, Henry, 104, 105
Emerson, George B., 83
Emerson, Ralph Waldo, 75
Endive, 111
Environmental Protection Agency, 239
Evelyn, John, 82

Everglades National Park, Florida, 217–218, 242
Executive Order 11987 (Carter administration), 240
Executive Order 13112 (Clinton administration), 246

Fairchild, David G., 126, *127,* 128–129, 147–149, 158, 214, 302n70
Fairchild Tropical Garden, Florida, 214
Fassett, Norman, 190–193
Fay, Joseph S., 86
Federal Horticultural Board, 153–154, 157, 162–163
Fermi National Accelerator Laboratory prairie, 251
Fernald, Charles H., 137
Fernow, Bernhard, 94
Ferris, Gordon, 227
Fessenden, Thomas G., 51
Fire blight, 69, 134
Fischer, William, 182, 184
Fisher, Carl, 215
Fisher, Richard T., 247
Flagg, Willard, 95
Flagg, Wilson, 83, 84
Flagler, Henry, 208–212
Flame thrower for insect control, 137, *140*
Flint Hills, Kansas, 252–254
Florida, *196;* early settlement, 198; historical biogeography, 197–198; latitude and climate, 207, 210–211; northern colonization of, 203–205; tropical agriculture in, 200
Florida East Coast Railway, 210–211, 219–221
Florida Exotic Pest Plant Council, 242
Florida Growers' and Shippers' League, 223
Florida State Plant Board, 223, 226
Forbush, Edward, 137
"Forest and Prairie Lands of the United States" (1857), *91*
French Revolution, 46
Frey, Lou, Jr., 242
Fruit: importance of, 53–54; improvement of, 54; nomenclature, 63–68
Fruits of America, 60
Furnas, Robert W., 93

Galloway, Beverly T., 126–127, 159, 212
Galston, Arthur, 239

Garden clubs, 235
Gardening as a hobby, 235–237
Genet, E. C., 103
Gifford, John C., 213
Gleason, Henry, 233
Grant, Jane, 318n14
Grant, Madison, 232
Grape: Alexander, 74, 77; Catawba, 74, 77; Concord, 75–79; *Vitis labrusca,* 13, 73; *Vitis vinifera,* 12, 22, 25–26, 29, 73, 78
Grape diseases, 78
Grass, 115–121; bluestem, 193–194, 252–254; brome, 126, 128; crested wheatgrass, 128; on Great Plains, 122; introductions, 122–123; Johnson grass, 117, 119, *120;* Kentucky bluegrass, 116, 118, 249–252; quackgrass, 118; in South, 122; Willard's bromus, 117
Gray, Asa, 87–88, 111, 133
Greeley, Horace, 76
Gypsy moth, 136–140, 145–146; control methods, *139, 140;* map of spread, *138*
Gypsy Moth Commission, Massachusetts, 137, 145–146

Hansen, Neils E., 126
Harlan, Jack R., 265
Harvard Forest, 247
Harvard University, landscape architecture at, 234
Harvesta Chemical Compounding Company, 221
Hasse, Clara, 223
Hawaii, Mediterranean fruit fly in, 151–152
Heade, Martin J., 209
Hedrick, U. P., 69
Hemings, Sally, 32
Herbicides: arsenical, 221–222; organochemical, 239–241
Herrick, Glenn, 227
Hessian fly, 34–49; developmental stages, *36;* name, 37–39; path of invasion, *35*
Hessians, 33
Hogg, James, 174
Holt, Joseph, 107
Horse, Arabian, 30
Horticultural freedom, 153
Horticultural societies, American, 53
Horticultural Society of London (Royal Horticultural Society), 55, 84

Horticulture, 1–2, 259–263; American, 2–4, 61–63; decline of, 235; at land-grant universities, 234
Horticulturist, 60, 76
Hosack, David, 55–56, 68, 167–168
Hough, Franklin, 86
Houseman, Jacob, 202–203
Hovey, Charles M., 58–60, *61*, 66, 69–72, 75
Hovey's Nursery, 59–60, 71–72
Howard, C. W., 116
Howard, Leland O., 141–142, 146
Howe, Gen. William, 34–35
Hudson, New York, *167*
Hudson River Valley, 166–167, 172
Hulbert, Lloyd C., 253–255
Hume, H. Harold, 226
Hunnewell, H. H., 171, 259
Hybridization, 65
Hyde Park (Hosack estate), 168
Hydrilla, 242

Idlewild (N. P. Willis estate), 171, *172*
Indian Key, Florida, 202
Industrializing Organisms, 265
Insecticides: arsenical, 137, 225; organochemical, 238
Invasive species, 245–246
Invasive Species Council, 246

Janzen, Daniel, 257
Japanese: and American diplomacy, 148–150; cherry trees, 129, 147–149, *150*; landscape gardening style, 185–186; plants, 86, 124; prejudice against, 148–149
Jefferson, Martha, 24, 26, 32
Jefferson, Thomas, 9–10, 23–24, 27–29, 32, 46, 48, 51; on climate, 25; gardening, 24, *25*, 26, 32; on race, 29–31
Jensen, Jens, 188–189
Jesup, Thomas, 201
Jordan, William R., III, 257
Juniper, 97, 256

Kalm, Peter, 16
Kansas State University, 253, 254, 255
Kinsey, Alfred, 162
Knapp, Seaman, 127, 128
Knight, Thomas Andrew, 54–55, 65
Know-Nothings (American Party), 76

Konza Prairie Research Natural Area, 252–254, *255*
Kosky, Francis, 173
Kudzu, 113, 247, *248*

Landscape gardening, 166–168; formal style of, 185; idea of repose in, 183, 186; Japanese style of, 185–186; racial basis for, 186; university programs in, 234. *See also* Naturalism, American
Larch, European, 86, 95, 97
Larkspur, 193
Lathrop, Barbour, 126
Leopold, Aldo, 190–193, 249–250
Lettuce, 111
Lilac, 182, 187
Lindley, John, 84
Linnaeus, Carl, 43
Locust, black, 94, 96, 182, 247, 256
Locust, honey, 183, 256, 257
Lombardy poplar, 85, *167*
Long Term Ecological Research Program (National Science Foundation), 255–257
Longenecker, William, 191, *192*, 249
Longworth, Nicholas, 59–60, 71–74
Lowell, Guy, 185
Lowell, John, 55–56, 66

Mack, Richard, 241
Madison, Rev. James, 25
Magazine of Horticulture (Hovey's Magazine), 58, 59, 75
Magnolia, 84, 169, 177
Maize, 13, 105
Mall, National. *See* Washington, D.C., National Mall
Mandarin, Florida, 204
Manischewitz's kosher wine, 79
Manning, Robert, 69
Marbois, Comte de, 26, 27
Marlatt, Charles L., *144*, 145–157, 161–164, 224, 226
Marsh, George P., 82
Martin, Paul S., 257
Mason, Charles, 106
Massachusetts Horticultural Society, 51, 56, 58, 60–61, 159, 228; Horticultural Hall, *62*
Mather, Cotton, 14
Mazzei, Filippo, 24, 26, 29
McFarland, J. Horace, 159, *160*, 161–162, 231

Means, John Hugh, 117
Mediterranean fruit fly, 131, *132*, 151–152, 224–227
Meehan, Thomas, 110
Melaleuca, 213, 242
Merchant Marine and Fisheries Committee, U.S. House, 243
Meyer, Frank, 128
Miami, Florida, 211
Miami Beach, Florida, 215
Miller, Philip, 70
Miller, Wilhelm, 189–190
Missouri Botanical Garden, 232
Mitchell, John, 14–19
Mitchill, Samuel L., 41, 48, 101, 199–200
Money-wort, 177
Montesquieu, Baron de, 15
Monticello orchard plan, *25*
Morgan, George, 33–34, 38–39, 44
Morton, J. Sterling, 93, 125
Morton Arboretum, 250
Mount Auburn Cemetery, 56, 57
Mount Vernon, Virginia, 85
Muddy River Improvement (Boston), 184
Muhlenberg, Henry, 117
Mulberry, 97, 102–103
Mulhern, Francis J., 240

National Arboretum (Washington, D.C.), 231, 233–234
National Convention for the Suppression of Insect Pests and Plant Diseases by Legislation, 143
National Mall. *See* Washington, D.C., National Mall
Nationalism, 85
Native, 6, 64–66, 78
Nativism, 76
Naturalism, American, 165, 182–183, 186–187, 258
Naturalization, 6, 63–65
Nature Conservancy, 251, 254, 255
Nebraska National Forest, 96
New York Botanical Garden, 232, 233
New York Horticultural Society, 55, 101, 199
Newell, Wilmon, 224, 226
Nolen, John, 190
North American Prairie Conference, 251
Notes on the State of Virginia, 9–10, 24–29

Noxious Weed Act, 240, 242
Nurserymen, 112–113

Office of Technology Assessment, 243–245
Olive, 12
Olmsted, Frederick Law, 173, 179–184, 188, 234
Olmsted, Frederick Law, Jr., 234
Orange. *See* Citrus
Ordway, Katherine, 254
Osage orange, 93, 96, 97, 253, 256, 257
Osborne, Herbert, 144
Owen, Ruth Bryan, 225

Pack, Charles L., 156
Packard, Stephen, 251
Paine, Thomas, 23, 46
Palatka, Florida, 219–221
Palm: coconut, 210; royal, 211–212
Palm Beach, Florida, 210
Pammel, Louis, 126
Parmentier, André, 103, 168
Parry, C. C., 120
Pascalis, Felix, 103
Patent Office Agricultural Division, 104–107, 111, 113
Pauly, Vincent A., 317n3
Pear, 68–69; Bartlett, 68; fire blight of, 69, 134; Kieffer, 69; Seckel, 68
Perrine, Henry, 199–203
Pest control, 134–136, 225; international, 147, 151–161, 228; interstate, 140–147, 151–152, 163
Pest controllers compared, 238–240; and ecologists, 241, 244, 245
Pests, 133–134, 137–138; cartoon, *156*
Pettingill, Amos, 318n14
Pilat, Ignaz, 176–177
Pimentel, David, 241, 322n41
Pine: eastern white, 95, 154–155; Scotch, 86, 95
Pine blister rust, 154–155
Pitkin, William, 151
Plant choice: in Boston, 179–184; in Central Park, 174–178; on Curtis Prairie, 194; at Cutler, 216; Downing on, 169–171; in Northeast, 83–84; in Palm Beach, 210, *211;* on prairie, 94–96; in Prospect Park, 182–183; in Saint Augustine, *209;* in south Florida, 213
Plant Quarantine Act, 153–154

Plant Quarantine and Control Administration, 163, 164, 224–228
Plant quarantine cartoon, *157*
Plant Quarantine 37, 156–162, 228
Plant succession: in Northeast, 83–84, 247; on prairie, *256*, 257
Platt, Charles, 184
Platt, Thomas, 212
Pleistocene extinctions, 258
Poinsett, Joel, 201–202
Poison ivy, 133, 193
Poiteau, Antoine, 65–66
Ponce de Leon Hotel (Saint Augustine), *209*
Powell, John Wesley, 90
Prairie: arboriculture on, 91–97; definitions of, 248; as Indian garden, 257; as naturalistic garden, 193–194, 258; planting a, *192*; problem of, 81, 88–91, 97–98, 249; re-wilding of, 258
Prairie burning: controlled, *250*, 252, 254–255; by Indians, 89–90, 257; by park visitors, 253
Prairie style in landscape gardening, 189–190
Prickly pear cactus, 177, 193
Princeton, New Jersey, 33–34, 37
Privy Council (England), 40–41, 46–47
Pyle, Robert, 235
Pyne, Stephen, 257

Racial differences, 15, 29–31
Rand, Edward S., Jr., 112–113
Raynal, Guilliame T. F., 9–10, 20–23, 28
Reasoner, Pliny, 211–212
Reed, Harrison, 204, 206
Rennselaer, Stephen van, 103
Rhododendron, 113
Rice, 12, 127
Riley, Charles V., 133
Ritcheson, Charles, 48
Rocky Mountain locust, 133, 145
Rose, 161
Royal Academy of Agriculture (France), 44
Royal Botanic Garden, Kew (England), 17, 113
Royal Palm Nurseries, Florida, 212, 219
Royal Palm State Park, Florida, 215
Royal poinciana, 210
Royal Poinciana Hotel (Palm Beach), 210, *211*

Rush, Richard, 101, 199
Russell, John L., 84–85, 110–111
Russian olive, 96, 189, 247, 256
Russian thistle, 113

Saint Augustine, Florida, 208–209
Saint Johns River, 219–221
San Jose scale, 141–143, 146
Sanders, Daniel C., 135
Sanford, Florida, 206–207
Sanford, Henry, 205–207
Sargent, Charles S., 83, 86, 90, 124, 159, 171, 179–185, 232
Saunders, William, 85
Schapiro's kosher wine, 79
Schmitz, Don C., 243–245
Scientific Committee on Problems of the Environment, 241
Scott, C. S., 147
Search and seizure, 163, *225*
Sectionalism, 106–107, 110
Seminole War, 200–203
Sequoia, giant, 84–87, *88*, 286n10
Shaler, Nathaniel S., 123
Shaw, E. Clay, Jr., 243
Shelterbelt Project, 96
Silk, 102–104
Silviculture. *See* Arboriculture
Simberloff, Daniel, 241, 244–245
Simpson, Charles T., 211–213, *214*
Sisal, 199, 203
Slaves: in Florida, 200; in Virginia, 29–31
Small, John K., 216–219
Smith, Adam, 41
Smith, Henry Nash, 93–94
Smith, John B., 141, 143
Soil Conservation Service, 247
Sorghum, 111
Soulé, Michael, 240
Soybean, 128
Sperry, Theodore, 190–193, 249
Sprague, Charles, 111, 185
Staten Island, New York, 35
Stiles, William A., 182, 307n43
Storm King, New York, 171
Stowe, Harriet Beecher, 204–205
Strawberry, 70–73; Hovey's Seedling, *59*, 71–72; Pineapple (Old Pine), 70–71; Wilson's Albany, 72
Strong, Lee, 228
Swaim, John, 203

Swingle, Walter T., 212
Systematic Seed and Plant Introduction, Section of, 126–128

Taft, William Howard, 148, 163
Tallgrass Prairie National Park, Kansas, 252–254
Taylor, Zachary, 202
Tea, 107, *108, 109,* 113
Temple, John, 44, 48
Temple, Stanley, 244
Thomson, Charles, 19, 20, 27
Thoreau, Henry David, 67, 75, 83
Thornton, Tamara, 69
Timber Culture Act, 92, 96
Tobacco, 12, 18, 30
Torrey, John, 117, 200
Toumey, James, 247
Tracy, S. M., 122
Treatise on the Theory and Practice of Landscape Gardening, 168, 169, 171
Tree culture. *See* Arboriculture
Trouvelot, E. L., 136
Tyler Townsend, C. H., 142

University of Wisconsin Arboretum, 190–191, *192,* 193, 249, *250*
U.S. government departments and bureaus. *See individual organizations*
U.S. Propagating Garden, 107, *110*

Van Fleet, Walter, 161
Van Mons, J. B., 65–66, 75
Vasey, George, 120, *121,* 122–123

Vaux, Calvert, 171, 173, 188
Virginia. See *Notes on the State of Virginia*
"Virginia flying weevil," 40, 42, 44

Wallace, Henry A., 228
Wallace, Henry C., 161–162
Wallace, Rev. Henry, 125
Washington, D.C., National Mall, 85, 107, 129, 150, 170, 185–186
Washington, George, 37, 39
Washington Monument, 85
Water hyacinth, 219–222, 241
Waugh, Frank, 186–187
Webber, Herbert J., 212
Webster, Francis, 143
Weed science, 239
Welch's grape juice, 78–79
Wellesley (Hunnewell estate), *260*
Wellesley College, 259
Wells, R. W., 89
Wheat, 13, 30, 40, 46–49, 127; durum, 128; stem rust on, 13, 134
White, Katherine S., 237
Wild Flowers of the United States, 237
Wilder, Marshall, 63
Willis, Nathaniel P., 171–172
Willits, Edwin, 142
Wilson, Ernest H., 128, 159
Wilson, James, 125–126
Windle, Phyllis, 244
Wood, Will R., 227

Zebra mussel, 243